World Scientific Series in Current Energy Issues Volume 4

Energy Storage

World Scientific Series in Current Energy Issues

Series Editor: Gerard M Crawley *(University of South Carolina & Marcus Enterprise LLC, USA)*

Published

Vol. 1 Fossil Fuels: Current Status and Future Directions
edited by Gerard M Crawley

Vol. 2 Solar Energy
edited by Gerard M Crawley

Vol. 3 Energy from the Nucleus: The Science and Engineering of Fission and Fusion
edited by Gerard M Crawley

Vol. 4 Energy Storage
edited by Gerard M Crawley

World Scientific Series in Current Energy Issues Volume 4

Energy Storage

Editor

Gerard M Crawley
Marcus Enterprise LLC, USA
&
Professor and Dean Emeritus
University of South Carolina, USA

World Scientific

NEW JERSEY • LONDON • SINGAPORE • BEIJING • SHANGHAI • HONG KONG • TAIPEI • CHENNAI • TOKYO

Published by

World Scientific Publishing Co. Pte. Ltd.
5 Toh Tuck Link, Singapore 596224
USA office: 27 Warren Street, Suite 401-402, Hackensack, NJ 07601
UK office: 57 Shelton Street, Covent Garden, London WC2H 9HE

Library of Congress Cataloging-in-Publication Data
Names: Crawley, Gerard M., editor.
Title: Energy storage / [edited by] Gerard M Crawley (University of South Carolina & Marcus Enterprise LLC, USA).
Other titles: Energy storage (World Scientific (Firm))
Description: [Hackensack] New Jersey : World Scientific, [2017] | Series: World Scientific series in current energy issues ; volume 4 | Includes bibliographical references and index.
Identifiers: LCCN 2016056678 | ISBN 9789813208957 (hc : alk. paper)
Subjects: LCSH: Energy storage.
Classification: LCC TJ165 .E497 2017 | DDC 621.31/26--dc23
LC record available at https://lccn.loc.gov/2016056678

British Library Cataloguing-in-Publication Data
A catalogue record for this book is available from the British Library.

Desk Editors: Herbert Moses/Amanda Yun

Typeset by Stallion Press
Email: enquiries@stallionpress.com

Printed in Singapore

Foreword to the World Scientific Series on Current Energy Issues

Sometime between 400,000 and a million years ago, an early humanoid species developed the mastery of fire and changed the course of our planet. Still, as recently as a few hundred years ago, the energy sources available to the human race remained surprisingly limited. In fact, until the early 19th century, the main energy sources for humanity were biomass (from crops and trees), their domesticated animals, and their own efforts.

Even after many millennia, the average per capita energy use in 1830 only reached about 20 Gigajoules (GJ) per year. By 2010, however, this number had increased dramatically to 80 GJ per year.[1] One reason for this notable shift in energy use is that the number of possible energy sources increased substantially during this period, starting with coal in about the 1850s and then successively adding oil and natural gas. By the middle of the 20th century, hydropower and nuclear fission were added to the mix. As we move into the 21st century, there has been a steady increase in other forms of energy such as wind and solar, although presently they represent a relatively small fraction of world energy use.

Despite the rise of a variety of energy sources, per capita energy use is not uniform around the world. There are enormous differences from country to country, pointing to a large disparity in wealth and opportunity, see Table 1. For example, in the United States the per capita energy use per year in 2011 was 312.8 million Btu[a] (MMBtu) and in Germany, 165.4 MMBtu. In China, however, per capita energy use was only 77.5 MMBtu, despite its impressive economic and technological gains. India, weighs in even lower at 19.7 MMBtu per person.[2] The general trends over the last decade suggest that countries with developed economies generally show

[a] *Note*: 1 GJ = 0.947 MMBtu.

Table 1: Primary Energy Use per Capita in Million Btu (MMBtu).[2]

Country	2007 (MMBtu)	2011 (MMBtu)	Percentage change
Canada	416.1	393.7	−5.4
United States	336.9	312.8	−7.2
Brazil	52.7	60.2	14.2
France	175.7	165.9	−5.6
Germany	167.8	165.4	−1.4
Russia	204.0	213.4	4.6
Nigeria	6.1	5.0	−18.0
Egypt	36.4	41.6	14.3
China	57.1	77.5	35.7
India	17.0	19.7	15.9
World	**72.2**	**74.9**	**3.7**

modest increases or even small decreases in energy use, but that developing economies, particularly China and India are experiencing rapidly increasing energy consumption per capita.

These changes, both in the kind of resource used and the growth of energy use in countries with developing economies, will have enormous effects in the near future, both economically and politically, as greater numbers of people compete for limited energy resources at a viable price. A growing demand for energy will have an impact on the distribution of other limited resources such as food and fresh water. All these lead to the conclusion that energy will be a pressing issue for the future of humanity.

Another important consideration is that all energy sources have disadvantages as well as advantages, risks as well as opportunities, both in the production of the resource and in its distribution and ultimate use. Coal, the oldest of the "new" energy sources, is still used extensively to produce electricity, despite its potential environmental and safety concerns in mining both underground and open cut mining. Burning coal releases sulfur and nitrogen oxides which in turn can lead to acid rain and a cascade of detrimental consequences. Coal production requires careful regulation and oversight to allow it to be used safely and without damaging the environment. Even a resource like wind energy using large wind turbines has its critics because of the potential for bird kill and noise pollution. Some critics also find large wind turbines an unsightly addition to the landscape, particularly when the wind farms are erected in pristine environments. Energy from nuclear fission, originally believed to be "too cheap to meter"[3] has not had the growth predicted because of the problem with long-term storage of the waste from nuclear reactors and because of the public perception regarding the danger of catastrophic accidents such as happened at Chernobyl in 1986 and at Fukushima in 2011.

Even more recently, the measured amount of CO_2, a greenhouse gas, in the global atmosphere has steadily increased and is now greater than 400 parts per million (ppm).[4] This has raised concern in the scientific community and has led the majority of climate scientists to conclude[5] that this increase in CO_2 will produce an increase in global temperatures. We will see a rise in ocean temperature, acidity, and sea level, all of which will have a profound impact on human life and ecosystems around the world. Relying primarily on fossil fuels far into the future may therefore prove precarious, since burning coal, oil, and natural gas will necessarily increase CO_2 levels. Certainly for the long term future, adopting a variety of alternative energy sources which do not produce CO_2 seems to be our best strategy.

The volumes in the *World Scientific Series on Current Energy Issues* explore different energy resources and issues related to the use of energy. The volumes are intended to be comprehensive, accurate, current, and international perspective. The authors of the various chapters are experts in their respective fields and provide reliable information that can be useful not only to scientists and engineers, but also to policy makers and the general public interested in learning about the essential concepts related to energy. The volumes will deal with the technical aspects of energy questions but will also include relevant discussion about economic and policy matters. The goal of the series is not polemical but rather is intended to provide information that will allow the reader to reach conclusions based on sound, scientific data.

The role of energy in our future is critical and will become increasingly urgent as world population increases and the global demand for energy turns ever upwards. Questions such as which energy sources to develop, how to store energy, and how to manage the environmental impact of energy use will take center stage in our future. The distribution and cost of energy will have powerful political and economic consequences and must also be addressed. How the world deals with these questions will make a crucial difference to the future of the earth and its inhabitants. Careful consideration of our energy use today will have lasting effects for tomorrow. We intend that the *World Scientific Series on Current Energy Issues* will make a valuable contribution to this discussion.

References

1. Our Finite World: World energy consumption since 1820 in charts. Available at: http://ourfiniteworld.com/2012/03/12/world-energyconsumption-since-1820-in-charts/. Accessed on February, 2015.
2. U.S. Energy Information Administration, Independent Statistics & Analysis. Available at: http://www.eia.gov/cfapps/ipdbproject/iedindex3.cfm?tid=44&pid=45&aid=2&cid=regions&syid=2005&eyid=2011&unit=MBTUPP. Accessed on March, 2015.
3. The quote is from a speech by Lewis Strauss, then Chairman of the United States Atomic Energy Commission, in 1954. There is some debate as to whether Strauss actually meant energy from nuclear fission or not.
4. NOAA Earth System Research Laboratory, Trends in Atmospheric Carbon Dioxide. Available at: http://www.esrl.noaa.gov/gmd/ccgg/trends/. Accessed on March, 2015.
5. IPCC, Intergovernmental Panel on Climate Change, Fifth Assessment report 2014. Available at: http://www.ipcc.ch/. Accessed on March, 2015.

About the Editor

Gerard "Gary" M Crawley is the President of Marcus Enterprises LLC based in North Carolina. Previously, Professor Crawley served as the Director of the Frontiers Engineering and Science Directorate of Science Foundation, Ireland, from 2004–2007. Prior to this, Professor Crawley served as the Dean of the College of Science and Mathematics at the University of South Carolina, USA, from 1998–2004. At Michigan State University, USA, he was Dean of the Graduate School from 1994–1998 and earlier Chair of the Department of Physics and Astronomy from 1988–1994. Professor Crawley served two terms at the US National Science Foundation, one as the Director of the Physics Division, 1987–1988, and earlier as a Program Officer in the Nuclear Physics Program. He has also served as the Chair of the Nuclear Physics Division of the American Physical Society in 1991–1992. Dr. Crawley was born in Scotland, and obtained his first degree from the University of Melbourne, Australia. He obtained his PhD in Physics from Princeton University in 1965. He is the author of over 150 articles in refereed journals and wrote the textbook *Energy* published in 1975. He is also the editor of the *World Scientific Handbook of Energy* published in February 2013. Currently, Professor Crawley is the editor of the *World Scientific Series on Current Energy Issues.* The first three volumes in the series were published in 2016. He is also the co-author of a book, *The Grant Writers Handbook,* which was published by Imperial College Press in 2015. He has previously consulted for the National Research Foundation of the UAE (2008–2011) and the National Center for Science and Technical Evaluation, Republic of Kazakhstan from 2008–2012.

Contents

Chapter 1

The Importance of Energy Storage

Anna Stoppato* and Alberto Benato†

Department of Industrial Engineering, Padova University
via Venezia, 35138 Padova, Italy
**anna.stoppato@unipd.it*
†alberto.benato@unipd.it

This chapter describes the role that energy storage can play in the present and in the short–medium term future energy scenario. Both stationary and automotive applications will be considered and the main features required by each of them for an energy storage system will be explained. A very brief description of the proven and most promising storage technologies will be given with the aim of providing an overview of the peculiarities of each one and consequently its better suited applications. Finally, the state-of-the-art, the opportunities and the barriers to the spread of energy storage systems will be summarized.

1 Introduction

The energy scenario has significantly changed in the last decade for a variety of factors. The first one is the increase in the amount of electrical capacity supplied by **variable and non-predictable renewable sources**. In recent years, the growing awareness at both the public and institutional levels of the "energy issue" has led to a series of initiatives aimed at promoting the use of renewable energy sources (RESs).[1] Definitely, their use often results in a lower environmental impact in terms of a reduction of both resources' consumption and emissions, with special regard to carbon dioxide, which is responsible for the greenhouse effect. Worldwide, the policy of incentives for renewable energy plants and for the reduction of greenhouse gas emissions,[2,3] and the development of less expensive and more efficient

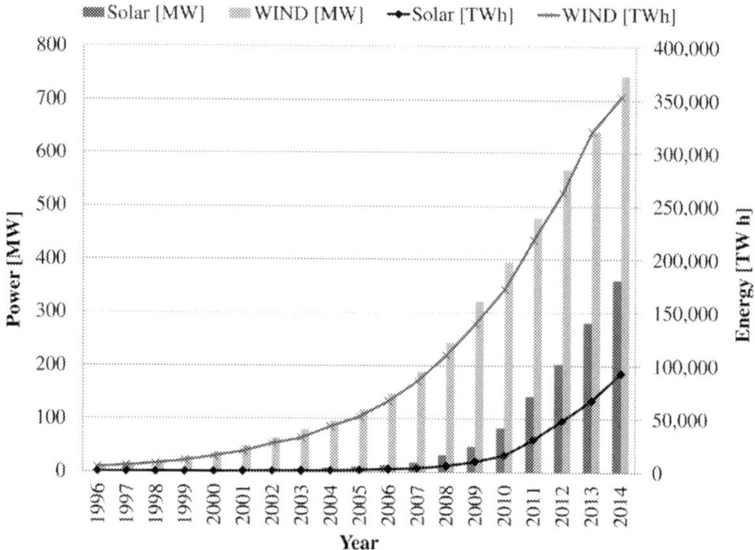

Fig. 1. Trend of capacity supply and energy production by wind and PV power plants.[4]

technologies has brought a significant increase in the number of such systems, especially wind power and photovoltaic plants, as shown in Fig. 1.

The comparison between the installed capacity and energy production allows us to evaluate the load factor of such plants, expressing the ratio between the energy actually produced and that theoretically producible if the plant would run throughout the year at its design power: the value is low and around 22% for wind power and 12% for solar. This is due to the extreme variability of the sun and of the wind over the year and even in a single day, so that the plant can operate at design load only for a limited number of hours per year. In addition, the low predictability of these variations makes the production scheduling more difficult, especially for wind power even in the very short term.

A second aspect is due to the progressive **deregulation** of the energy market carried out in many developed countries,[5,6] which aims to separate the activities of generation, transmission and distribution of electricity and where for every hour of the year the price of energy is determined by the intersection between supply and demand curves. The main goal of deregulation is the promotion of fair competition in the production and sale of electricity in order to reduce energy costs and increase the efficiency of the system. A first consequence has been the significant and rapid growth of the installed generation capacity[4]: in OECD countries in 2000, the power plants had a total installed capacity of 2,080 GW, of which 1,311 GW was from thermal power plants; in 2012, this installed capacity had become 2,777 GW (of which 1677.0 GW was from thermal plants) with a growth in 12 years of 33% (28% for the thermal segment only). But in some countries, this expansion has been even greater (in Italy, +65% and +37%, in the USA, +31% and +30%, respectively).

In many countries, the installed capacity increment was matched by a reduction in energy consumption mainly due to the global economic crisis of 2008 followed by a very slow recovery: for example, in Italy, the electricity demand in 2012 was equal to that in 2004, while in the USA, it was similar to that of 2006.

In almost all the developed countries, in a more or less marked way, the combined effects of the aforementioned factors has created a situation where the installed capacity is overabundant with respect to the users' peak demand, and where quite a high fraction of power is made available by variable and difficult to predict renewable sources. This condition definitely implies a significant saving of fossil fuels and the reduction of emissions, but introduces some critical elements into the market. In addition, the advent of generators powered by renewable sources has drastically changed the structure of the electricity network with the presence of a large number of small power production facilities spread over the country in the vicinity of the users and of the available sources, instead of a classic structure with "a few" large facilities concentrated in the industrial zones of the country.

It is important to remember that the power grid is a very complex system which transmits and distributes electricity generated from the production plants to users through a set of power lines, transformer stations, isolation and protection systems, and is subject to very stringent technical constraints, in particular:

- An instantaneous and continuous balance between the amount of energy released and that required by the network is necessary, taking into account the losses due to transformers, transport and distribution.
- The frequency and voltage must be kept within a very narrow range of values (60 Hz in USA and some other countries in America, 50 Hz in Europe and in many countries of the world), which is essential to protect the safety of the generation and end user facilities.
- It must always be ensured that the energy flow in each power line does not exceed the maximum permissible load on the power line itself.

The change of any one of the abovementioned parameters, even if minor and/or of very short duration, can rapidly induce a state of crisis into the entire local electrical system and subsequently, because of a "domino" effect, to a possible blackout of the entire network. For example, the sudden drop in power available from wind turbines caused by an unexpected reduction of the wind speed can cause stability problems to the network when the share of energy provided by these systems is significant.

But even in the absence of extreme events, during the normal hours when the wind is blowing or the sun is shining, a significant fraction of energy use is met by renewable energy and so thermal power plants are forced to stand idle or to work at partial load with low efficiency. Therefore, when thermal plants are called upon to operate, sometimes suddenly, they impose a high price on the market, which partly offsets the higher costs associated with frequent stops and part load operation. As an example, the quantity and the price of electricity sold on the Italian day-ahead market in two days of May 2013 are reported in Fig. 2,[7] showing two very different trends and average values. This great variability on the supply side has enhanced the role and the value of the markets for ancillary services, and has led to a major diffusion and increased importance of the capacity markets which have to ensure that supply will be available when it is needed.

This provides an additional incentive for owners of generating capacity (i.e. power plants or demand response providers) to make their capacity available to electricity markets where price signals alone would not.

Besides the great complexity of the electricity market itself, the general increase in fossil fuel costs plus the high variability and unpredictability of their trend is a further source of concern in the scheduling of thermal plant operation.

As a final issue, the transport sector is also undergoing many changes. The most important driver is the requirement to reduce pollutants, mainly particulates, in urban areas. This issue demands new transportation solutions: one of the most promising and studied is that of pure electric and hybrid vehicles, combining a traditional fossil fueled engine with an electric propulsion system (see Fig. 3, Ref. 8). These vehicles are powered by the energy stored in an on-board battery, which will be recharged at the so-called "charging-points" usually connected to the grid. With a proper operation strategy, these charging stations can be managed as users with flexible demand and are able to dampen the peaks and the gaps of energy supply.

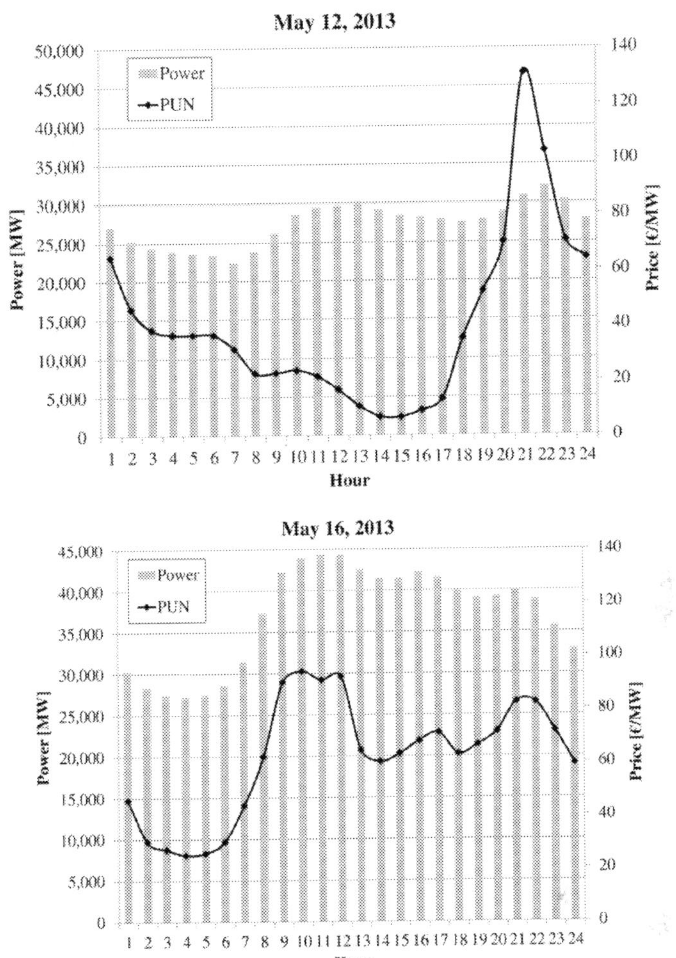

Fig. 2. Trend of electricity sold in the market and of its price on two different days of May 2013: Sunday 12 and Thursday 16.[7]

The combination of all these elements has led therefore to the need to rethink the arrangements for managing both the electricity network as a whole, as well as individual plants. The main target is to supply energy with high efficiency, low cost, high reliability, and low environmental impact.

As a last point, it is important to note that in developing countries or for communities looking for energy self-sufficiency, the exploitation of renewable sources is an opportunity to increase the number of people who have access to electricity with an adequate degree of availability and reliability. In

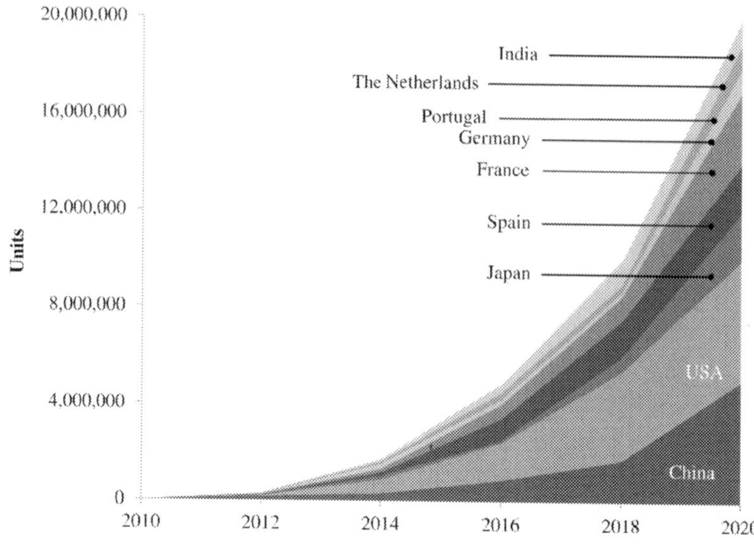

Fig. 3. Electric vehicle (EV) stock target.[8]

such countries, the number of renewable energy and/or hybrid power plants is increasing. In this case, the systems are not grid connected (stand-alone units) and it is even more important to arrange power units able to meet the demand in spite of the intermittency of variable energy sources.

The new situation needs to be addressed with adequate instruments. Possible areas of research and application are manifold and involve very different aspects:

- Thermal plant flexibility, partial load efficiency and start-up and shut-down speed.
- Emission control and removal, including carbon capture and storage technology.
- New technologies for the exploitation of low quality energy sources with high efficiency.
- Control systems and smart grids.
- Policies that are able to promote the use of renewable sources and energy efficiency without distorting the market.
- Energy storage, which permits the separation of energy production from its consumption.

In this volume, the attention will be on energy storage technologies.

2 Energy Storage: The Main Characteristics

As mentioned previously, energy storage is a challenge for the electricity system where the optimum efficiency can still be improved. Briefly, the aim is to separate the supply and demand for energy. However, each storage technology has specific features, which make it better suited for some applications than for others.

For this reason, before explaining the role of energy storage in actual and future markets, the most important characteristics of energy storage units are outlined below[9–14]:

(a) *Storage capacity* (C): This represents the maximum amount of energy that can be stored. For some technologies, due to the presence of a maximum allowable depth of discharge, it differs from the usable energy.
(b) *Charging and discharging rating power* (P): This is the nominal charge–discharge power, usually that of maximum efficiency. For some systems, the charge and discharge powers have the same value, while in other cases, they differ (for example if different devices are used for the two phases and/or for specific users' needs). Usually, the actual power can differ from the rating value and varies from a maximum to a minimum value.
(c) *Specific energy and specific power* (E_s *and* P_s): They quantify the density of energy or power, and are defined as E/V and P/V, where V is the volume of the storage. The higher these values, the lower the volume at constant energy or power. Sometimes, these values are supplied per mass unit.
(d) *Round trip efficiency* (η): This is the ratio of the discharged to the charged electricity. Its value is related to the losses both during the charge–discharge cycle itself and the self-discharge during the storage period. It usually differs from the cycle efficiency, considering only charge and discharge phases.
(e) *Rated discharge time* (T): This represents the duration of the discharge time at the rated discharge power starting from full storage, and can be defined as C/P. Obviously, the actual value of the rated discharge time depends on the actual discharge power and on the energy stored.
(f) *Response time* (t_r): This is the time between the request to change the operation and the system response.
(g) *Inversion time* (t_i): This is the time needed to pass from the charge to the discharge phase or *vice versa*. t_r and t_i represent the ability of a

system to vary the power and to quickly respond to the grid operation regulation signal.

(h) *Expected lifetime* (L): This value can be defined in terms of lifetime or as the number of charge–discharge cycles. For some technologies, this last value is strongly related to the charge–discharge history.

(i) *Reliability*: This gives an idea about the robustness of the storage units operation.

(j) *Environmental impact*: This point is very important and the public acceptability is closely related to it. The main impact can be during the building phase (for example, due to the use of materials in short supply or to the handling/transport of large amounts of material), during the operation (due to pollutant or electromagnetic emissions, or to the visual impact) or during the decommissioning (if potentially dangerous substances must be disposed of). The location of the storage system and its size influence this aspect.

(k) *Levelized unit electricity cost* ($LUEC$): This is the price at which the electricity should be sold in order to cover all the costs related to the building, operation and decommissioning costs and to assure a return on the investment. This value depends on many factors, such as the size, the location and the charge–discharge history of the storage unit, and also it is related to the market and the incentives policy. In any case, this value is fundamental in order to evaluate the feasibility of a storage solution.

(l) *Operation principle*: This aspect will be explored more fully in the following chapters and briefly in Sec. 4, but in synthesis, it is possible to distinguish among:

  (i) mechanical storage (pumped hydro, compressed air energy storage (CAES), flywheels);
  (ii) chemical storage (hydrogen, power-to-gas);
  (iii) electrochemical storage (batteries, supercapacitors);
  (iv) thermal storage;
  (v) electrical storage (magnetic superconductors).

As a last remark, for some technologies, the performance indices (C, P, η, T) are almost constant during the lifetime of the storage system, while for others, they decrease due to different deterioration phenomena.

Other aspects can be important for the selection of a storage technology, as, for example, its commercial maturity, the constraints required by the installation site, or the operational constraints (pressure, temperature), plus the safety problems.

3 Energy Storage: Its Role in the Energy Scenario

The traditional role of storage systems was to store energy when the demand, and consequently its cost, were low, typically at night, and to make the energy available during the hours of peak demand. In this way, the inflexible thermoelectric plants could work all day at almost constant power and an energy reserve for peak hours was assured. In these systems, the main requirement for energy storage was the ability to exchange power with a rated discharge time of several hours.

In the new scenario, the energy storage system is also required to provide, in a very short time, power to overcome the intermittence of renewable sources and contribute to the regulation of the mains voltage: in this case, the amount of energy that can be stored is less important, but the response speed is critical.

Therefore, depending on the application, different features are required.

The most evident application difference is between stationary and automotive storage.

3.1 *Stationary Applications*

For stationary applications, storage units can either be connected to the grid or work in isolated areas for stand-alone energy systems.[13,15–17]

In the first case, they can operate as independent units to serve the grid or may be connected to a RES plant or to an end-user to provide the needed support.

Generally, the so-called "energy performance" storage systems are able to provide power for many operation hours and having a low value of P/C, are suitable for **energy management applications** which include moving power over long timescales, and generally require continuous discharge ratings of several hours or more. Typically, one or a few cycles/day are required.

The traditional service of this kind to the grid is the **time-shift** or energy arbitrage. Storage systems are used to decouple synchronization between power generation and consumption. A typical application is load leveling, which implies storing up energy during off-peak hours (low energy cost) and using the stored energy during peak hours (high energy cost). This is convenient if the ratio between peak and base load prices is lower than the round trip efficiency of the system. In this case, the storage system is required to have high rated discharge times (from some hours to days) and capacity (from some MWh to about 10,000 MWh).

To the time shift service, the **peak shaving** service is often added, which helps decrease the number of shut-down/start-ups of traditional plants, the operation hours of more expensive/less efficient power plants, and the high line loss rates that occur during peak demand. Another service for the grid is the **not-programmable sources integration**, which helps boost the penetration of RES, decreases the energy losses from RES plants, compensates for the power fluctuations and provides a more regular and predictable power profile.

Deferral of grid investments and congestion relief is another benefit that energy storage can guarantee. Distribution systems must be sized for peak demand; as demand grows, new systems (both lines and substations) must be installed, often only to meet the peak demand for a few hours per year. New distribution lines may be difficult or expensive to build, and can be avoided or deferred by deploying distributed storage located near the load. Response in minutes to hours is required plus a rated discharge time of some hours.

The so-called power performances storage technologies, able to supply high power for relatively short periods, are suitable for providing **grid ancillary services**. One of the key challenges in grid management is maintaining reliability.

The "fast response" (in seconds to minutes from null to rated power) nature of these energy storage technologies makes them ideally suited to meet grid stability and reliability challenges. Among these services is the contribution to **primary frequency control**, which also requires a low inversion time to guarantee a high regulation band, and a good round trip efficiency. Some tens of cycles/day are usually required. Similar features are required of energy storage systems to contribute to **secondary and tertiary frequency control. Black start** is the restart of electricity supply after a major power system disturbance, and requires capacity and energy after a system failure restart and must provide a reference frequency for synchronization. It requires several minutes to over an hour of response time, plus several hours of discharge time. Usually, cycling is very low (about 1 cycle/year). For **contingency reserve**, stored energy is used for seconds to minutes to ensure service continuity when switching from one source of electricity to another. Discharge times in the range of up to about an hour are usually required. Far less cycling is required than for power quality applications.

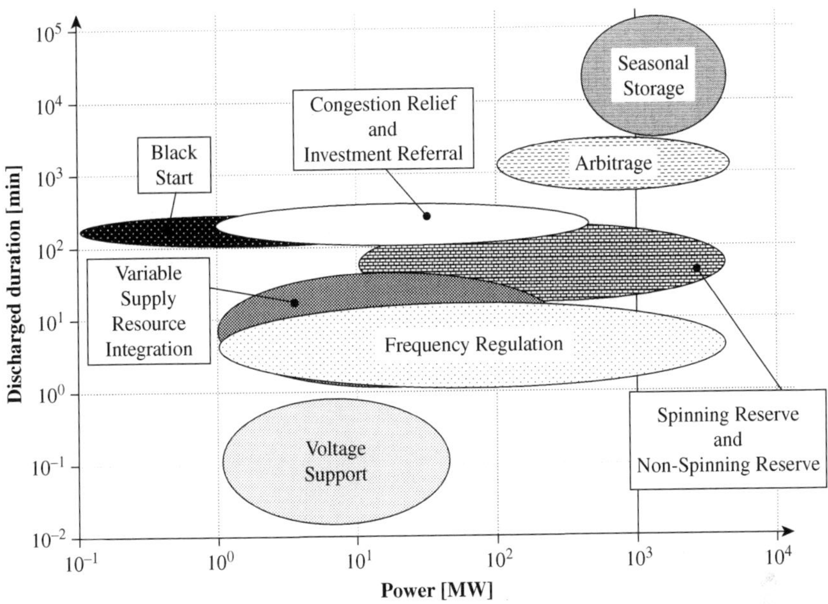

Fig. 4. Applications of energy storage systems for the grid support.

Another benefit that energy storage can provide to the grid is the **power quality control**: stored energy is only used for a few seconds or less to ensure the quality of power delivered. Power quality applications require rapid response — often within less than a second — and include transient stability and frequency and voltage regulation. As with the other applications, the timescales of discharge may vary; but this kind of services typically requires discharge times of up to about 10 min and nearly continuous cycling (hundreds of cycles/day).

Figure 4 summarizes the features required for energy storage for different applications to support the grid: energy performance applications are in the upper part of the diagram, while the power performance applications are towards the right side.

With regard to the services that a storage unit can directly supply to an RES plant, we can include the increase of its flexibility, the minimization of curtailment (forced stop or load reduction when electricity cannot be supplied to the grid due to an oversupply of generation) and displacements, the capacity firming, and the limitation of upstream perturbations. The features required for the storage system depend on the size of the RES plant, but usually a continuous discharge of several hours and a response of minutes are required.

The services to the end-users include the increase of reliability, the warranty of the continuity of service, the limitation of upstream disturbances, the end-user peak shaving, the compensation of reactive power and, if the user owns also a power generator, the maximization of the self-consumption of the produced energy (avoiding electricity exchange with the grid). These services become paramount for application in isolated areas relying on an intermittent renewable source. In this case, the key elements are high reliability, low response time, autonomy, the lowest possible self-discharge, and long life expectancy.

In conclusion, the presence of energy storage units provides benefits for:

- The traditional production facilities, which can work at nearly constant load, plan their production and limit the number of on/off switches.
- The renewable source plants, whose production can be entirely and profitably used.
- The grid, whose stability and reliability are enhanced.
- The users, who are ensured a safer and more reliable electric service.

3.2 *Automotive Applications*

The main constraint for energy storage units for EVs and plug-in hybrid electric vehicles (PHEVs) is the necessity to remain on board the vehicle.[8] Therefore, high volume and mass energy densities are paramount. Batteries are the most suitable technology for vehicles. For EVs they need to be designed to optimize their energy storage capacity, while for PHEVs they typically need to have higher power densities. Other important requirements are the rated discharge time, the fast charging, a high life expectancy, plus a low temperature sensitivity.

On the other hand, as more fully explained in Chapter 4, the charging of EVs can potentially be controlled, and provides a source of planned demand and demand response. Controlled charging can be timed to periods of greatest RES energy production, while charging rates can be controlled to provide contingency reserves or frequency regulation reserves. EVs could potentially provide the grid services discussed previously.

4 Energy Storage: The Main Technologies

In this section, only a very brief summary of the main characteristics of each technology will be given. The following chapters of the volume provide a more complete description of many of these storage technologies.[9,14,15]

4.1 *Pumped-Hydro Energy Storage (PHES)*

In brief, water is pumped into an upper reservoir and stored there; when energy is required by the grid or the price of electricity is high, water is released through one (or more) turbines to a lower reservoir and the electricity produced is sold. PHES is a proven technology, suitable for large-scale storage. It is very efficient and flexible in power, has a short response time, can ramp up to full production capacity within minutes providing a quick response for peak-load energy supply and is already used for both primary and secondary regulations. It is suitable for energy management applications.

On the other hand, it needs to be located in suitable geological sites, containing a geodetic head and natural upper and lower basins or at least the possibility of building artificial reservoirs. This requires relatively high initial costs. The environmental impact can also be non-negligible, in terms of land occupation and modification, disturbance of the aquatic life, and modification of the natural water flow.

4.2 *CAES*

In these storage systems, air is compressed during charging and then stored in an underground cavern or other pressure vessel. When electricity is required, the air is heated to avoid freezing and then expands in a turbine. If the heat generated during compression is stored and then used to preheat the air in order to increase the round trip efficiency, the process is called adiabatic. If external heat input by means of combustion is used to preheat the air, the process is called diabatic.

Despite the large interest for this technology, at present there are only two plants in operation around the world. CAES is basically suitable for medium and large energy storage for energy applications. Recently, many studies are related to its use in small systems. The use in direct connection to wind farms or other non-programmable RES plants, or for distribution grid support seems to be very promising. At present, the main drawback is the cost.

4.3 *Flywheels (FWs)*

FWs store energy in the form of kinetic energy. The storage unit is composed of a FW driven by an electric motor able to work either as a generator or as a motor and located inside a housing. If the motor provides a positive torque, the FW increases its rotation speed and energy is stored. When energy is needed from the FW, the electrical machine applies a negative torque and the stored energy is released.

This technology is already mature and is suitable for high power applications. In addition, FWs present a long cycling expected life and short response and inversion times. On the other side, mainly due to friction losses, the round trip efficiency strongly and speedily decreases during the operation. For this reason FWs are suitable only for short-term storage and, at present, are mainly used as voltage and frequency control, as support for wind farms or in transportation to increase the efficiency of trains, ferries or large EVs.

4.4 *Fuel Cell Hydrogen* (*FC-HES*)

Chemical energy storage is the transformation of electrical energy into chemical energy carriers (the so-called power-to-gas or P2G). At present, the most promising energy vector is hydrogen. Hydrogen is produced by means of water electrolysis which is a process consuming electricity. Then, hydrogen is stored as a liquid at cryogenic temperature, or as a gas at very high pressure or as solid in hydrides. Finally, the stored hydrogen may be used to produce electricity. The most common solution is by means of fuel cells. There are many different kinds of fuel cells, which mainly differ in the electrolyte used and the operating temperature. Note that the use of hydrogen in fuel cells produces only water and does not emit any pollutant or greenhouse gas.

This technology has a high energy density and the possibility of storing very large quantities of hydrogen for a long time. Hydrogen can also be transmitted from one location to another. These features make chemical storage suitable for energy management applications, even for seasonal storage. At the same time, electrolysis has a short response time. The main drawbacks are the excessive costs, the low round trip efficiency and the short lifetime expectancy.

4.5 *Electrochemical Batteries* (*EBES*)

Batteries, or accumulators, are based on a single device with the functions of energy storage and discharge of electricity. The basic element is an electrochemical cell having voltages from below 1 to 4 V; many cells can be put in series in order to reach higher voltages. Electricity is produced by an oxidation–reduction reaction where a flow of electrons is created from a chemical species (anode) to another one (cathode) in contact by means of an electrolyte. The reverse process can recharge the battery. Many different batteries are available on the market, and others are under study: they differ for the materials used for the anode, the cathode and the electrolyte, and in the design.

Batteries have a high technological maturity, high energy density, good round trip efficiency, and great modularity that permits them to be tailored to users' requirements. Their main inconvenience is their relatively low life for large-amplitude cycling. During their operation they do not emit pollutants or noise. However, their disposal can present a significant environmental impact due to the materials used for the electrodes and/or the electrolyte.

4.6 *Supercapacitors (ECES)*

Supercapacitors or electrochemical capacitors (ECs) or also electric double layer capacitors (EDLCs), store electrical energy in an electric field between two electrodes separated by a dielectric and immersed in a liquid electrolyte. The electrodes are characterized by a very large useful surface and the distance between the electrodes is very small (normally only a few Angstroms). The process is easily reversible.

ECs are suitable for high power applications since they have a very fast response time, high round trip efficiency, high power density, but low energy density, long expected lifetime and can guarantee a very high number of charge–discharge cycles. They are an interesting solution also for electric transportation both for brake energy recovery and for propulsion over very short stretches of roads without electric connection. Supercapacitors have not yet reached commercial maturity, but they are expected to improve their performances in the near future.

4.7 *Magnetic Superconductors (SMES)*

In magnetic superconductors, energy is stored in the magnetic field of one or more superconducting coils characterized by very low losses. To reach this condition, they must work at very low temperatures (near absolute zero).

At present, they have no commercial market, but are still in the research phase and are considered a promising technology. The main problem is the necessity of a cryogenic temperature with the related prohibitive cost and high energy requirement. This brings a low energy density and low round trip efficiency. However, they have very interesting characteristics, such as very fast delivery of high power at high cycle efficiency. For this reason, they are suitable for power applications, requiring continuous operation with many charge–discharge cycles.

4.8 *Thermal Storage (TES)*

Thermal energy storage (TES) includes many technologies where energy is stored in the form of heat. Heat can be stored as:

- *Sensible heat*: if storage is achieved by increasing or decreasing the temperature of a storage material. In this case, the amount of stored energy is proportional to the temperature difference.
- *Latent heat*: if storage is connected to a phase transition of the storage material, usually from solid to liquid and *vice versa*. In this case, the temperature remains constant and the stored energy depends on the latent heat of fusion of the material.
- *Thermochemical heat*: if heat is stored as chemical compounds created by an endothermic reaction and it is recovered again by recombining the compounds in an exothermic reaction. The stored energy is equivalent to the heat of the reaction.

These technologies are used for many heating and cooling applications. For example, where heat supply and demand are often not simultaneous, as for solar heating systems, or to utilize peak shaving of heating demand. Energy storage in these applications can be very long-term storage, even seasonal.

Nowadays, TES is also used for electric applications as support of renewable energy plants. For concentrated solar power plants, energy from the sun is stored by means of molten salt and then released, when needed, to steam which operates a Rankine steam power plant. Further support to the grid can be given by storing heat from electric boilers working as a fast balancing service, or helping limit the demand for electrical power from electric boilers where the need for heat is not continuous and varies in intensity.

As a last remark, recently many researchers are studying the possibility of combining different energy storage technologies in the same system in order to exploit the synergy among their different features: for example, the use of batteries together with FWs can increase the life of the batteries.

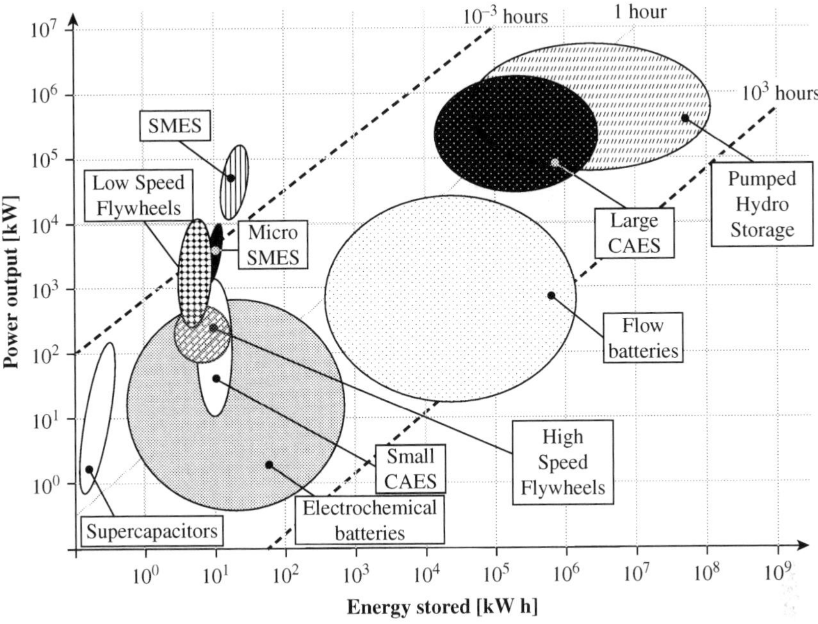

Fig. 5. Comparison among the main energy storage technologies in terms of capacity and rating power. The dotted lines represent discharge times.

In Fig. 5, the main storage technologies are summarized and it is possible to compare them taking into account the rating power and the discharging time. From this graph, it is possible to identify the technologies for power applications and the ones for energy applications. The power technologies are those with low discharging time and low rating power that fit well with the power quality applications. Some examples are batteries and FWs. The energy technologies are those with high discharging time and high rating power. These technologies fit well with applications like time shift, peak shaving and capacity reserve. Some examples are Pumped-Hydro Storage and CAES technologies.

Figure 6 shows the different maturity levels of the main energy storage technologies.[17] The present level (left end of each arrow representing a technology) and the expected level in 2030 (right end of the arrow) are reported. Some of them are still in a first research phase, others are already proven technologies. Nevertheless, improvements are expected for all of them. Note that for some of them, the expected development is really important.

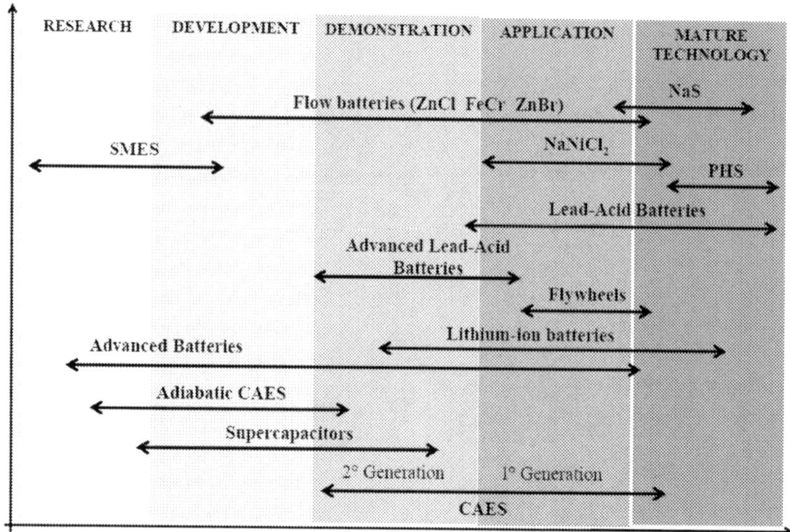

Fig. 6. State-of-the-art and forecast development of the most promising energy storage technologies.[17]

5 State-of-the-Art and Projects

At present (2015), the worldwide capacity of operating storage systems is estimated at nearly 145 GW and is equivalent to about 2% of the overall electric installed capacity[15,18] (in Europe, it is about 5%). Many of these systems were built between the late 1970s and 1980s when the increasing price of oil and natural gas drove the construction and operation of many coal and nuclear power plants, which need to work at a steady base load and are not very flexible.

Starting from the late 1990s, with the development and the large spread of Combined Cycle Gas Turbines, which are very efficient and flexible, the necessity for energy storage decreased until the last 10 years, as explained in the Introduction.

As summarized in Fig. 7, more than 99% of the present capacity is supplied by PHES systems.[19] For the remaining 1%, the major contribution is from sodium–sulfur batteries and CAES.

Batteries are mainly used for stand-alone applications: only two big grid-connected systems are installed, in Japan and Abu Dhabi, respectively. Also only one CAES plant in the USA and another one in Germany are operating.

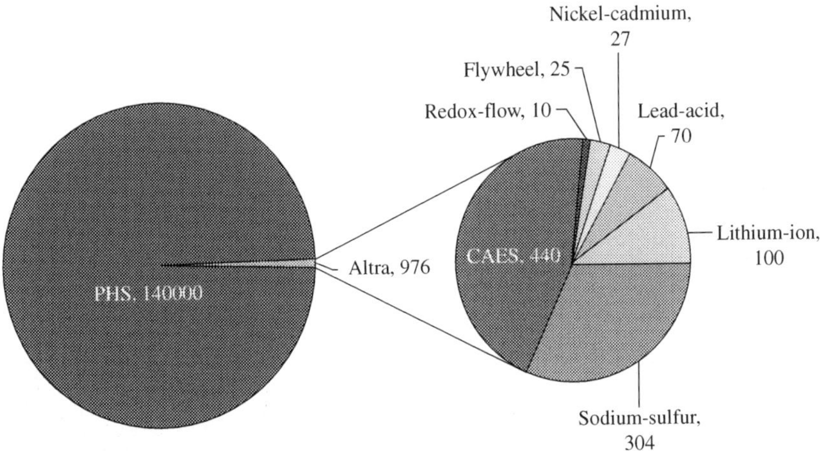

Fig. 7. Worldwide installed capacity of electric energy storage systems.[19]

The International Energy Agency (IEA) has estimated that additional 310 GW grid-connected electricity capacity would be needed in the United States, Europe, China and India to support electricity sector decarbonization.[20] The estimated worldwide installed capacity in 2050 is about 400 GW: the main contribution is expected from countries where there are (or are planned) many solar and wind power plants, or where geodetic heads available for PHES are present. Regarding this point, mountainous places are the most suitable, but there is an increasing interest, supported by a technological development for seawater pumped-energy storage systems, which exploit the geodetic head connected to high coasts and use the sea as the lower reservoir.

Even if it is possible to find many studies about energy storage all over the world, at present only a few official technology roadmaps are in force. In any case, most of the international and national administrations have put energy storage as a key objective of their respective work programs about energy.[12,15,19–24]

Japan, European Union and IEA elaborated the strategy about energy storage together with other energy technology roadmaps. In the United States, a federal map was not published, but some States have done so. California is the main example. In 2050, it is estimated that there will be an installed storage capacity of about 150 GW in the US, primarily achieved by means of the addition of new CAES plants.

At present, in China there is not any mention of energy storage in the Low Carbon Technology Development Roadmap, but in this country, there are nearly 50 storage demonstration projects in operation or in the planning phase. Among them, major applications include the support of wind power (53% of projects), distributed micro-grid projects (20%), and transmission and distribution grid support projects (7%).

Asian nations (Japan, Korea, China) dominate the market for new battery technologies because of their use in numerous consumer products (e.g. mobile phone, portable PC). Furthermore, the weakness of the Japanese grid contributes a lot to the development and commercialization of this storage technology. For the same reason, note that the only large sea-water pumped-hydro plant presently in operation is in Japan.

In developing countries, the main contribution to energy storage in the short–medium term is expected to be small-scale stand-alone systems: many projects of renewable energy or hybrid plants integrated with the use of batteries, hydrogen or small PHES have been presented. Often, these systems require a pumping station for access to water.

From the studies of many different agencies, it seems that in the short–medium term, batteries will be the winner for small-scale storage plants, while for large scale, probably CAES and Power to gas should be an attractive alternative to PHES, but they still need some improvements. It also calls attention to the importance that there is diversity in energy storage in order to exploit the suitable characteristics of each technology.

6 Barriers to Diffusion

As mentioned earlier, even if energy storage has a long history and much research and many improvements have been recently achieved, further efforts are needed in order to fulfill the new energy market needs. The main challenges for storage concern the technologies, the market and regulatory issues and the strategies.[15,25]

With regard to the **technologic** aspects, improvements are required to increase the capacity, the efficiency, the autonomy, the lifetime and the reliability of the existing technologies. For proven technologies, the improvements should mainly consist in the upgrading of the existing devices, while for new and developing ones, they will also involve new storage concepts or important modifications of the present design, such as the use of different materials (EBES, ECES, P2G) or working pressures and temperatures (P2G) or innovative cycles (CAES, FC HES).

In any case, the key point for each improvement is the reduction of the LUEC. Since the LUEC is also a function of the operating history of the storage unit, it is essential that the development paths are tailored to the specific applications.

Improvements are also required to the control and regulation systems of the grids where storage units are connected. Note that when the share of non-programmable RES electricity exceeds 20–25% energy storage could enable bi-directional energy flows in the grid, and this eventuality must be predicted.

For batteries used in the EVs, the major challenge is the increase of energy density and the reduction of charging times. This last point also requires development of the devices used in the recharging stations.

Great attention must also be paid to the environmental impact of the storage units: for many technologies, such as EBES, FC-HES and ECES, the major targets are the sustainable use of resources, the prevention of dangerous wastes and the possibility of recycling, for PHS the use of existing reservoirs and the minimization of the interference with the natural streams of water and of land use, for FWs, the main issue is noise control. In the analysis of the environmental impact of a storage unit, a life cycle approach must be used and the impact for each useful energy unit supplied must be evaluated.

Finally, improvements to the safety requirements for people living near the energy storage systems, and also the surrounding devices, have to be carried out: dangerous materials which can lead to explosions, toxic emissions, and corrosion are used in some technologies (EBES, SMES, ECES), failure of the rotating wheel is possible for FW, high pressure (CAES) or temperature (FC-HES) are required by some cycles, electrical hazards must be considered for all the technologies; for EVs also recharging security is a concern to be studied.

Note that the perceived environmental impact and safety of storage technologies is an important social barrier to the spread of energy storage, so this point needs particular attention.

A second important challenge for the diffusion of energy storage is the creation of a **market** able to incentivize the connection to the grid of storage capacity and the supplying of storage services.[22]

Since the value of energy arbitrage alone is not sufficient to justify the building of new plants, the market must recognize and pay for the help that an additional offered capacity gives to the grid in terms of flexibility, service safety and reliability (market for the ancillary services). It should be paramount that the price paid by the final users is not increased. So, a preliminary detailed cost–benefits analysis is very important.

To attract investment in fast response energy storage technologies, the market must be willing to pay for the value of the speed and accuracy that energy storage provides to the grid, reducing the overall need for, and cost of, regulation services. So the payment can be composed of two different terms: the first one based on the speed and amount of energy transferred by the resource in response to a control signal, and the second one based on the capacity that a unit makes available to provide regulation.

Another interesting approach can be the employment of a regulation dispatch algorithm that selects fast response resources before slow response resources in order to minimize the total amount of regulation capacity required in the balancing area.

The difficulty in defining the rules for the proper support of energy storage systems is increased by the complexity due to the different functions that the same energy storage unit can assume in-service to the grid and to the different features that its function requires. When a unit is in the direct service of an RES plant or of final users, the possible presence of different owners and stakeholders makes the regulation more complicated.

Another important issue is that the market rules must be clear and well defined in the short–medium term since the costs for a storage unit are often very high and investors need to plan their investments with a certain degree of confidence.

This is paramount for Europe where a common balancing market must be built up.

In the United Stated, there is already a quite favorable environment for energy storage due to the well-developed ancillary services market: energy storage is allowed to participate and provide services that account for both its qualities (e.g. fast response) and shortcomings (e.g. energy-limited supply).[26–28]

Finally, it is important to note that the presence of energy storage also involves other markets such as the gas market (P2G), local districting heating markets (TES), and the transportation market (automotive sector). A common development strategy must be implemented.

It is clear that some **strategic** issues also have to be faced. The development of energy storage systems is linked to the progress of the whole energy system. Therefore, it is important that a systemic approach is employed, where technical, regulatory, market and political aspects are combined. Some points are particularly related to the spread of energy storage systems:

- The regulatory framework for the reduction of CO_2 emissions, which can strongly encourage the growth of RES.
- The penetration of EVs, which is connected to the evolution of the automotive industry and to the motivation for change in the behavior of users.
- The development of smart grids, which support the diffusion of small-scale energy storage and also of EVs as storage units.
- The upgrade of the Transmission and Distribution grids with the construction of new cables, connecting areas where large amounts of RES are available to areas where electricity is needed. This includes both new long cables in large geographic areas and an increase of the number of interconnections between different smart energy grids.

It is clear that strong public engagement is needed, which depends on investment priorities. Public investments are required to develop new projects and to help the construction of large test case facilities to validate the effective features of the storage technologies.

It is important, at this development stage, that technology diversity is encouraged and promoted. A mix of all solutions is needed, tailored for each region and system architecture.

In educating the public, it can be helpful to emphasize the value of energy storage (and EVs' use) in reducing CO_2 emissions, fossil resource consumption, pollution, and noise. Energy storage can also promote the development of new industries with new job creation.

With regards to safety concerns, it is urgent that security standards and operation protocols are defined for all the technologies to simplify system procurement, installation and operation.

7 Conclusions

This chapter has introduced the topic of this volume by emphasizing the importance of Energy Storage. One main conclusion is the crucial role that energy storage plays in the present energy scenario in increasing the penetration of RESs without decreasing the reliability and availability of the system. Another important point is the necessity of further improvements in the technologies and in the market rules in order to make storage systems more competitive. Finally, it is important to emphasize that there is no single storage technology that is better than the others for every application: for this reason, all the options must be developed and promoted considering their specific features.

In the following chapters, the main technologies will be described in more detail.

Glossary

C	storage capacity
E_s	specific energy
L	expected lifetime
P	charging/discharging rating power
P_s	specific power
T	rated discharge time
V	volume
t_i	inversion time
t_r	response time
η	round trip efficiency

Acronyms

CAES	compressed air energy storage
EBES	electrochemical battery energy storage
ECES	supercapacitors energy storage
EDCL	electric double layer capacitor
EV	electric vehicle
FC-HES	fuel cell hydrogen energy storage
FW	flywheel
LUEC	levelized unit electricity cost
P2G	power-to-gas
PHES	pumped-hydro energy storage
PHEV	plug-in hybrid electric vehicle
RES	renewable energy source
SMES	superconductors magnetic energy system
TES	thermal energy storage

References

1. EC (European Commission), Directive 2009/28/EC of the European Parliament and of the Council of 23 April 2009 on the promotion of the use of energy from renewable sources and amending and subsequently repealing Directives 2001/77/EC and 2003/30/EC.

2. EC (European Commission), Directive 2009/29/EC of the European Parliament and of the Council of 23 April 2009 amending Directive 2003/87/EC so as to improve and extend the greenhouse gas emission allowance trading scheme of the Community.
3. United Nations, Kyoto Protocol to the United Nations Framework Convention on Climate Change (1998).
4. BP, *Statistical Review of World Energy* (2015). Available at: www.bp.com/statisticalreview.
5. EC (European Commission), Directive 2009/72/EC of the European Parliament and of the Council of 13th July 2009 Concerning Common Rules for the Internal Market in Electricity and Repealing Directive 2003/54/EC.
6. FERC (Federal Energy Regulatory Commission), *Wholesale Competition in Regions with Organized Electric Markets*, Order no. 719 (2008).
7. Available at: www.mercatoelettrico.org/en.
8. IEA, Global EV Outlook (2013).
9. J. K. Kaldellis and D. Zafirakis, Optimum energy storage techniques for the improvement of renewable energy sources-based electricity generation economic efficiency, *Energy* **32** (2007), pp. 2295–2305.
10. X. Luo, J. Wang, M. Dooner and J. Clarke, Overview of current development in electrical energy storage technologies and the application potential in power system operation, *Appl. Energ.* **137** (2015), pp. 511–536.
11. S. Koohi-Kamali, V. V. Tyagi, N. A. Rahim, N. L. Panwar and H. Mokhlis, Emergence of energy storage technologies as the solution for reliable operation of smart power systems. A review, *Renew. Sust. Energ. Rev.* **25** (2013), pp. 135–165.
12. F. Díaz-González, A. Sumper, O. Gomis-Bellmunt and R. Villafáfila-Robles, A review of energy storage technologies for wind power applications, *Renew. Sust. Energ. Rev.* **16** (2012), pp. 2154–2170.
13. P. Denholm, E. Ela, B. Kirby and M. Milligan, The role of energy storage with renewable electricity generation, NREL/TP-6A2-47187 (2010).
14. H. Ibrahim, A. Ilinca and J. Perron, Energy storage systems — characteristics and comparisons, *Renew. Sust. Energ. Rev.* **12** (2008), pp. 1221–1250.
15. EASE/EERA, European energy storage technology development roadmap towards 2030, *Technical Annex* (2013). Available at: http://www.ease-storage.eu/Technical_Documents.html.
16. W. F. Pickard, Q. A. Shen and N. J. Hansing, Parking the power: Strategies and physical limitations for bulk energy storage in supply-demand matching on a grid whose input power is provided by intermittent sources, *Renew. Sust. Energ. Rev.* **13** (2009), pp. 1934–1945.
17. EPRI, Functional Requirements for Electric Energy Storage, Application on the Power System Grid (EPRI, Palo Alto, CA, 2011).
18. IEA, *Energy and Climate Change*, World Energy Outlook Special Report (2015).
19. G. Damato, Energy Storage Value Propositions (STRATGEN Consulting, 2011).
20. IEA, Technology Roadmap: Energy Storage (OECD/IEA, Paris, 2014).

21. European Commission Directorate-General for Energy, Working Paper: The Future Role and Challenges of Energy Storage.
22. EPRI, Electric Energy Storage Technology Options: A White Paper Primer on Applications, Costs and Benefits (EPRI, Palo Alto, CA, 2010).
23. EC (European Commission), Communication from the Commission to the European Parliament, The Council, The European Economic and Social Committee and the Committee of the Regions, Energy Roadmap 2050 (2011).
24. Imperial College London, Strategic Assessment of the Role and Value of Energy Storage Systems in the UK Low Carbon Energy Future (2012).
25. Available at: http://ec.europa.eu/energy/infrastructure/doc/energy-storage/2013/energy_storage.pdf.
26. FERC (Federal Energy Regulatory Commission), Transmission Planning and Cost Allocation by Transmission Owning and Operating Public Utilities, Order no. 1000 (2012).
27. FERC (Federal Energy Regulatory Commission), Frequency Regulation Compensation in the Organized Wholesale Power Markets, Order no. 755 (2011).
28. FERC (Federal Energy Regulatory Commission), *Third-Party Provision of Ancillary Services*, Accounting and Financial Reporting for New Electric Storage Technologies, Order no. 784 (2013).

Chapter 2

Pumped-Storage Hydropower Plants: The New Generation

Giovanna Cavazzini[*,‡], Juan I. Pérez-Díaz[†,§], Francisco Blázquez[†], Carlos Platero[†], Jesús Fraile-Ardanuy[†], José A. Sánchez[†] and Manuel Chazarra[†]

[*]*University of Padova, Italy*
[†]*Technical University of Madrid, Spain*
[‡]*giovanna.cavazzini@unipd.it*
[§]*ji.perez@upm.es*

This chapter describes some current trends and future challenges related to pumped-hydro energy storage (PHES) with special emphasis on the mechanical aspects of hydraulic machinery, power electronics devices used for variable speed operation, and utilities' operation strategies. After a brief introduction and historical background, the new generation of PHES is presented with particular focus on those equipped with variable speed technology. Typical configurations of pumped-storage hydropower plants (PSHPs) are also briefly described. The next section focuses on reversible pump-turbines, discussing their operating limits and presenting the state-of-the-art of the research on their unstable behavior. The operating principle and some basic aspects of the electrical machines most widely used in PSHPs are next described. Power electronics devices typically used in PSHPs for variable speed operation are described in detail, along with some recent developments on variable speed drives. Finally, utilities' operation strategies are reviewed in detail, and some future challenges to make the best possible use of PHES assets according to their new role are identified.

1 Introduction

Hydropower is undoubtedly one of the most mature energy technologies producing about 3,500 TWh of electrical energy in 2010 (16.3% of the world's electricity), greater than that of all the other renewable energies

(RENs) combined (3.6%), but much smaller than that of fossil fuel plants (67.2%).[1]

Besides the positive effects on climate mitigation, hydropower also provides other significant advantages: it promotes price stability because, unlike petroleum fuel and natural gas, it is not subject to market fluctuations; it reduces environment vulnerability to floods; it contributes fresh water storage for drinking and irrigation; it makes a significant contribution to development by bringing electricity, roads, industry, and commerce to communities. This will benefit future generations as hydropower projects are long-term investments with an average life span of 50–100 years.

Despite numerous examples of excellent, sustainable, and safe exploitation of water resources by hydropower, large hydro projects, which include a dam and a reservoir, have encountered substantial opposition in the latter part of the last century because of their environmental and social effects (landscape, wildlife, biodiversity, population settlement, health and water quality, etc.). This opposition was one of the main factors in the slowdown of hydroelectricity generation between the late 1990s and the early 2000s.[1,2]

However, eliminating large hydropower projects from REN programs will not reduce the demand for electric power, which will be partially satisfied by thermal fossil fuel plants, thereby increasing global greenhouse gas emissions. For instance, it was demonstrated that two pumped-hydro energy storage (PHES) units combined with a thermal generation unit make it possible to reduce the excess emissions of the thermal unit by 60%.[3] Therefore, a renewed interest in large pumped-storage hydropower plants (PSHPs) is emerging globally due both to further increases in the share of RENs to electricity production and to the ability to support an increased exploitation of other renewable but intermittent energy sources such as wind and solar power, because of storage capacity.

At present, the electricity grid is highly centralized with a complex system of energy production and transmission characterized typically by long distances between power plants and end-users and by a limited use of storage. The installed generating capacity was only about 127.9 GW in 2010 (2.5% of the world installed capacity).[4] To ensure the security of the power system, a continuous balance between demand and supply should be guaranteed and this actually limits penetration in the grid of intermittent RENs, whose energy production is fluctuating, unpredictable, and delocalized. The development of a significant energy storage capacity is, therefore, a necessary component to increase the deployment of RENs not only in isolated grids, but also in interconnected grid systems, as demonstrated by several analyses carried out on a national scale.[5,6]

For this reason, the European Union (EU) is carrying out a Climate and Energy policy, defined in the Strategic European Technology Plan (SET-Plan),[7] one goal of which is to study more in depth the benefits of storage applications. In such a context, several studies[4,8,9] have been carried out to analyze the current status of the wide range of available technologies (mechanical, electromagnetic, chemical, thermal) in terms of technology maturity, efficiency, energy storage capacity, power discharged capacity, application size, cost of investment, life time, and environmental impact.

All these analyses identify PHES as being the most cost-efficient large-scale storage technology currently available, with an efficiency range of 75–85% and competitive costs (600–1,000 €/kW). In Europe, this technology represents 99% of the on-grid electricity storage[4] with more than 7,400 MW of new PSHPs proposed and a total investment cost of over 6 billion €.[10] In spite of this boost provided by the increasing need for storage capacity, one of the major limits to the further development of PHES is the lack of suitable locations for the construction of new facilities. To overcome this problem, analyses to identify areas that could be quite easily modified in order to construct the reservoirs of the PSHPs were carried out[11] and the possibility of exploiting the sea as a lower reservoir (seawater PSHPs) and of excavating underground reservoirs was considered.[12,13] However, to meet the need for increased storage capacity, besides the installation of new PHES sites, it would be necessary to adapt and exploit existing hydropower plants, as was achieved in France during the 1970s and 1980s to support nuclear power reactors or as recently proposed in Greece to support REN penetration. To reach this goal, the existing hydropower and PHES plants should not only optimize turbine performance by means of innovative design criteria (see Sec. 2) so as to increase the corresponding storage efficiency and to achieve the required greater flexibility, but also modify their operation strategies in order to maximize the revenue from the day-ahead market and from the regulation services (see Sec. 5). The resulting storage capacity will also favor a different economic approach to the investment in REN production based not only on the incentive mechanisms, but also on the techno-economic optimization of the plant operation. This would be aimed at maximizing returns by storing excess electricity from "green" production during low demand periods and selling it to the system during the peak demand periods, thereby providing the required grid stability by means of a fast response time.

1.1 *The New Generation of Pumped-Storage Hydropower Plants*

A PSHP converts grid-interconnected electricity to hydraulic potential energy (so-called "charging"), by pumping the water from a lower reservoir to an upper one during the off-peak periods, and then converting it back during the peak periods ("discharging") by exploiting the available hydraulic potential energy between the reservoirs similar to a conventional hydropower plant.

These plants require very specific site conditions to be viable, including proper ground conformation, a difference in elevation between the reservoirs and water availability. For these reasons, the earliest PSHPs were built in the Alpine regions of Switzerland and Austria whose topography together with the presence of hydro resources was suitable for PHES.

The first PSHPs owned by state utilities, were built to supply energy during peak periods, allowing the base-load power plants to operate at high efficiency, and to provide balancing, frequency stability, and black starts.[a] The period from the 1960s to the late 1980s was characterized by a significant development of these plants mainly due to the corresponding deployment of nuclear power plants whose great inertia was compensated by the PSHPs flexibility (Fig. 1).[10]

In the 1990s, the reduced growth of nuclear plants together with the increasing difficulty to identify suitable locations limited the further development of new PSHPs. For this reason, in 2006 the average percentage of PHES installed capacity was only about 6% of the full generation installed capacity in the majority of countries with the exception of Luxemburg (67%). This percentage was greater than 10% only for those countries characterized by a significant availability of hydro sources (Croatia) or by a significant percentage of installed nuclear power capacity (Latvia, Japan, and the Slovak Republic). The USA and Japan still maintained the world highest installed PHES capacity with 20,815 MW and 24,575 MW, respectively, whereas in Europe, the largest number (23) of PSHPs were concentrated in Germany.

In recent years, after the liberalization of the market, the increasing interest in RENs has again turned public attention towards PHES as a mature and large-scale energy storage technology to support green energy

[a] "Black start" is the process of restoring a power station to operation without relying on the external electric power transmission network (in case of total or partial shut-down of the transmission system).

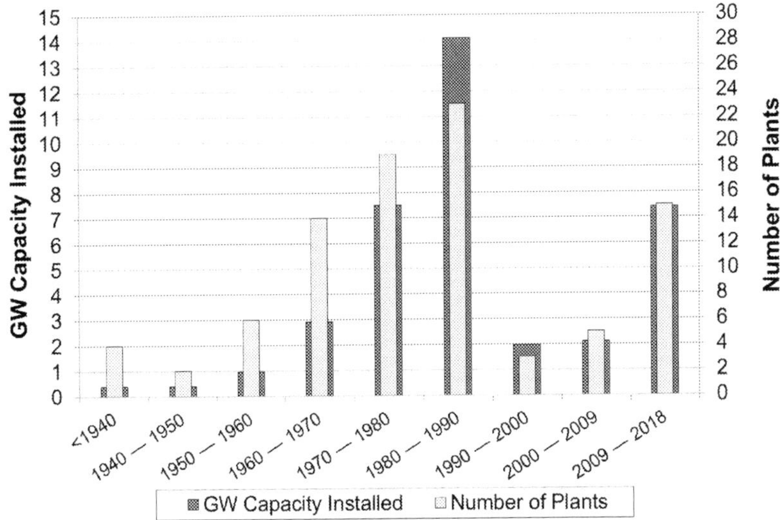

Fig. 1. Development of PSHPs in the EU.

production and to provide grid stability. For these reasons, several new PSHPs have been planned in Europe for a total power capacity of 7,426 MW[14] and some of them will adopt the variable speed reversible pump-turbines breakthrough technology (Sec. 1.3.4.2), which will improve the pump-turbine efficiency over a wide range of operating conditions and improve the capability of grid regulation of the PSHP.[15,16]

Due to the great interest in such a technology, several new PSHPs have been constructed or are under construction in Europe and around the world. Table 1 presents an overview of this new generation whereas Tables 2 and 3 present technical details of the most important PSHPs under construction equipped with the variable speed technology.

Two of the largest PSHPs under construction in Europe that are equipped with this technology are Nant de Drance and Linthal, located in the south-west and the north-east of Switzerland, respectively. In both these power plants, the use of the variable speed technology was justified by the wide head variation: in the power station of Nant de Drance, equipped with six variable speed units with a unit output of 157 MW (rated speed = 428.6 rpm; speed range = ± 7%), the gross head varies between 250 and 390 m; in the power station of Linthal, equipped with four variable speed units with a unit output of 250 MW (rated speed = 500 rpm; speed range = ± 6%), the gross head varies between 560 and 724 m; in the power station of

Table 1: The new generation of PSHPs.

PSHP	Head (m)	Power (MW/ Unit)	Runner diameter (m)	Speed (rpm)	Country
Afourer I	600	175		750.00	Morocco
Afourer II		60		500.00	Morocco
Alqueva I	71.1	129		136.40	Portugal
Alqueva II	72	134		136.40	Portugal
Baixo Sabor Jusante	35	18	3.948	150.00	Portugal
Baixo Sabor Montante	100	77	4.112	214.29	Portugal
Beni Haroun	680	90	2.205	750.00	Algeria
Cruachan	396	106			UK
Dniester	147.5	324	7.300		
Dos Amigos	38.1	28			USA
Feldsee	548	73	1.919	1000.0	Austria
Foz Tua	99	125	4.837	187.50	Portugal
Grand Maison	955	152.5		600.00	France
Grimsel II	400	90		750.00	Switzerland
Guangzhou	535	306		500.00	China
Hintermuhr	518	74	1.870	1000.0	Austria
Hohhot	503/585	306		500.00	China
Huizhou	630	300		500.00	China
Kozjak	713.2	220		600.00	Slovenia
Kops II (ternary)	723.1/818.2	162	2.800	500	Austria
La Muela II	520	850	—	600.00	Spain
Lam Ta Khong	360	260		428.60	Tahiland
Lang Yashan	153	166	4.700	230.77	China
Limberg II	288/436	240	3.920	428.57	Austria
Pedreira		20		120.00	Brasil
Salamonde II		207		166.70	Portugal
Shahe		51		300.00	China
Tierfehd (Nestil)	1066	142	2.263	600.00	Switzerland
Tongbal	289	306	4.802	300.00	China
Venda Nova II	410	191.6		600.00	Portugal
Vianden M11	295	200	4.286	333.33	Luxemburg
Yang Yang		258		600.00	South Korea
Yecheon		408		400.00	South Korea
Yixing	420	262	4.394	375.00	China
Zarnowiec	128	188	6.008	166.67	Poland
Zanghewan		255		333.33	China

Goldisthal, equipped with two fixed units and two variable speed unit, the gross head varies between 279 and 334 m; in the power station of Frades II, the speed range is $+2\%/-7\%$ and the gross head varies between 414 and 432 m reports the project features of other variable speed PSHPs still under construction.

Table 2: The new generation of PSHPs equipped with the variable speed technology.

PSHP	Nant de Drance 2019	Linthal 2015	Tehri 2016	Goldisthal	Frades II
Rotational synchronous speed (rpm)	428.6	500.0	230.77	333	375
Speed range	±7%	±6%	±6%	+4%/ − 10%	+2%/ − 7%
Head variation (m)	250/390	560/724	830/740	279.2/334.0	413.64/431.8
Nominal output per unit (MW)	157	250	255	265	383
Maximum pump discharge per unit (m^3/s)	56	—	—	80	—
Maximum turbine discharge per unit (m^3/s)	60	—	—	103	100
Generator Mode (MVA)	175	280	278	—	419.5
Motor mode (MW)	172	250	—	—	372
Runner diameter (m)	6.009	4.230	—	4.593	4.52

Table 3: Under construction variable speed PSHPs.

PSHP	Venda Nova III	Omarugawa III	AVCE	Grimsel III
Rotational synchronous speed (rpm)	—	600	600	—
Speed range	—	±4%	±4%	—
Max head (m)	—	688/720	521	580
Nominal output per unit (MW)	380	310	185	200
Generator mode (MVA)	420	319	195	—
Motor mode (MW)	380	310	185.25	—
Maximum turbine discharge per unit (m^3/s)	—	—	40	—
Maximum pump discharge per unit (m^3/s)	—	44.1	34	—

It is interesting to highlight that in most of these plants, design priority was given to flexibility in the pumping mode, maximizing the pump operating range, whereas in other cases, such as the PSHP of Tehri under construction in India (four variable speed units of 255 MW; rated speed = 230.77 rpm; speed range = ±6%) the hydraulic design was optimized to increase the global plant efficiency. These different design strategies, aimed at maximizing the plant revenue, are due to the different electricity market regulations and in particular to the existence (or not) of a remunerative regulation market that could significantly affect the definition of the plant management strategies (Sec. 5).

As regards the power converter technology, significant advances have been made in recent years, as, for example, the 100 MVA variable speed frequency converter provided by ABB for the PSHP of Grimsel 2 (Switzerland), enabling a wider speed variation in pumping mode (±12%) but not in generating mode. However, increasing the voltage range and reducing cost, size and losses still remain the main challenges to develop a ±100% power converter.

Even if these plants represent the starting point of a new generation of PSHPs, the possibility in terms of speed variation is limited to about ±10% due to the lack of full variable speed design criteria, allowing the pump-turbine to operate at a high efficiency in a wide range of rotation velocities, and to the lack of a full power converter technology, enabling a full-range variable speed turbining and pumping (Tables 2 and 3).

As regards the mechanical equipment, computational fluid dynamics (CFDs) has allowed significant advances in the understanding of the reasons for the unstable behavior of the pump-turbines that prevent PHES units from operating at low load (see Sec. 2). However, innovative design criteria, allowing electricity production in the whole operating range (0–100% of the peak power), still have to be developed and certainly represents a future challenge for the development of a new generation of pump-turbines.

1.2 *Unconventional Pumped-Storage Hydropower Plants*

In recent years, new storage concepts have been proposed and developed on the basis of the possible exploitation of unconventional lower reservoirs. Two of the most interesting proposals are undoubtedly seawater and underground PHES, exploiting sea and underground caverns, respectively.

Since one of the main problems limiting the significant spread of PHES is the identification of suitable sites, the idea of exploiting unconventional lower reservoirs represents a method for further exploitation of this storage technique. Moreover, these plants are characterized by lower civil construction costs due to the need of constructing a single reservoir instead of the two required by conventional PSHPs. However, in spite of the recognized advantages of these new storage concepts, there are also critical barriers

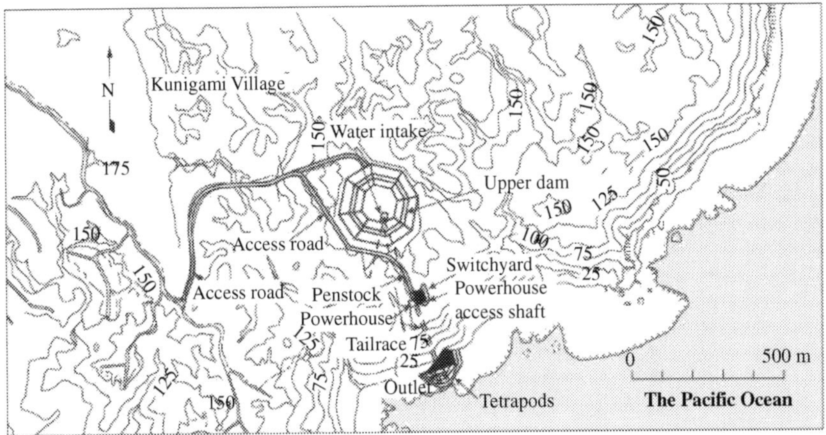

Fig. 2. Plan view of the seawater PSHP in Okinawa (Japan).[17]

(technological, economic, regulatory, environmental, social acceptance, etc.) limiting their intensive deployment.

(a) Seawater PHES

The Japanese were pioneers in building the first demonstration plant in the northern part of Okinawa Island. This project, whose construction was carried out from 1991 to 1999, commenced operation in 1999 and was tested for 5 years in order to demonstrate the feasibility of a seawater pumped-storage power generation technology (Fig. 2). The upper reservoir, located 500 m away from the seashore at an elevation of 150 m, is clearly identifiable because of its octagonal shape. The outlet of the tailrace was protected from the waves by tetra-pods. Table 4 reports the main specifications of the Okinawa power plant, whereas much more details can be found in other references.[17,18]

The most interesting aspects of the construction of this seawater PSHP are undoubtedly the challenging problems that have been faced due to the need to pump up seawater and to mitigate the environmental impact of this type of plant on the ocean eco-system.

Studies were carried out to define structural features of the mechanical equipment required for seawater applications and measures were taken to prevent corrosion and the adhesion of marine organisms.[17] A binary set was preferred to other configurations (see Sec. 1.3) and the assembly was designed in order to easily remove the runner for maintenance work. In

Table 4: Main specifications of the Okinawa project.

Okinawa Yanbaru power plant	Specification
Power plant	30 MW
Max. output	
Max. discharge	26 mc/s
Effective head	136 m
Upper regulating pond	Excavated type,
Type	
	Rubber sheet-lined
Max. embankment height	25 m
Crest circumference	848 m
Max. width	251.5 m
Total storage capacity	0.59 × 106 mc
Max. depth	22.8 m
Waterway	Inside dia. 24 m
Penstock	Length 314 m
Tailrace	Inside dia. 27 m
	Length 205 m

the zones of expected water stagnation, the pump-turbine was coated with anti-pollution dirt-prevention paint.

The most significant environmental impacts were identified by a study committee, *viz.*:

- Outflow of muddy water from the construction area into the gullies and sea area near the river mouth.
- Reduction of habitat area due to land changes.
- Noise and vibration from heavy equipment.
- Damages to small animals from construction vehicles and accidents due to falling down into roadside gutters.

Proper countermeasures were carried out to significantly reduce these factors, such as water chemical treatments, low-noise machinery, animal intrusion prevention nets, etc.[18]

As a consequence of the positive results obtained by the Okinawa plant, other possible seawater projects have been considered and feasibility studies have been proposed in the last few years.[19–21]

(b) Underground PHES

The first idea of exploiting a disused mine as an underground reservoir dated from 1960[22] and it was developed by several studies and technical reports but not accompanied by functioning pilot projects.[12] This storage concept presents several advantages in comparison with conventional PHES, including the higher probability of social acceptance and the larger number of potential sites. From a technical point of view, even though the construction of an underground storage reservoir is possible, the main limit is the need for appropriate rock, especially at reservoir depths.

An extremely interesting techno-economical analysis of the possible construction of underground PHES in abandoned coal mines in the Ruhr valley (Germany) was presented in 2013.[23] In this area, the geological conditions were found not to be optimal and the project might be uneconomical with the exception of the use of tubular underground drift grids for the intake of water.

As regards the investment costs, the profitability depends upon the realizable head. In case of low heads, the extension of the lower reservoir, in order to increase the accumulation capacity, dominates the cost in comparison with other expenses. However, the relative cost decreases with increasing heads.

Apart from the higher construction cost of the lower reservoir, higher maintenance and repair costs, as well as lower service life, are also expected.[23] However, the possibility of placing the lower reservoir directly below the upper reservoir certainly reduces the length of the headrace and/or tailrace and hence the corresponding costs.

An interesting **unconventional pumped-hydro project**, proposed in Estonia,[24] is that of Muuga whose completion is expected in 2020. Table 5 reports some details of the proposed project.[24]

The peculiarity of this project is that it combines two different unconventional reservoirs: the sea as the upper reservoir and underground chambers, resulting from granite excavations, as the lower reservoir.

As for underground PHES, the total volume of the lower underground reservoir defines the number of hours of plant operation at the nominal capacity, which is 12 h for the Muuga project. The technological and environmental issues from the use of seawater will require the adoption of proper countermeasures based on the Okinawa project experience. More details about civil works and technical solutions can be found in the description of the project.[24]

To conclude this section, it is worth mentioning a proposal of unconventional PSHP, presented by De Boer *et al.*[25] about an **inverse offshore pump accumulation station** (IOPAC), which is a seawater PSHP combined with an offshore wind power plant.

Table 5: Main specification of Muuga project.

Muuga Plant	
Max. plant output	500 MW
Max. unit output	1 × 100 MW pump-turbine
	2 × 175 MW pump-turbine
	1 × 50 MW Francis turbine
Average operation height	500 m
Max. discharge	110 mc/s
Diameter of delivery conduit	7.0 m
Max. water velocity	3.1 m/s
Water consumption	
(by 12-h operation full capacity)	4.75 Mmc

The idea is to station the plant on an artificial island, called the "Energy Island", consisting of a ring of dikes enclosing a seawater reservoir, approximately 6 km long and 4 km wide. According to this proposed scenario, the reservoir would be dredged to a depth of 50 m below sea level with a capacity of 30 GWh and a maximum generation capacity between 2,000 and 2,500 MW. The island will be built from the sand, obtained from deepening the reservoir.

When the wind power exceeds the demand, the station pumps the water out of the reservoir. Otherwise, the floodgates are opened and the seawater flows into the turbines, producing the requested power difference.

A possible site location has been found on the Netherland coast, where the presence of a layer of clay on a right depth (50 m below the sea level) and with a good thickness (40 m) should prevent the groundwater entering in the inner reservoir by percolating through the substrata. Bentonite walls, already feasible to a depth of 60 m, should prevent seawater leakage. An in-depth analysis of the technical, economical, and ecological aspects of this project is still underway.

1.3 *Power Plant Configurations*

Several machine configurations have been used throughout the history of pumped storage. These configurations differ in the number of hydraulic and electric machines used. In general, they can be classified as:

- *Binary set*: One pump-turbine and one electrical machine (motor/generator).
- *Ternary set*: One turbine, one pump, and one electrical machine (motor/generator).

- *Quaternary set*: One turbine driving one generator and one pump driven by one motor.

Each configuration has its own advantages and disadvantages. In what follows, these configurations will be described.

1.3.1 *Binary set*

This is, by far, the most used scheme. Without doubt, this is because it is the cheapest one. The most common configuration uses a single-stage pump turbine coupled to a synchronous electrical machine directly connected to the grid.

Single-stage pump-turbines can be used with heads from 10 m up to 700 m (Fig. 3(a)).[26] For larger heads, multistage pump-turbines should be used (Fig. 3(b)). Multistage pump-turbines can be used with heads from 700 m up to 1200 m.[26]

However, this increase in head range comes at a price: multistage pump turbines do not usually have wicket gates[27] and therefore are unable to contribute to load frequency control. In the 1980s, a partial solution to this drawback was proposed and tested: *viz.*, double-stage pump-turbines with two adjustable sets of wicket gates, one for each runner.[26] An example

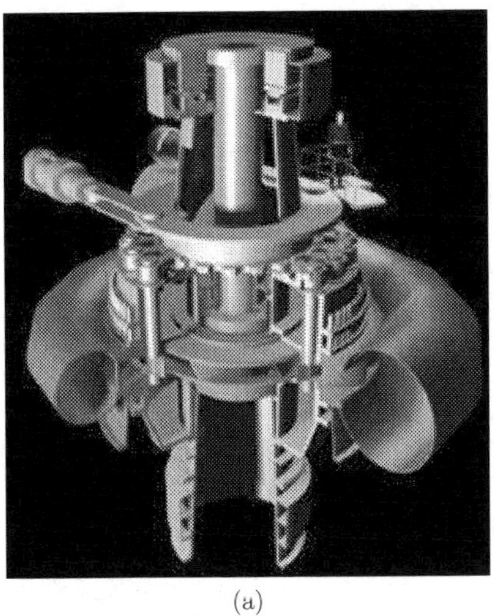

(a)

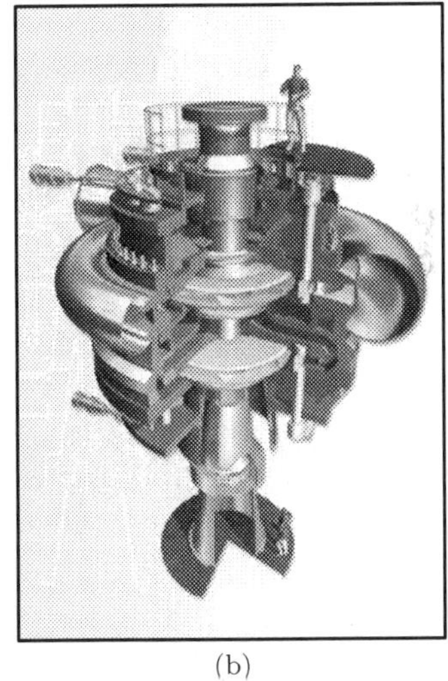

(b)

Fig. 3. Scheme of a single stage pump-turbine (a) and of a multistage pump-turbine (b).[26]

is the regulated double-stage reversible pump-turbine recently supplied by Alstom for the Yang Yang PSHP.[26]

Cycle efficiency, η_{cycle}, can be obtained from:

$$P_b = \frac{\gamma Q'_n (H_0 + H'_r)}{\eta_b \eta_g}, \tag{1}$$

$$P_t = \gamma \eta_t \eta_g Q_n (H_0 - H_r), \tag{2}$$

$$\eta_{\text{cycle}} = \frac{P_t / Q_n}{P_b / Q'_n} = \eta_t \eta_b \eta_g^2 \frac{H_0 - H_r}{H_0 + H'_r}, \tag{3}$$

where γ is the water specific weight; Q_n and Q'_n are, respectively, the rated flow in generating and pumping modes; H_0 is gross head; H'_r and H_r are the friction head losses between upper and lower reservoirs with Q'_n and Q_n, respectively; η is the average efficiency of turbine, pump, or electrical machine depending on subindex: t, b, and g, respectively.

There are many examples of this type of PSHPs; La Muela II PSHP[28] is a recent one.

1.3.2 *Ternary set*

It is composed of a turbine, an electrical motor/generator, and a pump coupled altogether on the same shaft. The electrical motor/generator is, usually, a synchronous machine (SYM). Unlike the binary set, in which the pump-turbine design is the result of compromises between the two operating modes, in a ternary set both turbine and pump designs are optimized.

Prior to 1960s, this was the preferred scheme with a horizontal shaft and a Francis turbine. Nowadays, this configuration is used only when single stage pump-turbines are not appropriate (i.e. for too large heads) and, therefore, the turbine is a Pelton turbine. Although both horizontal and vertical shaft configurations could theoretically be used in these cases, vertical ones allow installing the pumps below water level in the lower reservoir and the Pelton turbines above the water level. In order to reduce shaft length, the Pelton turbine can operate inside a compressed air chamber that provides atmospheric conditions in the Pelton runner outlet.

Unit operation in generating mode is similar to conventional Pelton units operation if the pump is decoupled (through the clutch) from the turbine-generator set. Since these turbines can use deflectors in the water jets, water hammer overpressures can be properly controlled. Pump start-up is carried out with the help of the turbine; once the motor is connected and synchronized to the grid, the nozzles are closed.

Cycle efficiency can be derived from Eqs. (1) to (3).

1.3.3 *Quaternary set*

Quaternary configurations have two different powerhouses; one for pump units and the other for turbine units; therefore, pumps and turbines are not mechanically coupled. Operation in generating mode is similar to that of the ternary set configuration, without the need of compressed air in the turbine chamber. Cycle efficiency and grid support capabilities are also similar to that of the ternary set configuration.

Pump start-up can be carried out '*back to back*' with a turbine or with the help of a frequency converter. Both possibilities imply an increase in the cost since proper electrical connections or an electronic converter are needed.

An example of this configuration can be found in the *Gorona del viento* wind-hydro power plant.[29]

1.3.4 *Solutions for load-frequency control in consumption mode*

In outline, load-frequency control can be provided in consumption mode[b] by means of:

(1) Hydraulic short-circuit operation
(2) Variable speed operation

1.3.4.1 Hydraulic short-circuit operation

The so-called hydraulic short-circuit operation allows providing load-frequency control by simultaneously pumping at the rated power and controlling the turbine power generation. This approach has been only implemented in a ternary set configuration PSHP (Kops II).[30]

Nevertheless, theoretically it can be also implemented in binary set configurations with two or more reversible units, as well as in quaternary set configurations with at least one pump and one turbine. The power regulation range in consumption mode depends on the power regulation

---

[b]The term "consumption mode" is intentionally used to include those cases where both turbines and pumps are running and the plant net power supply (generation minus consumption) is negative.

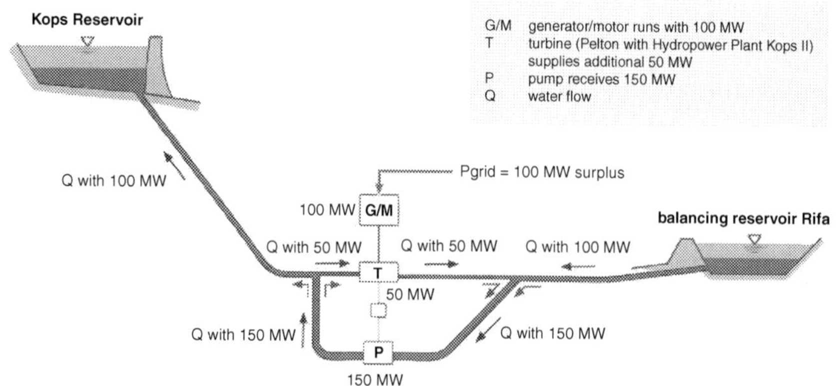

Fig. 4. Kops II hydraulic short-circuit operation.[31]

range of the turbines in operation. In Fig. 4, an example of this operating mode is shown.

The "excess" water that reaches the upper reservoir during load-frequency control operation can be used later for power generation. The cycle efficiency corresponding to the excess water depends on the operating point of the turbine during load-frequency control operation, and is therefore variable. Let F be the fraction of pump rated flow diverted for load-frequency control purposes (in per unit values) to either other unit (binary set configuration) or to the turbine coupled to the pump within the same ternary unit, the cycle efficiency corresponding to the water that reaches the upper reservoir during load-frequency control operation in consumption mode, η^{LFC}, can be derived from:

$$P_b = \frac{\gamma Q'_n (H_0 + H'_r)}{\eta_b \eta_g} - \gamma \eta_t \eta_g (F Q'_n) \cdot (H_0 + H'_r), \tag{4}$$

$$P_t = \gamma \eta_t \eta_g Q_n (H_0 - H_r), \tag{5}$$

$$\eta^{\mathrm{LFC}} = \frac{P_t / Q_n}{P_b / ((1 - F) Q'_n)} = \eta_t \eta_b \eta_g^2 \frac{H_0 - H_r}{H_0 + H'_r} \frac{(1 - F)}{1 - F \eta_t \eta_b \eta_g^2}, \tag{6}$$

where, H'_r and H_r are the friction head losses between upper and lower reservoirs with $(1 - F)Q'_n$ and Q_n, respectively; see Eqs. (1)–(3) for the rest of notation.

The overall cycle efficiency of the plant will therefore depend on:

- The time during which the plant provides load-frequency control
- The turbine operating point during load-frequency control operation
- The time during which the plant operates in a "classical" way

A new ternary set PSHP, specifically designed for hydraulic short-circuit operation, is currently under construction in Switzerland.[32]

1.3.4.2 Variable speed operation

The most common solution for providing load-frequency control in pumping mode is the use of variable speed drives.[15,16]

Figure 5 shows a comparison of operating ranges in terms of head and discharge variation between a fixed and a variable speed pump-turbine with

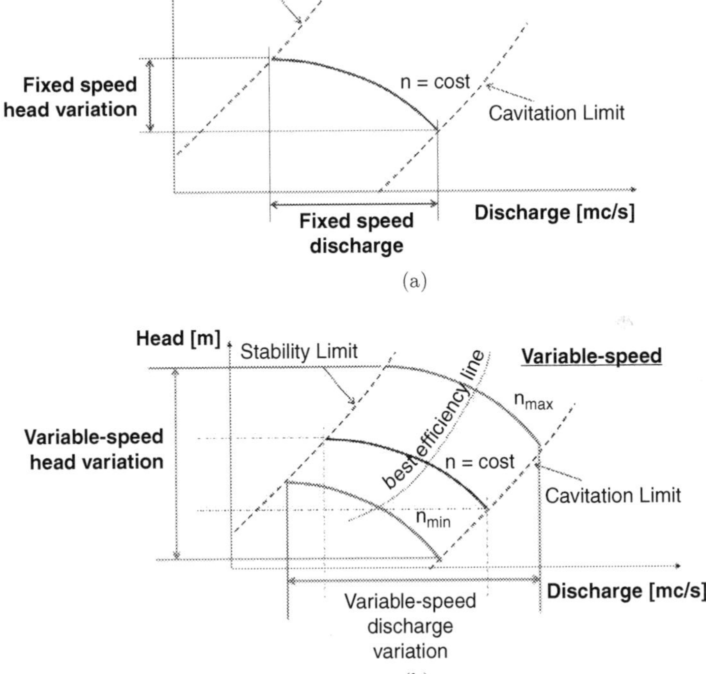

Fig. 5. Comparison of operating domains between fixed (a) and variable speed (b) pump-turbines (pumping mode).[33]

reference to the pumping mode. The possibility of varying the speed significantly increases the operating range of the machine, which is limited, on the one hand, by cavitation problems (broken line on the right) and on the other by an instability operating area (broken line on the left). The latter is a typical behavior of the pump-turbine at part load that should be avoided during the plant operation since it may lead to self-excited vibrations of the hydraulic system (Sec. 2).

From an electrical point of view (Sec. 3), the possibility of speed variation is obtained by means of an appropriate frequency converter.

The power regulation range of a variable speed pump strongly depends on the following factors[34]:

- The stable operation range of the pump
- The minimum power necessary to provide at least the static head
- The rated head and head variation

The cycle efficiency of a variable speed PSHP can be calculated from Eqs. (1) to (3). The overall cycle efficiency of the plant will depend on:

- The time during which the plant provides load-frequency control.
- The turbine and pump operating points during load-frequency control operation in generating and pumping modes.
- The time during which the plant operates in a "classical" way.

1.3.5 *Start-up and shut-down times*

The new generation of PSHPs are characterized by good connection properties, depending on the power plant configuration.

Updated values of start-up and shut-down times have been recently presented by Voith Hydro in 2013,[35,36] (Fig. 6) considering four operating conditions:

- Standstill with the runner filled with water (ST)
- Pump Mode — full load (PU)
- Turbine Mode — full load (TU)
- Synchronous Condenser (SC)

2 Mechanical Equipment

The most common mechanical equipment adopted in the new generation of PSHPs is pump-turbines, which are generally preferred to other technical

A - Reversible PT

B - Ternary set (HTC + Pelton Turbine)

	MODE CHANGE	A	B
1	Standstill → TU – Mode	90	60
2	Standstill → PU – Mode	340	120
3	Standstill → SC – Mode	120	60
4	TU – Mode → SC - Mode	80	20
5	SC – Mode → TU - Mode	70	20
6	SC – Mode → PU - Mode	70	30
7	PU – Mode → SC - Mode	140	30
8	TU – Mode → PU - Mode	420	30
9	PU – Mode → TU - Mode	190	30
10	TU – Mode → Standstill	200	110
11	PU – Mode → Standstill	160	50
12	SC – Mode → Standstill	200	100

Fig. 6. Comparison of start-up and shut-down times (s) between a reversible pump-turbine and a ternary set.[35]

arrangements, such as the combination Francis turbine/pump or Pelton turbine/pump, due to their cost-effectiveness.

Although pumped storage may solve several problems in the grid, fast and frequent changes between pumping and generating modes are required, extending the operation of the machine at off-design conditions. However, the design of a pump-turbine is the results of compromises between contradictory targets, such as pump and turbine performance, regulation capacity, efficiency, and cavitation behavior.

Moreover, in the design of reversible pump-turbines, great attention is generally paid to the behavior in pumping mode, due to the greater sensitivity of the decelerated flow field to boundary layer detachments and flow separation, which causes recirculation and hydraulic losses. This design approach causes the onset of unstable behavior at off-design conditions, which is not acceptable and may lead to self-excited vibrations of the hydraulic system

Figure 7 shows the evolution of the discharge factor $Q_{11} = Q(\frac{D_{11}}{D})^2\sqrt{\frac{H}{H_{11}}}$ [l/s] vs. the speed factor $n_{11} = n\frac{D}{D_{11}}\sqrt{\frac{H_{11}}{H}}$ [rpm] (where

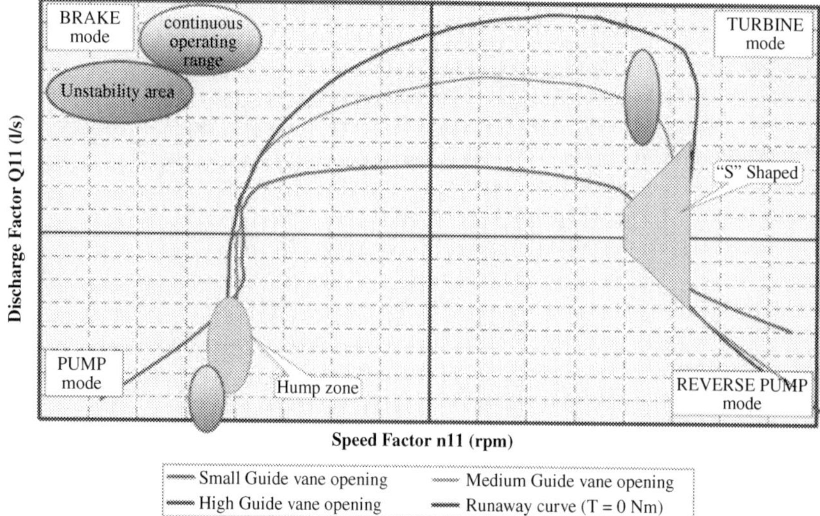

Fig. 7. Typical four quadrant characteristics of a high head reversible pump-turbine.[37]

H_{11} and D_{11} are equal to 1 m) for three different guide vane openings in the so-called four quadrant characteristics,[37] where H is the head, Q is the flow rate and D is the runner diameter.

Two main features of unstable behavior of pump-turbines are identifiable:

- One occurring in the generating mode at low load off-design operation close to runaway conditions (S-shape of the turbine characteristic).
- The other one occurring in pumping mode at part load (saddle-type pump instability of head curve — "hump" zone).

In pumping mode, the unstable behavior is associated with a positive slope of the head-flow curve $\frac{dH}{dQ} > 0$ (Fig. 8) and is associated with unsteady flow patterns developing inside the machine, accompanied by pressure fluctuations that may lead to self-excited vibrations of the hydraulic system with severe consequences for the PSHP.

If the hump zone is located below the highest operating head, start-up of the machine at higher heads becomes impossible due to the head drop and hence it is necessary that the saddle-type head drop has sufficient margin against the highest regular operating head.

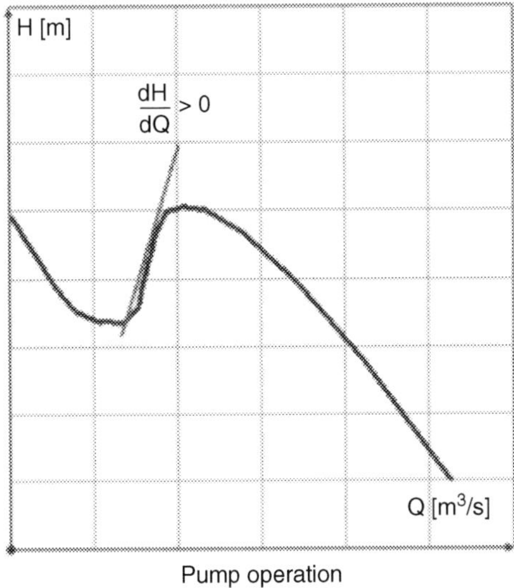

Fig. 8. Hump zone in pumping mode (Ref. 38).

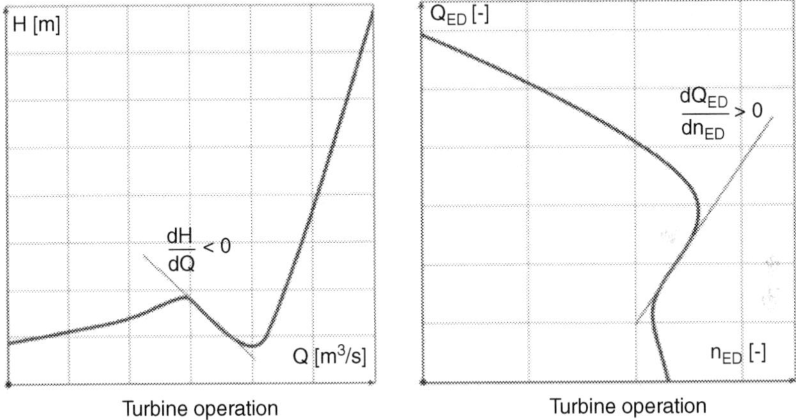

Fig. 9. S-Shape in turbine mode at a constant wicket gate angle (Ref. 38).

In turbine mode, the criterion for instability is expressed by a negative slope of the head-flow curve $\frac{dH}{dQ} < 0$ or by a positive slope of the dimensionless flow-speed curve $\frac{dQ_{ED}}{dn_{ED}} > 0$ (Fig. 9), where $n_{ED} = nD/\sqrt{gh}$ and $Q_{ED} = Q/D^2\sqrt{gh}$[39] (Fig. 9).

This instability, associated with fluctuations of the head and discharge flow rates in the system, is highly undesirable during start-up and synchronization, since faster start-up and switch-over times are extremely important nowadays. As a consequence of the unstable interaction of the machine and hydraulic systems, additional efforts are required during the pump-turbine start-up, as the misalignment of a few wicket gates.

So, even though the ability to vary the speed has significantly increased the operating range of pump-turbines, unstable operating conditions still represent a limit for the exploitation of the full-working range of pump-turbines. For these reasons, in recent years, several studies have been carried out on these topics and a summary is presented in the two following sections.

2.1 *Instability in Turbine Mode: S-Shaped Region*

This instability is associated with fluctuations of speed, torque, head, and flow rate that negatively affect the start-up and synchronization of the pump-turbine with the grid as well as the turbine brake.

At the start-up, the machine operates in the no-load condition in which the energy input (the hydraulic energy) of the turbine is completely dissipated by the energy losses. Due to the severe torque fluctuations caused by the instability development, in this operating area the pump-turbine could be affected by sudden changes in working mode (from turbine to pump and *vice versa*) as well as by significant fluctuations of the head and flow rate with possible self-excited vibrations[40] or water hammers which stress not only the mechanical equipment but the whole power plant.[41]

The main reason for the S-Shape instability is the design criterion of the pump-turbine being much more focused on the behavior in pumping mode,[39] due to the greater sensitivity of the decelerated flow field to boundary layer detachments and flow separation. For this reason, pump-turbines present some geometry differences with Francis turbines, such as the inlet blade angle being much smaller than that in a Francis turbine. These geometry differences affect the stable behavior of pump-turbines in turbine mode, determining the appearance of the S-Shape region in the pump-turbine characteristics (Fig. 10).

Four geometrical changes positively affect the pump-turbine behavior at part load but not its performance[39]: the increase of the inlet blade angle, the increase in the radius of curvature of the pressure side of the leading edge in turbine mode (to reduce the incidence conditions and the corresponding detachment at off-design conditions), the decrease of the inlet radius and the increase of the blade length.

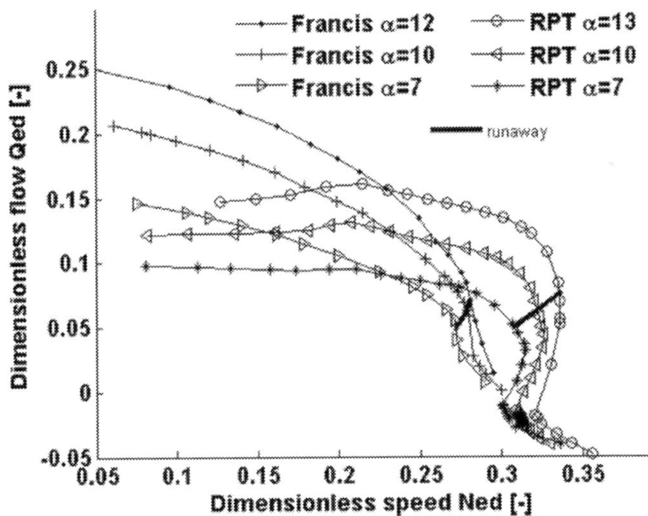

Fig. 10. Pump-turbine vs. Francis characteristics for different wicket gate angles.[39]

The pump-turbine performance at different flow rates has been numerically analyzed,[42] showing that reverse flows inside the runner were identified both at high and partial loads.

Steady and unsteady state numerical simulations were carried out on the entire model of a pump-turbine,[43] bringing the machine from a stable operating condition to runaway and turbine brake conditions. The analysis revealed the onset of flow instabilities inside the runner blade channels at the runaway condition.

These instabilities were amplified in the S-shape region, characterized by stall cells totally blocking some runner blade channels. In agreement with experimental studies,[44] the stall was demonstrated to rotate with a frequency equal to 70% of the runner rotating frequency.

At lower discharge conditions, the increased intensity of the rotating stall affected not only the runner but also the rotor-stator gap and the flow channels of stay vanes and wicket gates. This rotating stall, which blocks the flow rate passage in some runner channels, produces an increase of the head, causing a change of the pump-turbine characteristics in turbine mode. The reverse flow developing in some runner channels could also favor a change of pump-turbine working mode: from turbine to reverse pump mode.

To reduce this blocking action and to eliminate the S-shape characteristic, the use of a misaligned guide vanes technique has been proposed.[45,46]

However, this technique also resulted in increasing pressure fluctuations and unbalanced radial forces on the runner, leading to an unstable behavior at no-load condition.

The influence of the wicket gate angle on the unstable behavior in the S-Shape region has been analyzed by Widmer *et al.*,[47] and its development has been identified only at large wicket gate angles.

With a dedicated design, it is possible to significantly reduce the unstable behavior at no-load condition with a consequent gain in time during the start-up sequence.[37]

2.2 *Instability in Pump Mode: The Saddle Zone*

The saddle-type pump instability is undoubtedly the most challenging problem to face in order to significantly increase the operating range of pump-turbines in pumping mode, even in the case of a variable speed pump-turbine.[16]

This unstable pump operating zone is characterized by a head drop associated with an increase of the hydraulic losses in the runner, in the stator parts or in both. At off-design conditions, the diffuser and the draft tube do not work properly and give awkward boundary conditions to the runner, together with a strong fluid-dynamical interaction between runner and stator parts. Flow features, such as flow separations and re-circulations, occur in an unsteady manner and guide vanes may experience strong vibrations. For this reason, the unstable behavior does not simply determine a head drop due to the increased hydraulic losses but also it is accompanied by high-cycle fatigue stress that may result in crack propagation and in the failure of shear pin or guide vanes stem. The root causes of damage can range from misalignment during shear pin assembly on the guide vane activation mechanism, which causes anomalous loading, to the strong excitation due to the Rotor–Stator Interaction (RSI). It is certain that the complexity of the pump-turbine structures favors the RSI, that is a fluid-dynamical interaction between the rotating element of the pump-turbine (runner) and the stator parts of the pump-turbine (stator vanes, return channel, draft tube, adduction).

Among the phenomena developing inside the machine, the onset of a fluid-dynamical instability as a consequence of the RSI still represents a crucial research point and authors are still debating the reasons for its onset and development. Several experimental and numerical analyses have been carried out on this interaction to identify a possible connection between unsteady flows and pressure fluctuations developing inside centrifugal pumps with different runner/diffuser geometries and operating conditions.

Even though experimental analyses[48–55] have allowed the identification and characterization of the development of unsteady pressure pulsations in the hump zone of the pump-turbine operating range, a significant boost to the understanding of this instability was provided by CFD, whose capability of modeling the flow through the entire machine in a single CFD simulation has significantly increased in the last few years.

Numerical analyses on different configurations of pump-turbines allowed an improvement of the understanding of these unsteady phenomena, highlighting on one hand the existence of pulsating reverse flows inside the runner channels, and on the other, the existence in the stator parts of related re-circulating phenomena partially or totally blocking the flow in channels/regions downstream or upstream of the runner.

The pulsating onset of reserve flow cells in the runner moving along the blade length and from one channel to another has been identified by means of simulations.[56] This unsteady behavior in the runner was associated with a perturbation of the diffuser flow field, characterized by an unsteady flow rate migration between passages and by unsteady flow jets. This flow rate migration in its turn was caused by vortexes developing in the first part of the return channel and partially or totally blocking the flow coming out from the diffuser. This strong RSI resulted not only in an asymmetrical distribution of the flow coming out from the runner but also in the development of rotating pressure pulsations, causing severe vibrations and an unstable pump-turbine behavior in terms of fluctuating head and flow rate.

Some transient numerical analyses carried out for a pump-turbine highlighted the dependence of the flow behavior in the head drop with the specific speed of the pump-turbine.[38] For high values of the pump-turbine, the unstable operating conditions resulted in the onset of a fully-developed pre-rotation at the runner inlet shroud which extends beyond the pump leading edge. At a lower specific speed, the head drop was due to a strong interaction between the runner pre-rotation (previously identified) and a stall developing in the stator.

2.3 *Future Challenges*

Even though CFD has allowed us to obtain interesting information on the unstable behavior of pump-turbines, to reduce the instability problems and to significantly enlarge the working range of pump-turbine, many more details about the influence of the pump-turbine geometry on the flow patterns and on the dissipation mechanisms are needed.

The characteristics and flow mechanisms that determine the onset and the development of these unsteady phenomena and the effects of the geometry on their onset and stabilization still represent a challenging issue which merits further research.

CFD has certainly improved the understanding of these unsteady phenomena. However, in spite of the significant advances of the last few years, the investigation of the unstable behavior of a pump-turbine is still extremely time-consuming and requires a huge amount of computational effort. It has been demonstrated that steady state calculations, even if carried out with a fine mesh, failed to correctly predict the head drop.[38] So, to provide detailed knowledge of the unstable behavior of pump-turbine, further improvements in CFD are also necessary in order to model turbulent flow and performance at a wider range of discharge rates more accurately and faster.

3 Electrical Machinery

Most electric energy is produced by rotating electrical generators that are driven by prime movers, called turbines. Depending on the type of prime mover, the most common power plants can be classified as[57]:

- Steam turbine thermal plants (nuclear, coal, oil, gas, biomass, solar, geothermal)
- Combustion thermal plants (gas, diesel, combined cycle)
- Hydraulic turbine plants (reservoir, run-of-the-river, tidal)
- Wind turbine plants

The prime mover supplies rotational mechanical energy to the generator, which in turn converts it into electrical energy that is injected into the grid (Fig. 11). These machines generate alternating current (AC) with constant frequency and voltage.

For PSHPs, different types of electrical machines can be used depending on the desired application. For fixed speed applications and large units, conventional SYMs are commonly used. For variable speed applications and small units (less than 60 MW), the SYM is linked to the grid by a static frequency converter. For larger variable speed units, this solution is not justified economically and doubly fed induction machines (DFIMs) are the chosen solution.

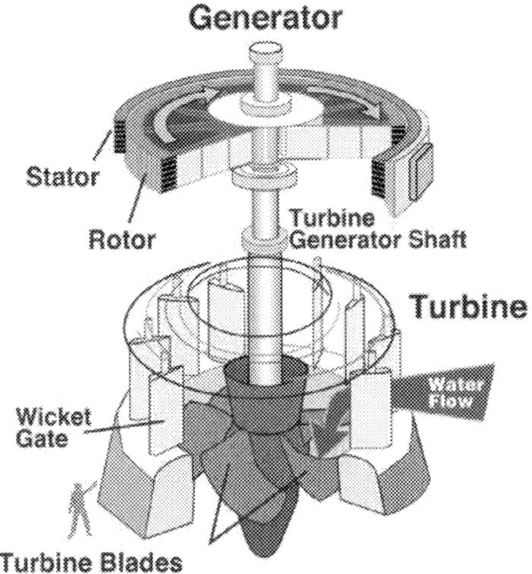

Fig. 11. Turbine and electric generator.[58]

3.1 *Conventional SYM*

The bulk of electric energy is produced by three-phase SYMs, with power ratings of several hundred MVA; the biggest machines have a rating up to 1,500 MVA.

SYMs are characterized by a uniformly slotted stator laminated core that houses a three-phase AC winding and a direct current (DC) excited rotor.[59,60]

SYMs used in hydro power plants are built with salient-pole concentrated-excitation rotors.

DC excitation power on the rotor can be transmitted by:

- Copper slip-rings and brushes
- Brushless excitation systems

A controlled rectifier, with a nominal power around 3% of generator rated power controls the DC excitation currents according to the needs of generator voltage and frequency stability.

3.1.1 *Operating principle*

A DC current is applied to the rotor winding, which produces a rotor magnetic field. The rotor is then driven by the prime mover (i.e. water) producing a rotating magnetic field. This rotating magnetic field induces a three-phase set of voltages within the stator windings of the generator.

These machines are synchronous in the sense that the electrical frequency produced is *synchronized* with the mechanical speed of the generator by Eq. (7):

$$n = \frac{60 f_1}{p}, \tag{7}$$

where n is the mechanical speed, f_1 is the grid frequency, and p the number of pole pairs.

For small power applications, it is also possible to use permanent magnet synchronous machines (PMSYMs) which use permanent magnets in the rotor. These machines are simpler and cheaper than the conventional SYMs but they have some limitations. In PMSYMs, the output voltage is directly proportional to the rotor speed and since the air gap flux is not controllable by adjusting the DC voltage, its terminal voltage cannot be easily regulated without using power electronics devices.

3.1.2 *Conventional synchronous generation regulation*

In order to connect a synchronous generator to the electric grid, four voltage conditions must be satisfied:

- Voltage must have the same phase sequence as the grid voltages.
- Voltage must have the same frequency as the grid.
- Voltage must have the same amplitude at its terminals as the grid voltage.
- Voltage must be in phase with the grid voltage.

When the generator is connected to a large grid, its output voltage and frequency (and rotational speed) are locked to the system values.

The equivalent circuit of the SYM connected to an infinite bus[60] is represented by an internal electromotive force (EMF), E_0, and a series reactance X_s (Fig. 12). This reactance X_s is called **synchronous reactance** and is constant during normal steady-state conditions.

Immediately after coupling to the grid, the machine is neither feeding power to nor absorbing power from the grid. The water going through the turbine is the same as before the coupling action and it is just enough to drive the rotor and compensate for the losses in the turbine and the generator.

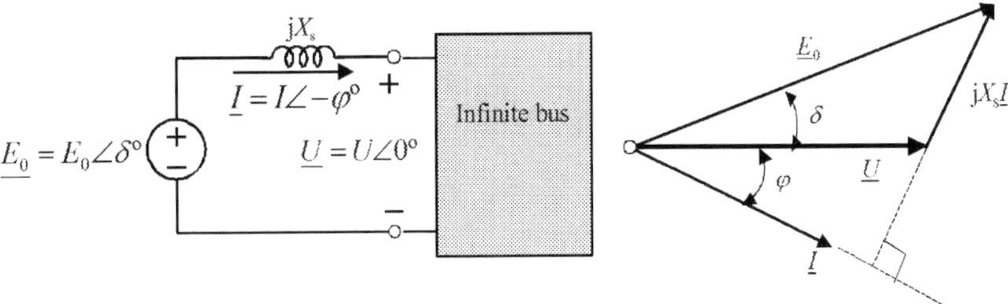

Fig. 12. The equivalent circuit of an SYM connected to an infinite bus.[60]

If more water is fed into the turbines, the internal EMF, E_0, leads the terminal voltage. The amplitude of the internal EMF, E_0, is a function of the DC field current and can be controlled by the operator. In Fig. 12, the internal EMF, E_0, leads the system voltage U by an angle δ, which is known as the **power angle**. As a consequence, current and active power is fed into the grid.

The expression for the current is given by:

$$I = \frac{E_0 - U}{jX_s}, \tag{8}$$

where I is the current phasor, E_0 is the internal EMF phasor, U is the system voltage phasor (also called terminal voltage) and X_s is the machine reactance.

The three-phase complex power supplied to the power grid is:

$$S = 3UI^* = P + jQ. \tag{9}$$

The expression for the active and reactive power is found by substituting I from Eq. (8) in Eq. (9):

$$P = \frac{3E_0U}{X_s}\sin\delta = P_{\max}\sin\delta, \tag{10}$$

$$Q = 3\frac{E_0U\cos\delta - U^2}{X_s}. \tag{11}$$

A closer look at the active power equation (Eq. (10)) shows that the sign of the active power is determined only by the power angle.

- $\delta > 0 \rightarrow P > 0$, the machine works as a generator.
- $\delta = 0 \rightarrow P = 0$, no active power exchange with the grid.
- $\delta < 0 \rightarrow P < 0$, the machine works as a motor.

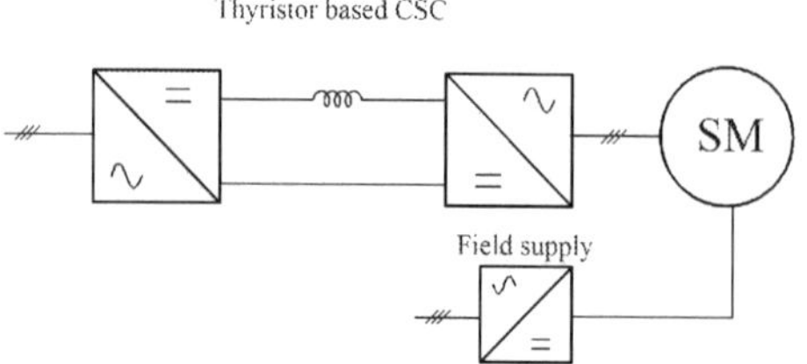

Fig. 13. Full-scale thyristor-based current source converter driving an SYM.[59]

A similar approach can be obtained by analyzing the reactive power equation (Eq. (11)):

- $|\mathrm{E}| \cos(\delta) > |U| \rightarrow Q > 0$, the machine supplies reactive power to the grid and is overexcited.
- $|\mathrm{E}| \cos(\delta) = |\Upsilon| \rightarrow Q = 0$, no reactive power exchange with the grid.
- $|\mathrm{E}| \cos(\delta) < |U| \rightarrow Q < 0$, the machine absorbs reactive power from the grid and is under-excited.

3.1.3 *Variable speed operation using SYMs*

Some of the first applications of variable speed operation of pump-turbines using SYMs employed full-scale converters based on the configuration shown in Fig. 13.

The use of a full-scale converter has an important drawback: the converter losses closely offset the gain in turbine efficiency, leaving little net improvement of efficiency. There would be a gain in capacity at low heads and speeds, but not sufficient to offset the substantial cost of the power electronics converters.[61] Therefore, only a few PSHPs with high demands on power regulation in pumping mode have used this topology for continuous operation.

3.2 *Induction Machines*

3.2.1 *Operating principle*

The induction machine (IM) is the most commonly used industrial motor. There are two general types of IMs: the **squirrel-cage** type and the **wound-rotor** machine. Both machines have a stator structure similar to that of the SYM, consisting of a hollow cylinder of laminated sheet steel in which are punched longitudinal slots. A symmetrical three-phase winding is laid in these slots which, when connected to a suitable voltage source, produces a traveling magnetomotive force (MMF) wave in the air gap, rotating at synchronous speed given in Eq. (7).

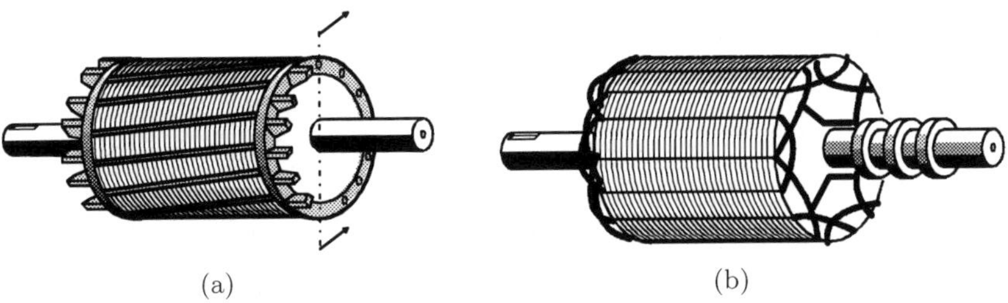

(a) (b)

Fig. 14. Squirrel-cage (a) and wound rotor (b) of an IM.[60]

The squirrel-cage type of rotor (Fig. 14) is made up of sheet steel laminas keyed to the shaft and having slots punched in the periphery. The rotor conductors in most machines are made of aluminum alloy either molded or extruded in place in the slots.

The wound-rotor IM (Fig. 14) has a three-phase insulated winding. This winding is usually wye-connected with the terminals brought out to three slip rings on the shaft. Graphite brushes connected to the slip rings provide external access to the rotor winding.

Application of a three-phase balanced set of currents in the three-phase stator windings, produces a rotating magnetic field of constant magnitude in the uniform air gap. This field links the short-circuited rotor windings and the relative motion induces short-circuit currents in them, which move about the rotor in exact synchronism with the rotating magnetic field. Induced currents react in opposition to the flux linkages producing them, resulting herein a torque on the rotor in the direction of the rotating field.[59,60] This torque causes the rotor to revolve so as to reduce the rate of change of flux linkages, and so the magnitude of the induced current and the rotor frequency. If the rotor were to revolve at exactly synchronous speed, there would be no changing flux linkages about the rotor coils and no torque would be produced. However, the real IMs have friction losses requiring some electromagnetic torque, and the system stabilizes with the rotor revolving at slightly less than synchronous speed. A mechanical shaft load will cause the rotor to decelerate and the rotor current to increase, automatically increasing the torque produced and stabilizing the system at a slightly reduced speed.

The difference in speed between rotor and the rotating magnetic field is termed *slip* which is equal to:

$$s = \frac{n_1 - n}{n_1}, \tag{12}$$

where n_1 is the rotating magnetic field speed (also called synchronous speed and given by Eq. (7)) and n is the rotational rotor speed.

An IM operates as a generator when its stator is connected to the electric grid with fixed frequency and voltage, f, U and it is driven by a prime mover above the synchronous speed, given by:

$$n > \frac{60f}{p}. \tag{13}$$

The electric grid provides reactive power to magnetize the machine and increasing the rotational speed n (above $60f/p$) will increase the active power delivered from the machine to the power grid.

There exist many variable speed IMs using this cage-rotor configuration, for example, in early wind turbine generators. The main drawback of this configuration is its stiffness, as these machines are stable only until n reaches the value:

$$n_{\max} > \frac{60f}{p}(1 + |S_k|), \tag{14}$$

where S_k is the critical slip, which decreases with power and is below 8% for IMs in the hundreds of kilowatts range.[62]

Alternatively, reactive power may be provided by parallel (plus series) capacitors to operate in stand-alone mode. The main drawback of this configuration is that, even if the speed is kept constant through prime mover speed control, the output voltage and frequency vary with the load.

3.3 *DFIM*

For most applications of PSHPs, only a limited controllable speed range is needed during normal operation. This allows for obtaining variable speed operation by utilizing the concept of a DFIM and a power electronic converter with reduced converter rating compared to the total machine rating. This topology has been preferred in most large-scale implementations to limit the converter ratings. With this concept, the industry has been able to build units with total ratings in the range of several hundred of MVAs.[63–65] In addition, the reactive power exchange with the grid can be easily controlled. This can be used for voltage control in the grid and contribute to improving the stability and the operating conditions in the rest of the power system.

When the first commercial, large scale, implementation of variable speed on PSHPs was investigated, the power electronic converters had to be based on thyristors to achieve sufficient ratings. Since the required frequency for the rotor circuit in the DFIG is given by the deviation from synchronous speed, it is usually limited to few Hz. Thus, configurations with **cyclo-converters** have been considered suitable solutions that can be produced with a rugged design for high capacity and low losses[65] (see next section for more details).

As the voltage and current ratings of gate-controlled switches like Gate Turn Off Thyristors (GTOs), Insulated Gate Bipolar Transistor (IGBTs), etc. have increased, topologies based on **back-to-back voltage source converters (VSCs)** have become relevant for feeding the rotor windings of the DFIMs, and are being used in some of the most recent pumped-storage facilities.[66,67]

4 Power Electronics Equipment

The power available in the electronic converter determines the use of one technology or another. Although some manufacturers provide new developments in power converters of 100 MVA,[68] most of them are located around 60 MVA.

An SYM is used when the generation unit does not exceed 60 MVA. It has two main advantages:

- It is a conventional and economical machine.
- The full converter allows starting the pump in a fully controlled mode, avoiding electrical transients.

Its use is particularly interesting in islanded electrical systems, where at the same time the PSHP can be controlled to contribute to the frequency control of the system and also to control grid voltage or reactive power flow. This can reduce the dependency on power generation based on fossil fuels.[69]

A DFIM is the solution of choice when the generation unit exceeds 60 MVA. As explained in the previous section, the electronic converter active power is proportional to the rotor slip, which is less than 0.25 (per unit value) in conventional DFIM. Even though the converter must be oversized because the DFIM consumes reactive power, DFIM units of up to 400 MVA,[70] can be developed with converters of around 60 MVA.

Power generation companies currently consider 60 MVA as an approximate limit for selecting either SYM or DFIM. This limit strongly depends on the cost of the power converter and consequently it will go up with the development of new high power converters. Nowadays, there are two types available:

- Cyclo-converter (direct converter) for high-power devices but with reactive power consumption on the grid side.
- DC-link converter for reduced power but with separate active and reactive power control possibility.

4.1 *Cyclo-converter*

The cyclo-converter is a frequency converter which is based on the three phase — full-wave controlled rectifier. This topology allows higher power development. It has been used in PSHPs by the Japanese, who are pioneers in these applications.[71]

The cyclo-converter is constituted by an association of three rectifiers, as detailed below (Fig. 15). This type of converter is the only one in which there is a direct conversion from AC to AC power; for this reason, it is also called direct converter.

The output frequency (f2) is lower than the input frequency (f1), so it does not require a high-speed semiconductor; thyristors are used instead. Its field of application is in high power and low speed drives, which needs a low frequency power supply. Voltage amplitude and frequency output can be adjusted, from a fixed voltage and frequency input.[72]

As shown in Fig. 15, the cyclo-converter is composed of three double Graetz bridges, so at least 36 thyristors must be used in cases in which only one thyristor is needed to carry the rated current. High current cyclo-converters can be made with several thyristors connected in parallel; hence the total number of thyristors is a multiple of 36.

Since thyristors are line commutated, the control system is relatively simple but the output frequency is lower than 0.4 times the grid frequency.

It is a simple, robust, and cheap technology but it cannot be used in units with SYMs.

The main disadvantages of cyclo-converters are:

- Reactive power consumption by thyristors and the absence of a DC link, causes the reactive consumption required for operating DFIM.
- The inability to force the turn off of the thyristor results in a large ripple in the current exchanged with the grid.

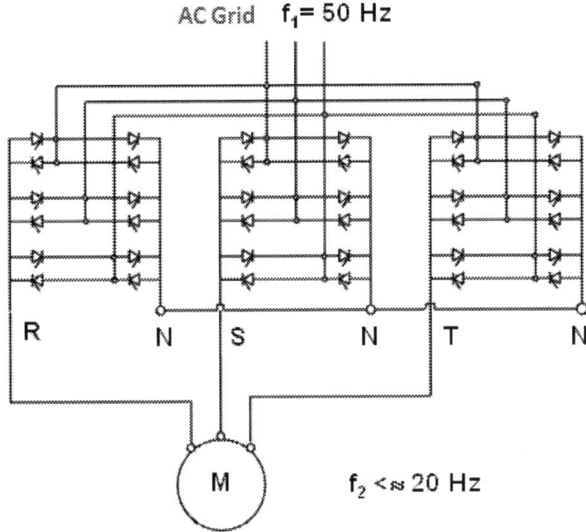

Fig. 15. Cyclo-converter detailed power diagram.

- In order to avoid the transmission of harmonics to the grid, high power harmonic filters must be installed, for which more space must be provided.
- There is a risk of loss of commutation of the thyristors when a grid fault appears. This can cause the DFIM to shut down.

4.2 *DC-link Converter*

This consists of a rectifier stage (AC/DC), a DC-link, and an inverter stage (DC/AC) and it allows operation of the electrical machine with a better efficiency at voltages and frequencies different from the grid ones.

The DC-link provides a decoupling between the frequency and amplitude of the grid voltage and the applied AC machine. The most common is that the DC-link is achieved through a certain capacitance. The capacitance value is chosen to minimize voltage ripple produced by electronic converters and switches.

The simplified diagram in Fig. 16 shows a general configuration of a VSC.

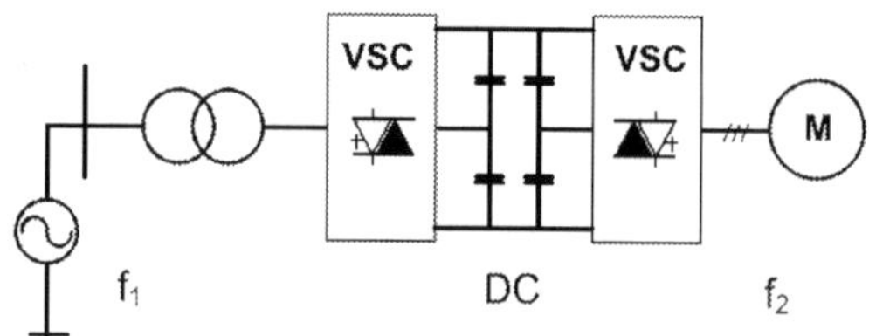

Fig. 16. Simplified scheme of a VSC.

Both rectifier and inverter stages are developed with semiconductors whose shut-down can be forced; this is a more complex power electronics system compared to a direct converter. Although more expensive, this nevertheless greatly improves the performance of the DC-link converter with respect to the cyclo-converter:

- The DC-link converter can be used for both DFIM and SYM.
- The separation of the rectifier and the inverter via a DC link allows controlling P and Q independently.[73]

The use of VSC converter has a number of additional advantages:

- **Low harmonic currents on the grid side.** The installation of expensive filters is not necessary. It produces a reduction in the cost and space requirements.
- **Improvements in the ability to start-up and brake.** Machine start-up and braking are possible without a separate converter, by means of the short circuit connection of the stator windings. The ability for shut-down control of semiconductor devices used in DC-link converters cannot be achieved by thyristors.

DC-link converters have been made possible by the development of semiconductor devices with the ability for shut-down control that cannot be achieved with thyristors. There are three semiconductors with these features:

- GTO
- IGBT
- Insulated Gate Commutated Thyristors (GCT)

Table 6 presents a comparison of the performance of these semiconductors. The development of high-power converters and medium voltage (MV) drives started in the mid 1980s when 4,500 V GTO became commercially available.[74] The GTO was the standard for the MV drives until the advent of high-power IGBTs and GCTs in the late 1990s.[75,76] These switching

Table 6: Comparison of different semiconductors used in DC link converters.

	GTO	IGBT	IGCT
Switching capacity	Low frequency	High frequency Integrated gate driver	High frequency Integrated gate driver
Losses	Low conduction	Low switching High conduction	Low switching Low conduction
Converter design	Complex clamping circuit	Suitable for series/parallel connection at high voltages	Suitable for series/parallel connection at high voltages

devices have rapidly progressed into the main areas of high-power electronics due to their superior switching characteristics, reduced power losses, easy of gate control and snubberless[c] operation.

The MV drives cover power ratings from 0.4 to 40 MW at the medium voltage level of 2.3 to 13.8 kV. The power rating can be extended to 100 MW, where synchronous motor drives with load commutated inverters are often used summarizes the characteristics of these semiconductors.

4.3 *Control Schemes*

One important motivation for variable speed operation of pump-turbines has been the possibility of improving the efficiency of the pump-turbine, since the speed corresponding to maximum efficiency is different for pumping and generating modes, and changes with the water head.

Even more important is the possibility for power control in pumping mode, since traditional pump-turbines with SYMs connected directly to the grid will operate at constant speed and therefore constant power in pumping mode.

From the power system point of view, this possibility is also one of the most important benefits obtained by variable speed operation of PSHPs. The power electronic drive system can also be used to increase the response time for power control by utilizing the inertia of the pump-turbine and the electrical machine, both in generating and pumping mode. The fast response allows for compensation of power fluctuations and the damping of power oscillations, and thus can improve the stability of the power system.

[c]A **snubber** is a device which helps reduce the stress experienced by semiconductors during commutation.

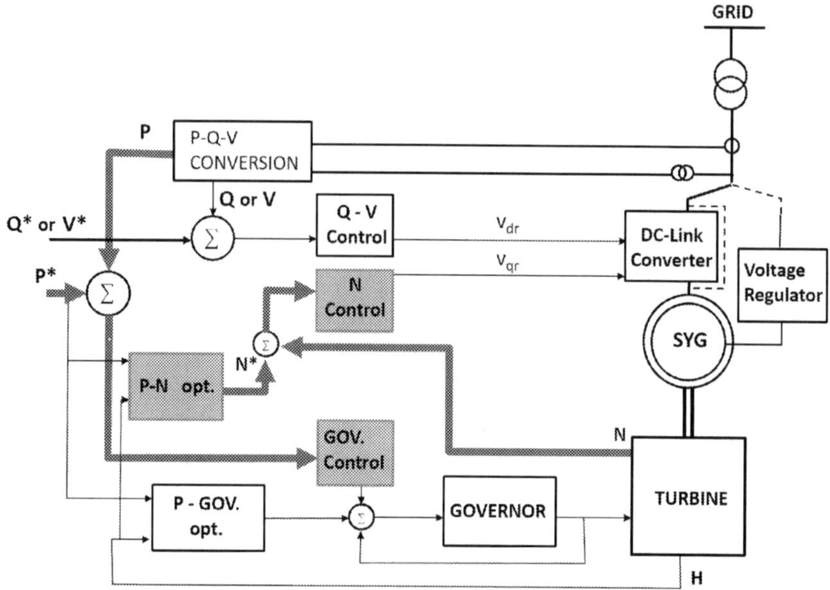

Fig. 17. Scheme of P–Q control system (SYM in generating mode).

In Figs. 17–20, the control schemes typically used for active and reactive power control (P–Q control) in both SYM and DFIM connected pump-turbines are shown.

4.4 *Current Trends and New Developments*

As mentioned above, the cyclo-converter has been the most used converter in the early development of PSHPs. For example, the Ohkawachi Power Station, in Japan, uses this technology.[65] It had a complicated architecture called 6-pulse cyclo-converter composed of 36 thyristors. Its main drawback was the high value of the total harmonic distortion (THD); in the rotor current it can reaches values of 15.69 %.[77]

The second generation of PSHPs began to use DC-link converters also called back to back converters. The simplest topology of back to back converters is the two levels VSC (Fig. 16). They are composed of 12 IGBT or IGCT switching devices, with a free-wheeling diode in parallel with each one. By means of a pulse wide modulation (PWM) or a space vector modulation (SVM), a switching signal is generated. In this device, the THD factor in the rotor current decreases to 4.42%.[77]

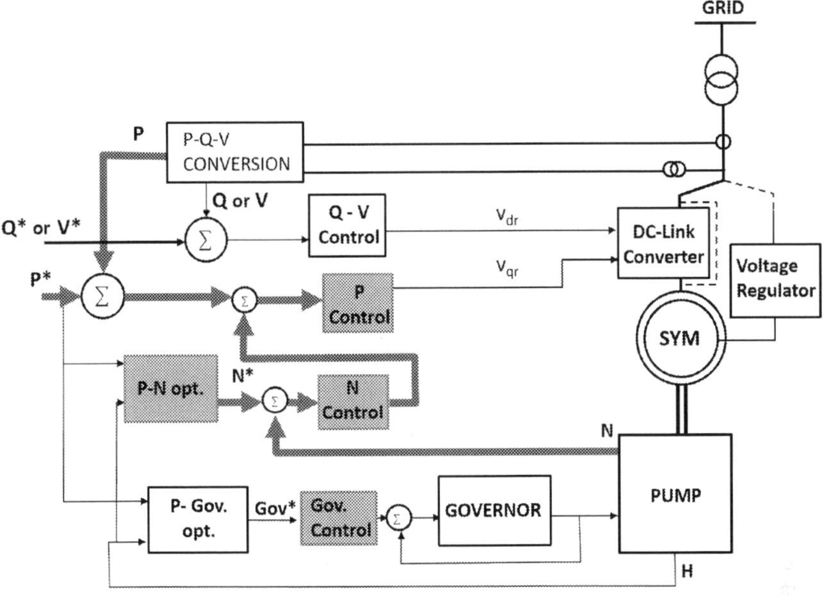

Fig. 18. Scheme for P–Q system (SYM in pumping mode).

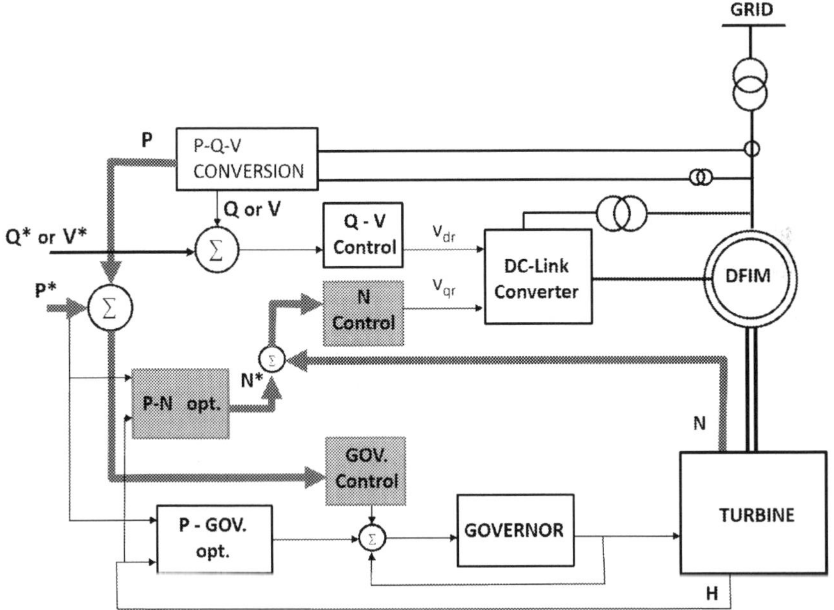

Fig. 19. Scheme of P–Q control (DFIM in generating mode).

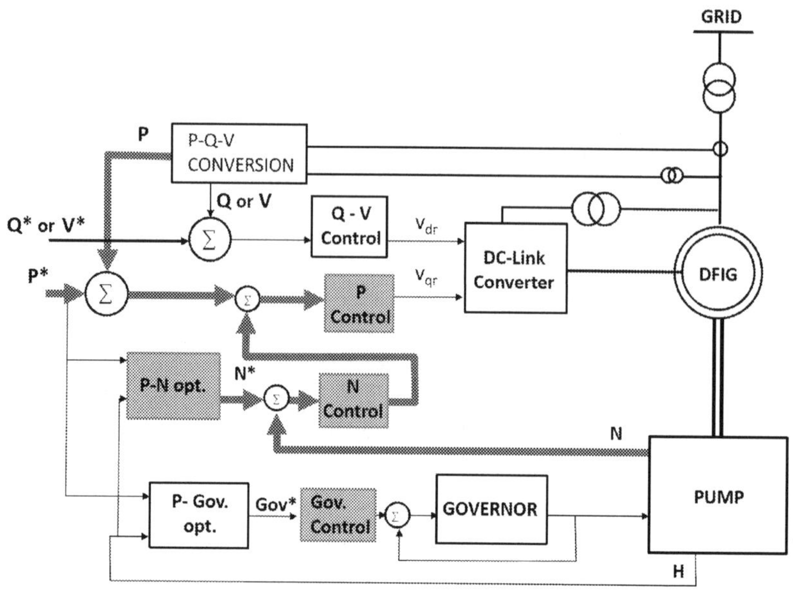

Fig. 20. Scheme of P–Q control system (DFIM in pumping mode).

With a non-negligible complication of the back to back topology, it is possible to improve the THD factor. For example, with a 3-level diode clamped VSC composed of 24 IGBT or IGCT, the THD factor in rotor current decreases down to 4.19%.[77] Industry has developed VSC converters of 5 and 7 levels but their designs are quite complicated.

Nowadays, the research in power electronics converters for variable speed PSHPs is focused on the multilevel modular converter (MMC). This topology is receiving special attention for being used in transforming devices DC/AC and AC/DC as part of the links in high voltage direct current (HVDC).

It is composed of a large number of modules in series. Each module, called an H-bridge, is capable of providing positive, negative, or zero voltage at its output terminals. Therefore, each branch is capable of providing $(2n+1)$ voltage levels where n is the number of modules in a branch. The main advantages of this topology are:

- It requires fewer components than the diode-clamped and circuit for the same number of levels.
- Optimized circuit layout and packing is possible because each level has the same modular structure.

- Almost sinusoidal output voltage.
- Low THD. It can be reduced to 1.07%.[77]

5 Pumped-Storage Hydropower Plants Operation Strategies

Since the early 1990s, electricity markets all over the world have gradually experienced a continuous process of deregulation. These new market schemes have served as a framework for the development of planning tools that assist the power generation companies to better utilize their energy resources.

5.1 *Short-term Operation Strategies*

As liberalized electricity markets evolved and increased in complexity, the marginal approach and peak-shaving method traditionally used to determine the short-term operation schedule of a PSHP[78,79] have been gradually replaced by more sophisticated scheduling tools or models.

Nevertheless, some authors have used a marginal approach to optimally schedule the energy and reserves for the corresponding markets. In 2004, Lu *et al.*[80] obtained an analytical condition depending on the energy and reserve prices, as well as on the unit cycle efficiency, that should be fulfilled in order for the bids to be profitable. In 2010, the above-mentioned condition was revised in order to consider unit start-up and shut-down costs.[81]

The consideration of uncertainty in energy prices in short-term power generation scheduling was a "hot topic" during the last decade. The bidding risk has been modeled by including a weighted penalty term in the objective function as a function of the energy and reserve price variances.[82] By varying the weight value, a set of trade-off solutions can be obtained for which the expected profit cannot increase without increasing risk; the so-called efficient frontier.[83]

The use of the variance as a risk measure has been criticized for its symmetry with respect to the expected value of the revenue.[84] Other indicators, such as the value at risk (VaR) or the conditional value at risk (CVaR), are considered more efficient than the variance as a risk measure. Nevertheless, some authors have recently used it to consider the risk in pumped-hydro scheduling problems.[85]

The MILP-based model proposed by Kazempour *et al.*[85] is aimed at maximizing the weekly profit of a price-taker PSHP in a multimarket framework. The uncertainty associated with the hourly prices of energy, spinning reserve and regulation markets, is considered through the variance of the corresponding forecast errors.

The influence of each risk penalty factor on the expected profit was analyzed separately. The risk penalty factor corresponding to the spinning reserve prices turned out to have the most significant impact on the expected profit, mainly due to the high percentage of the profit that is expected to be obtained from the spinning reserve market.

The CVaR has been used as a risk measure within a day-ahead power generation scheduling problem of a hydrothermal system with pumped-storage units by Dicorato *et al.*[86]

Ugedo *et al.*[87] proposed a multistage stochastic MILP-based model to calculate the bids of a price-maker power company for the day-ahead, the secondary reserve and the first intraday markets. The power company is assumed to own thermal, hydro, and pumped-storage units. The influence of the power company on the prices of the above-mentioned markets is considered through a set of scenarios of residual demand curves (RDCs).

Pumped-hydro units are modeled by means of a single energy balance equation; no decisions on the operation mode (generating/pumping) or the on/off status of the units are considered. Previous formulation was later improved by including binary variables in order to select the operation mode as well as the on/off status of the units.[88]

It is worth mentioning that a significant change in the strategy of the pumped-hydro portfolio was observed by Ugedo and Lobato,[88] as a consequence of considering the intraday market when bidding in the day-ahead and secondary reserve markets. As shown in their paper, the consideration of the reserve market in a deterministic case, yields a 5% increase in the daily revenue, in agreement to some extent with the conclusions drawn in previous works.[79,85]

In this sense, it is interesting to mention the works presented by Pinto *et al.*[89] and Connolly *et al.*[90] Pinto *et al.*[89] used a MILP-based deterministic model to calculate the bids of a price-taker PSHP for the day-ahead and spinning reserve markets. The optimal operation schedules obtained in the paper led in all analyzed cases to negative revenue in the day-ahead market, whereas that obtained in the spinning reserve market turned out to be positive and more than one order of magnitude bigger than the previous one, in absolute value.

Connolly *et al.*[90] compare different price arbitrage strategies in a series of electricity markets. No ancillary service market is considered in the paper. The operation of a PSHP with a 6-h charge–discharge cycle, following each strategy, is simulated during a one-year period, using data from different electricity markets. The results of the simulations show that the expected profit of the PSHP differ considerably from one electricity market to another; the maximum expected profit ranges from 60 to 70 M€/year, whereas the minimum expected profit is roughly 3 M€/year.

From the results of the above-mentioned simulations, authors determined the most profitable strategy and simulated the operation of the PSHP, in accordance with said strategy, during a five-year period, using data from different electricity markets. The results show that even for the same electricity market, the expected profit may vary considerably from one year to another and that even with a low investment cost and a low interest rate, a PSHP is a risky investment in most electricity markets.

The results obtained by Connolly *et al.*,[90] provide evidence that the traditional price arbitrage strategy is far from being economically feasible for a PSHP in a liberalized market context, and point to the importance of the reserve markets to guarantee feasibility. Nevertheless, in order to maximize the profits that a PSHP can obtain from the reserve markets, it is necessary to deal with the uncertainty associated to the effective power delivery request (i.e. the real time use of the committed power reserves). This remains a challenging task.

As far as we know, the influence of the so-called power delivery request on the expected revenues of a PSHP has been only preliminary analyzed by Varkani *et al.*,[91] who proposed a mixed integer nonlinear (MINLP)-based model to obtain the power offers of a wind power plant and a PSHP for the day-ahead, spinning, and regulating reserve markets. Hourly prices of each market are considered in a deterministic way, whereas the uncertainty associated with the wind power production is modeled through a set of scenarios. Different probabilities of power delivery request were considered in the paper. The results included in the paper show that a significant increase in revenues could be obtained as a result of the coordinated operation. In addition, the higher the probability of a power delivery request in both the spinning and regulation reserve markets, the bigger the added value obtained as a result of the coordinated operation.

Similar conclusions regarding the benefits of coordinated operation of wind and pumped-hydro assets were obtained by García-González *et al.*[92] In that paper, a MILP-based model is used to calculate the power bids of a wind farm and a PSHP for the day-ahead market considering the uncertainty associated to both hourly prices and wind power, by means of a scenario tree. The results show that the benefits of the coordinated operation increase as the PSHP size increases; nevertheless, the rate of increase decreases as the PSHP size approaches that of the wind farm.

Such an operation strategy has received special attention in island power systems, such as the Greek and Canary islands, where the fuel costs are usually higher,[93] and where in most cases the electricity market is organized in a centralized way.[94]

Tuohy and O'Malley[95] analyzed the benefits of a PSHP on the operation of the Irish electric power system considering different levels of wind power penetration. For these purpose, the authors use a stochastic MILP-based unit commitment model with hourly time resolution and 36-h rolling planning horizon, updated every 3 h as new wind and load forecasts become available. The objective function of the model is aimed at minimizing the costs for meeting the system demand and spinning and replacement reserves. The results presented in the paper indicate that the investment in the proposed PSHP would be justified for a wind installed capacity greater than or equal to 8.5–9 GW (48–51% of energy from wind), assuming a PSHP lifetime of 40 years.

With respect to the PSHP operation, results show that for a wind installed capacity higher than 6 GW (34% of energy from wind), there would be more days in which the PSHP would pump at peak hours and generate at off-peak ones, as wind energy would be more frequently the marginal technology to meet the system demand.

As demonstrated by Tuohy and O'Malley,[95] it is very important to take into account wind stochasticity to analyze the impact of a PSHP in centralized markets since, otherwise, wind curtailments would be underestimated and, as stated by the authors, the more uncertain the wind energy is, the more beneficial is the flexibility of the PSHP.

Pezic and Moray Cedrés[96] proposed a deterministic MILP-based minimization-cost model to calculate the optimal short-term operation schedule of the world's first Megawatt-level REN system in El Hierro Island (Spain).[29] The objective function consists in minimizing the sum of thermal generation costs (fuel and start-up) and six "artificial" cost terms related to the energy stored in the upper reservoir and wind curtailments, among others.

Given the high wind penetration (higher than maximum historical power demand), both inertia and spinning reserve requirements are considered as constraints in the scheduling model. The power generation schedules turned out to be quite different when considering inertia and spinning reserve requirements, the most noteworthy difference being the continuous simultaneous use of both pumps and turbines to minimize wind power curtailments and meet the inertia and reserve requirements.

5.2 *Long-term Operation Strategies*

All the papers discussed in the previous section are focused on short-term scheduling, and therefore either an end of day or week target of stored energy,[85] or an end of day or week marginal *water value* function[88] is assumed to be known in advance. The use of one or another approach depends on the characteristics of the PSHP under study. For closed-loop PSHPs, a target value is normally used. Traditionally, the target value is determined in such a way that the PSHP can make as much profit as possible from the price arbitrage. For daily-cycle PSHPs, this usually means beginning and ending every day (0:00 am) with the upper reservoir empty. For weekly-cycle PSHPs, the end of week target strongly depends on the difference between Sunday and Monday energy prices.

Nevertheless, considering the ongoing shift from the traditional price arbitrage to an operation more focused on the ancillary services markets, further studies on the determination of the end of day and week targets of closed-loop PSHPs are necessary.

For open-loop PSHPs, both weekly storage targets and marginal water value functions have been used recently in the literature.[88,97] These targets or water value functions are usually obtained as a result of a long-term scheduling problem. The most widely used techniques for long-term hydropower generation scheduling are stochastic dynamic programming (SDP)[98] and stochastic dual dynamic programming (SDDP).[99] The use of one or another technique depends mainly on the "size" of the hydro system under study. The former is subject to the so-called *curse of dimensionality*[100] and therefore, is limited to reduced size hydro systems with one or a few reservoirs.[101] The latter combines the ability of SDP to solve multistage stochastic programming problems, with the use of Benders cuts to approximate the marginal water value functions.[102]

As in the short-term, electricity market considerations have been gradually introduced into long-term hydro scheduling models during the past decades.[103] Nevertheless, given the large computational requirements of these models, it has occurred at a slower pace, and still today remains a challenging task.

The long-term hydro scheduling model presented by Lohndorf *et al.*[104] is, to authors' knowledge, the one where uncertainty in hourly energy prices has been considered in greatest detail.

Another noteworthy long-term scheduling model is the one presented by Helseth *et al.*[105] In that paper, a SDDP-based model is proposed for the long-term scheduling of a hydrothermal system with pumped-storage and wind power. As far as we know, the paper by Helseth *et al.*[105] is the first work where the start-up costs of pumps and thermal generating units are considered within a long-term generation scheduling model.

Considering the new reserve-driven operation paradigm, it seems rational to think that the above-mentioned targets or water values, could vary substantially.

In this regard, it is worth mentioning the study presented in 2012 by Abgottspon and Andersson,[106] where the participation of a large open-loop PSHP in the secondary load-frequency control market was considered within a long-term scheduling model. The authors compared the expected profit at each feasible state with and without the possibility to offer secondary control. The differences observed in the comparison turned out to be of little importance; a 1–2% increase in the expected profit might be attained when secondary control is considered.

The results obtained in the case study regarding the weak sensitivity of the long-term scheduling to the secondary control offers should not be generalized. On the contrary, they should encourage researchers to carry out similar analyses in other PSHPs and electricity markets. The structure of the secondary control market in the Swiss power system might introduce a certain "rigidity" into the PSHP operation, which may not occur in other reserve markets with hourly auctions.

5.3 *Current Trends and Future Challenges in Pumped-Storage Hydropower Plants Operation Strategies*

Current trends in PSHPs operation strategies can be summarized as follows:

— Traditional operation strategies based on price arbitrage and/or peak shaving appear to be no longer economically feasible.

— Ancillary services markets, particularly those related to balancing generation and demand, emerge as the main source of revenue for PSHPs in liberalized market contexts. Analogously, support for the integration of non-dispatchable energies in the electric power system is an essential

role of PSHPs in centralized market contexts, such as those currently in place in many island power systems.
— Consistently, special attention is being given to accurately model the uncertainty in energy and ancillary services prices, as well as in wind and/or solar power production, for the power generation scheduling of PSHPs in both liberalized and centralized market contexts.

Future challenges that would be interesting to deal with in the near future can be summarized as follows:

— Modeling the uncertainty in the power delivery requested in real time for ancillary services purposes, in the short-term, could contribute significantly to increasing the participation of PSHPs in such services, and therefore the penetration of intermittent energy into the electric power system, and thus improving the PSHPs economic feasibility.
— The determination of the long-term guidelines for open-loop PSHPs should be "revisited", considering the new ancillary services or balancing-based operation paradigm. Analogously, the suitability of the end of day or week storage targets traditionally used for closed-loop PSHPs operating in liberalized electricity markets should be reevaluated.
— Taking into account the ongoing PSHPs projects all over the world, studies of the operation strategy of more flexible PSHPs, such as those described in previous sections of this chapter, become mandatory.

References

1. IEA (International Energy Agency), Technology roadmap. Hydropower, 2012. Available at: https://www.iea.org/publications/freepublications/publication/2012_Hydropower_Roadmap.pdf. Accessed on March 26, 2014.
2. IEA (International Energy Agency), *Key World Energy Statistics*, 2012.
3. M. E. Nazari, N. M. Ardehali and S. Jafari, Pumped-storage unit commitment with considerations for energy demand, economics, and environmental constraints, *Energy* **35** (2010), pp. 4092–4101.
4. EPRI (Electric Power Research Institute), Electric energy storage technology options: A white paper primer on applications, costs and benefits, Palo Alto, CA, 1020676, 2010.
5. J. S. Anagnostopoulos and D. E. Papantonis, Study of pumped storage schemes to support high RES penetration in the electric power system in Greece, *Energy*, **45** (2012), pp. 416–423.
6. B. Steffen, Prospects for pumped-hydro storage in Germany, *Energy Pol.* **45** (2012), pp. 420–429.

7. EC (European Commission), Technology map of the European strategic energy technology plan (SET-Plan) — Part I: Technology descriptions, JRC-SETIS work group, EUR 24979 EN, 2011.
8. R. Loisel, Power system flexibility with electricity storage technologies: A technical-economic assessment of a large-scale storage facility, *Int. J. Elec. Power* **42**(1) (2012), pp. 542–552.
9. B. Rangoni, A contribution on electricity storage: The case of hydro-pumped storage appraisal and commissioning in Italy and Spain, *Utilities Policy* **23** (2012), pp. 31–39.
10. J. P. Deane, B. P. O Gallachoir and E. J. McKeogh, Techno-economic review of existing and new pumped hydro energy storage plant, *Renew. Sustainable Energy Rev.* **14** (2010), pp. 1293–1302.
11. D. Connolly, S. MacLaughlin and M. Leahy, Development of a computer program to locate potential sites for pumped hydroelectric energy storage, *Energy* **35** (2010), pp. 375–381.
12. W. F. Pickard, The history, present state, and future prospects of underground pumped hydro for massive energy storage, *Proc. IEEE* **100**(2) (2011), pp. 473– 483.
13. N. Uddin, Geotechnical issues in the creation of underground reservoirs for massive energy storage, *Proc. IEEE* **100**(2) (2012), pp. 484–492.
14. EC (European Commission), European energy pocket book 2009. Available at: http://bookshop.europa.eu/en/eu-energy-and-transport-in-figures-pbKOAB08001/. Accessed on April 29.
15. G. D. Ciocan, O. Teller and F. Czerwinski, Variable speed pump-turbines technology, *Universitatea Politehnica din Bucuresti Scientific Bulletin, Series D.* **74**(1) (2012), pp. 33–42.
16. J. M. Henry, J. B. Houdeline, S. Ruiz and T. Kunz, How reversible pump-turbines can support grid variability — the variable speed approach, HYDRO 2012 Innovative Approaches to Global Challenges, Bilbao, Spain, October 29–31, 2012.
17. T. Fujihara, H. Imano and K. Oshima, Development of pump turbine for seawater storage power plant, *Hitachi Rev.* **47**(5) (1998), pp. 199–202.
18. T. Hino and A. Lejeune, Pumped storage hydropower developments, *Comprehensive Renewable Energy* **6** (2012), pp. 405–434.
19. A. Pina, C. S. Ioakimidis and P. Ferrao, Economic modeling of a seawater pumped-storage system in the context of São Miguel, in *IEEE International Conference on Sustainable Energy Technologies*, pp. 707–712 (2008).
20. E. McLean and D. Kearney, An evaluation of seawater pumped hydro storage for regulating the export of renewable energy to the national grid, *Energy Procedia* **46** (2014), pp. 152–160.
21. J. Alterach, A. Danelli, M. Meghella, G. Cavazzini and A. Stoppato, Seawater pumped hydro plant with a variable speed reversible pump-turbine: A case study for the Italian Islands, submitted for HYDRO 2014.
22. R. D. Harza, Hydro and pumped storage for peaking, *Power Eng.* **64**(10) (1960), pp. 79–82.

23. R. Madlener and J. M. Specht, An exploratory economic analysis of underground pumped-storage hydro power plants in abandoned coal mines, FCN Working Paper No. 2/2013, February 2013. Available at: http://ssrn.com/abstract=2350106.
24. Project ENE 1001, Brief description of the Muuga seawater-pumped hydro accumulation power plant, 2010. Available at: http://energiasalv.ee/wp-content/uploads/2012/07/Muuga_HAJ_17_02_2010_ENG.pdf.
25. W. W. De Boer, F. J. Verheij, D. Zwemmer and R. Das, The energy island — An inverse pump accumulation station, in EWEC 2007, European Wind Energy Conference (2007).
26. Alstom, Hydro pumped storage power plant, 2011. Available at: http://www.alstom.com/Global/Power/Resources/Documents/Brochures/hydro-pumped-storage-power-plant.pdf. Accessed on March 18, 2014.
27. L. Cuesta and E. Vallarino, *Aprovechamientos Hidroeléctricos*, (ed.) (Colegio de Ingenieros de Caminos, Canales y Puertos, Madrid, 2000).
28. J. Navarro and J. C. Elipe, La Muela II pumped storage plant: Electromechanical and construction project, in *Proc. HYDRO Conference* (2012).
29. J. Merino, C. Veganzones, J. A. Sánchez, S. Martínez and C. A. Platero, Power system stability of a small sized isolated network supplied by a combined wind- pumped storage generation system: A case study in the Canary Islands, *Energies* **5** (2012), pp. 2351–2369.
30. A. Mittereger and G. Penninger, Austrian pumped storage power stations supply peak demands, *World Pumps* **500** (2008), pp. 16, 18, 20–21.
31. Voralberger Illwerke AG, Kopswerk II, Vorarlberger Illwerke Aktiengesellschaft, Bregenz Austria. Available (in German) at: http://www.kopswerk2.at/downloads/090505_Vorstand_OK_KOWII_Prospekt.pdf. Accessed on March 18, 2014.
32. F. Lippold and N. Hellstern, Hongrin-Léman hydroelectric pumped storage plant, Veytaux II powerhouse — developing a new generation of multistage storage pumps, in *Proc. HYDRO Conference* (2012).
33. G. Ardizzon, G. Cavazzini and G. Pavesi, A new generation of small hydro and pumped-hydro power plants: Advances and future challenges, *Renew. Sustainable Energy Rev.* **31** (2014), pp. 746–761.
34. E. Kopf and S. Brausewetter, Control strategies for variable speed machines — design and optimization criteria, in *Proc. HYDRO Conference* (2005).
35. M. Geise, Pump Turbines — State-of-the-art, in *Proc. I Latin American Hydro Power and Systems Meeting*, Campinas, Sao Paolo, Brasil (2013). Available at: http://www.latiniahr.org/meeting/apresentacoes/geise.pdf. Accessed on March 18, 2014.
36. J. Koutnik, *Hydro Power Plants*, AGCS Expert Days 2013, Munich.
37. J. B. Houdeline, J. Liu, S. Lavigne, Y. Laurant and L. Balarac, Start-up improvement in turbine mode for high head PSP machine, *IOP Conf. Series: EES* **15**(4) (2012), pp. 1–10.

38. Ch. Gentner, M. Sallaberger, Ch. Widmer, O. Barun and T. Staubli, Numerical and experimental analysis of instability phenomena in pump turbines, *IOP Conf. Series: EES* **15** (2012), pp. 1–8.
39. G. Olimstad, T. Nielsen and B. Borresen, Stability limits of reversible pump turbines in turbine mode of operation and measurements of unstable characteristics, *J Fluid. Eng.* **134** (2012), p. 111202.
40. J. X. Zhou, B. W. Karney, M. Hu and J. C. Xu, Analytical study on possible self- excited oscillation in S-shaped regions of pump-turbines, *PI Mech. Eng. A — J. Pow.* **225** (2011), pp. 1132–1142.
41. S. Pejovic, Q. F. Zhang, B. Karney and A. Gajic, Analysis of pump-turbine S instability and reverse waterhammer incidents in hydropower systems, 4th International Meeting on Cavitation and Dynamic Problems in Hydraulic Machinery and Systems, Belgrade, Serbia, 2014.
42. R. Barrio, J. Fernandez, E. Blanco, J. Parrondo and A. Marcos, Performance characteristics and internal flow patterns in a reversible running pump turbine, *PI Mech. Eng. C — J. Mech. Eng. Sci.* **226**(3) (2012), pp. 695–708.
43. U. Seidel, J. Koutnik and G. Martin, S-curve characteristic of pump-turbines, HYDRO 2012 Innovative Approaches to Global Challenges, Bilbao, Spain, October 29–31, 2012.
44. V. Hasmatuchi, M. Farhat, S. Roth, F. Botero and F. Avellan, Experimental evidence of rotating stall in a pump-turbine at off- design conditions in generating mode, *J. Fluid. Eng.* **133** (2011), p. 051104.
45. L. Wang, J. Yin, L. Jiao, D. Wu and D. Qin, Numerical Investigation on the S characteristics of a reduced pump turbine model, *Sci. Technol. Sc.* **54**(5) (2011), pp. 1259–1266.
46. H. Sun, R. Xiao, W. Liu and F. Wang, Analysis of S characteristics and pressure pulsation in a pump-turbine with misaligned guide vanes, *J. Fluid. Eng.* **153** (2013), pp. 0511011–0511016.
47. C. Widmer, T. Staubli and N. Ledergerber, Unstable characteristic and rotating stall in turbine brake operation of pump turbines, *J. Fluid. Eng.*, **133** (2011), pp. 1–9.
48. J. Gonzales, C. Santolaria, E. Blanco and J. Fernandez, Numerical simulation of the dynamics effects due to impeller-volute interaction in a centrifugal pump, *J. Fluid. Eng.* **124** (2002), pp. 348–355.
49. T. Sano, Y. Yoshida, Y. Tsujimoto, Y. Nakamura and T. Matsushima, Numerical study of rotating stall in a pump vaned diffuser, *J. Fluid. Eng.* **124** (2002), pp. 363–370.
50. S.-S. Hong and S.-H. Kang, Flow at the centrifugal pump impeller exit with the circumferential distortion of the outlet static pressure, *J. Fluid. Eng.* **126**(1) (2004), pp. 81–86.
51. K. Majidi, Numerical study of unsteady flow in a centrifugal pump, *J. Turbomach.* **127**(2) (2005), pp. 363–371.
52. S. Guo and Y. Maruta, Experimental investigations on pressure fluctuations and vibration of the impeller in a centrifugal pump with vaned diffusers. *JSME Int. J., Ser. B* **48**(1) (2005), pp. 136–143.

53. C. G. Rodriguez, E. Egusquiza and I. F. Santos, Frequencies in the vibration induced by the rotor stator interaction in a centrifugal pump turbine, *J Fluid. Eng.* **129**(11) (2007), pp. 1428–1435.
54. G. Pavesi, G. Cavazzini and G. Ardizzon, Time–frequency characterization of the unsteady phenomena in a centrifugal pump, *Int. J. Heat Fluid Fl.* **29** (2008), pp. 1527–1540.
55. J. F. Gülich, *Centrifugal Pumps* (2nd edition) (Springer, Berlin, Heidelberg, New York, 2010).
56. G. Cavazzini, G. Pavesi and G. Ardizzon, Pressure instabilities in a vaned centrifugal pump, *PI Mech. Eng. A — J. Pow.* **225**(7) (2011), pp. 930–939.
57. L. Knapen, A. U. H. Yasar and D. Janssens (eds.), *Data Science and Simulation in Transportation Research* (Chapter 13) (IGI-Global, Hershey, 2013).
58. USGS (United States Geological Survey), Hydroelectric power: How it works. Available at: http://ga.water.usgs.gov/edu/hyhowworks.html. Accessed on March 24, 2014.
59. A. E. Fitzgerald, Ch. Kingsley and S. Umans, *Electric Machinery* (6th edition) (McGraw-Hill, London, 2003).
60. J. Fraile Mora, *Máquinas Eléctricas* (6th edition) (McGraw-Hill, London, 2008).
61. W. B. Gish, J. R. Schurz, B. Milano and F. R. Scheilf, An adjustable speed synchronous machine for hydroelectric power applications, *IEEE Trans. Power App. Syst.* **100**(5) (1981), pp. 2171–2176.
62. I. Boldea, *Synchronus Generators* (CRC Taylor & Francis, London, 2006).
63. J. K. Lung, Y. Lu, W. L. Hung and W. S. Kao, Modeling and dynamic simulations of doubly fed adjustable speed pumped storage units, *IEEE Trans. Energy Convers.* **22**(2) (2007), pp. 250–258.
64. J. M. Merino and A. López, ABB Varspeed generators boosts efficiency and operating flexibility of hydropower plant, *ABB Rev.* **3** (1996), pp. 33–38.
65. T. Kuwabara, A. Shibuya and H. Furuta, Design and dynamic response characteristics of 400 MW adjustable speed pumped storage unit for Ohkawachi power station, *IEEE Trans. Energy Convers.* **11**(2) (1996), pp. 376–384.
66. A. Hodder, J. J. Simond and A. Schwery, Double-Fed asnychronous motor-generator equipped with a 3-level VSI cascade, in *Proc. 39th IEEE IAS Annual Meeting* (2004).
67. A. Hämmerli and B. Odergard, AC excitation with ANPC-ANPC converter technology tailored to the needs of AC excitation equipment for pump storage plants, *ABB Rev.* **3** (2008), pp. 40–43.
68. H. Schlunegger and A. Thöni, 100 MW full-size converter in the Grimsel 2 pumped-storage plant, in *Proc. HYDRO Conference* (2013).
69. J. A. Suul, K. Uhlen and T. Undeland, Wind power integration in isolated grids enabled by variable speed pumped storage hydropower plant, in *Proc. IEEE International Conference on Sustainable Energy Technologies* (2008).
70. Mitsubishi Electric Corporation, Pumped storage power station with Adjustable speed pumped storage technology (2008). Available at: http://www.teriin.org/events/docs/present_japan/sess4/yokota_part1-2-malco.pdf. Accessed on March 18, 2014.

71. O. Nagura, M. Higuchi, K. Tani, and T. Oyake, Hitachi's adjustable speed pumped-storage systems contributing to prevention of global warming, *Hitachi Rev.* 59(3) (2010), pp. 99–105. Available at: http://www.hitachi.com/rev/field/powersystems/2133182_43332.html. Accessed on March 18, 2014.
72. J. M. Merino, Convertidores de Frecuencia Para Motores de Corriente Alterna (Mcgraw-Hill, New York, 1997).
73. N. Mohan, T. M. Undeland and W. P. Robbins, *Power Electronics: Converters, Applications, and Design* (John Wiley & Sons, New York, 1989).
74. S. Rizzo and N. Zargari, Medium voltage drives: what does the future hold?, in *Proc. International Power Electronics and Motional Control Conference* (2004).
75. H. Brunner, M. Hierholzer, T. Laska and A. Porst, Progress in development of the 3.5kV high voltage IGBT/diode chipset and 1200A module applications, in *Proc. IEEE International Symposium on Power Semiconductor Devices and IC's* (1997).
76. P. K. Steimer, H. E. Gruning, J. Werninger, E. Carroll, S. Klaka and S. Linder, IGCT — A new emerging technology for high power low cost inverters, in *Proc. IEEE Industry Applications Society Annual Meeting* (1997).
77. O. H. Abdalla and M. Han, Power electronics converters for variable speed pump storage, *IJPEDS* **3**(1) (2013), pp. 74–82.
78. A. J. Wood and B. F. Wollenberg, *Power Generation Operation and Control* (2nd edition) (John Wiley & Sons, New York, 1996).
79. R. Deb, Operating hydroelectric plants and pumped storage units in a competitive environment, *The Electricity Journal* **13**(3) (2000), pp. 24–32.
80. N. Lu, J. H. Chow and A. Desrochers, Pumped-storage hydro turbine bidding strategies in a competitive electricity market, *IEEE Trans. Power Syst.* **19**(2) (2004), pp. 834–841.
81. P. Kanakasabapathy and K. Shanti Swarup, Bidding strategy for pumped-storage plant in a pool-based electricity market, *Energy Convers. Manage.* **51**(3) (2010), pp. 572–579.
82. E. Ni, P. B. Luh and S. Rourke, Optimal integrated generation bidding and scheduling with risk management under a deregulated power market, *IEEE Trans. Power Syst.* **19**(1) (2004), pp. 600–609.
83. A. J. Conejo, F. J. Nogales, J. M. Arroyo and R. García-Bertrand, Risk-constrained self-scheduling of a thermal power producer, *IEEE Trans. Power Syst.* **19**(3) (2004), pp. 1569–1574.
84. L. Hongling, J. Chuanwen and Z. Yan, A review on risk-constrained hydropower scheduling in deregulated market, *Renew. Sustainable Energy Rev.* **12**(5) (2008), pp. 1465–1475.
85. S. J. Kazempour, M. P. Moghaddam, M. R. Haghifam and G. R. Yousefi, Risk- constrained dynamic self-scheduling of a pumped-storage plant in the energy and ancillary service markets, *Energy Convers. Manage.* **50**(5) (2009), pp. 1368– 1375.

86. M. Dicorato, G. Forte, M. Trovato and E. Caruso, Risk-constrained profit maximization in day-ahead electricity market, *IEEE Trans. Power Syst.* **24**(3) (2009), pp. 1107–1114.
87. A. Ugedo, E. Lobato, A. Franco, L. Rouco, J. Fernández-Caro and J. Chofre, Strategic bidding in sequential electricity markets, *IET Gener. Transm. Distrib.* **153**(4) (2006), pp. 431–442.
88. A. Ugedo and E. Lobato, Validation of a strategic bidding model within the Spanish sequential electricity market, in *Proc. IEEE 11th International Conference on Probabilistic Methods Applied to Power Systems* (2010).
89. J. Pinto, J. de Sousa and M. Ventim Neves, The value of a pumping-hydro generator in a system with increasing integration of wind power, in *Proc. 8th International Conference on the European Energy Market* (2011).
90. D. Conolly, H. Lund, P. Finn, B. V. Mathiesen and M. Leahy, Practical operation strategies for pumped hydroelectric energy storage (PHES) utilizing electricity price arbitrage, *Energy Pol.* **39**(7) (2011), pp. 4189–4196.
91. A. K. Varkani, A. Daraeepour and H. Monsef, A new self-scheduling strategy for integrated operation of wind and pumped-storage power plants in power markets, *Appl. Energy* **88**(12) (2011), pp. 5002–5012.
92. J. García-González, R. Moraga, R. Matres and A. Mateo, Stochastic joint optimization of wind generation and pumped-storage units in an electricity market, *IEEE Trans. Power Syst.* **23**(2) (2008), pp. 460–467.
93. T. L. Jensen, Renewable energy on small islands, Copenhagen, Denmark (1998). Available at: http://www.inforse.dk/doc/renewable_energy_on_small_islands.pdf. Accessed on March 19, 2014.
94. STORIES Project, Deliverable 3.4: Scheme for market organization of autonomous electricity systems (2012). Available at: http://www.stories project.eu/docs/STORIES_Deliverable3_4FINAL 2_.pdf. Accessed on March 19, 2014.
95. A. Tuohy and M. O'Malley, Pumped storage in systems with very high wind penetration, *Energy Pol.* **39**(3) (2011), pp. 1965–1974.
96. M. Pezic, M. and V. Moray Cedrés, Unit commitment in fully renewable, hydro- wind energy systems, in *Proc. International Conference on the European Energy Market* (2013).
97. A. Borghetti, C. D'Ambrosio, A. Lodi and S. Martello, An MILP approach for short-term hydro scheduling and unit commitment with head-dependent reservoir, *IEEE Trans. Power Syst.* **23**(3) (2008), pp. 1115–1124.
98. J. R. Stedinger, B. F. Sule and D. P. Loucks, Stochastic dynamic programming models for reservoir operation optimization, *Water Resour. Res.* **20**(11) (1984), pp. 1449–1505.
99. M. V. F. Pereira and L. M. V. G. Pinto, Stochastic optimization of a multireservoir hydroelectric system: A decomposition approach, *Water Resour. Res.* **21**(6) (1985), pp. 779–792.
100. S. Dreyfus and A. Law, *The Art and Theory of Dynamic Programming* (Academic Press, Cambridge, 1977).

101. J. W. Labadie, Optimal operation of multireservoir systems: State-of-the-art review, *J. Water Res. Pl — ASCE* **130**(2) (2004), pp. 93–111.
102. A. Gjelsvik, B. Mo and A. Haugstad, "Long- and medium-term operations planning and stochastic modelling in hydro-dominated power systems based on stochastic dual dynamic programming", in *Handbook of Power Systems I, Energy Systems*, eds. S. Rebennack *et al.* (Springer, Berlin, 2010), pp. 33–55.
103. O. B. Fosso, A. Gjelsvik, A. Haugstad, B. Mo and I. Wangensteen, Generation scheduling in a deregulated system: The Norwegian case, *IEEE Trans. Power Syst.* **14**(1) (1999), pp. 75–81.
104. N. Löhndorf, D. Wozabal and S. Minner, Optimizing trading decisions for hydro storage systems using approximate dual dynamic programming, *Oper. Res.* **61**(4) (2013), pp. 810–823.
105. A. Helseth, A. Gjelsvik, B. Mo and U. Linnet, A model for optimal scheduling of hydro thermal systems including pumped-storage and wind power, *IET Gener. Transm. Distrib.* **7**(12) (2013), pp. 1–9.
106. H. Abgottspon and G. Andersson, Approach of integrating ancillary services into a medium-term hydro optimization, in *Proc. XII Symposium of Specialists in Electrical Operational and Expansion Planning* (2012).

Chapter 3

Compressed Air Energy Storage

Jihong Wang[*,†,‡], Xing Luo[*,§], Christopher Krupke[*,¶] and Mark Dooner[*,‖]

[*]*School of Engineering, University of Warwick, Coventry CV4 7AL, UK*
[†]*School of Electrical and Electronic Engineering, Huazhong University of Science and Technology, Wuhan, P.R.China*
[‡]*jihong.wang@warwick.ac.uk*
[§]*xing.luo@warwick.ac.uk*
[¶]*c.krupke@warwick.ac.uk*
[‖]*m.dooner.1@warwick.ac.uk*

With the rapid increase of power generation from renewable energy sources, electrical power networks face a great challenge in maintaining operation stability and reliability. Various solutions are currently under investigation, which include energy storage (ES). Compared with all the ES technologies under consideration, compressed air energy storage (CAES) has the power rating and scale comparable to pumped-hydro ES. This distinguishes CAES from other ES technologies in grid scale applications. This chapter explains how CAES works and what its advantages and disadvantages are. The recent developments in CAES technology are reviewed and the application potential of CAES in supporting power network operation is analyzed. The research challenges and needs are also highlighted in this chapter.

1 Introduction

Compressed air energy storage (CAES) refers to the energy stored in the form of high pressure compressed air and consumed in a different form of energy converted from the compressed air. When CAES is used to serve power network operation, air is compressed to high pressure using compressors during the periods of low electric energy demand and then the stored

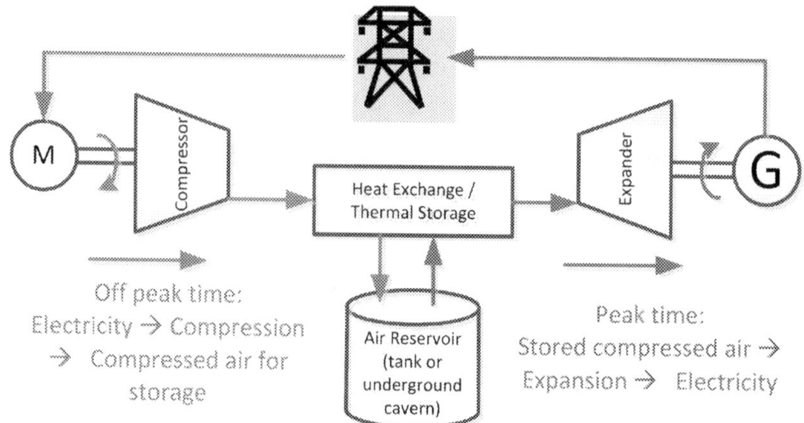

Fig. 1. A CAES process connected to the power network.

compressed air is released to drive an expander for electricity generation to meet high load demand during the peak time period[1,2] as illustrated in Fig. 1. A CAES system can work for a range of power scales from small (kW) to large (>100 MW) and for storage time durations from short (minutes) to long (days), depending on the volume of the storage reservoir used. CAES is often combined with thermal energy storage (ES) to store the heat generated in the compression process and then reuse it during the expansion stage, which improves the system round-trip efficiency.

CAES technology development and deployment started in the middle of the 20th century. In 1949, Laval obtained the patent on using air to store energy inside an underground air storage cavern, which marked the beginning of the new era of CAES exploration.[1,3] The world's first utility-scale CAES plant was installed and commissioned for operation by Brown Boveri (today "Asea Brown Boveri, ABB") in Huntorf, Germany, in 1978.[3] It had a rated generation capacity of 290 MW, for providing load following service and meeting the peak demand while maintaining the constant capacity factor of the nuclear power plant.[3,4] Due the relatively low cost of CAES compared to oil and gas through the 1980s–1990s, the development and applications of CAES technology continued to stay attractive. In 1991, the second large-scale CAES plant commenced operation in McIntosh, Alabama, USA.[4] This plant has a generation capacity of 110 MW, with a storage capacity of 2,700 MWh, and is capable of continuously delivering its full power output for up to 26 h; the plant is used for off-peak power storage, peak power generation, and spinning reserve services.[4,5]

Research and development (R&D) of CAES has continued and has become more active in recent years. In 2010, RWE Power, General Electric, Züblin and German Aerospace Centre (Deutsches Zentrum für Luft- und Raumfahrt e.V — DLR) started working on the world's first large-scale advanced adiabatic (AA-CAES) demonstration plant project, aiming to avoid using any fossil fuel in this CAES plant by using the heat generated in the compression stage at the expansion process.[6] However, this project was put on hold due to the lack of clarity about its economic and business viability. Meanwhile, some R&D work in small-scale CAES attempted to use CAES to replace chemical batteries in some applications, such as uninterruptible power supplies (UPS), plus stand-alone and backup power systems. The US-based LightSail Energy Ltd. patented and developed a CAES technology which came very close to achieving isothermal compression and expansion (I-CAES) and which captures the heat from the compression process by spraying water (with water drop sizes at the nano-scale) for efficient heat absorption and storage. The stored heat is then added to the compressed air during expansion using the same technology.[7] Flowbattery (ex Energetix Group) in the UK launched its small-scale (less than 50 kW/unit) compressed air UPS products in 2009.[1,8] Liquid air energy storage (LAES) can be considered as a variant of CAES since there are similarities in the operation mechanism between CAES and LAES; many important components needed in building a LAES system are also required by CAES. The UK-based start-up company, Highview Power Storage, built a pilot LAES demonstration plant with 300 kW power rating and 2.5 MWh storage capacity. The LAES plant was in operation associated with an 80 MW biomass plant from 2010 to 2014.[9] In 2015, the LAES plant was transferred to the University of Birmingham campus in the UK and is used for education and research, serving as a "living laboratory".[10] Highview and their partner, Viridor, were awarded funding from the British Government Department of Energy and Climate Change (DECC), to build a 5 MW LAES technology system. It is planned to be operational in 2016.[9]

With CAES technology development, various concepts with different strengths and weaknesses are proposed and demonstrated in different applications. Figure 2 shows one way of classifying different CAES technological concepts,[11] which is dependent on the management of the thermal energy in the CAES process,[12] including (i) diabatic — the heat generated from compression is released to the ambient via a cooling system and an additional heat source is required for the expansion process to preheat the compressed air before it enters the expander. This process has an efficiency of around

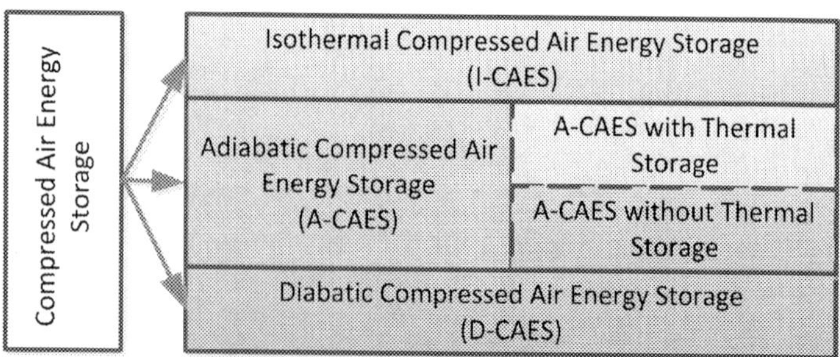

Fig. 2. A way of classifying CAES technologies.

54%[11]; (ii) adiabatic — the heat generated from the compression process is captured and stored and is then used to heat the compressed air prior to expansion. The energy efficiency for the adiabatic approach has not yet achieved the optimal level and the technology development aims to bring the cycle efficiency to over 70%[6,11]; (iii) isothermal — in an isothermal process, the process is well controlled to maintain the temperature with the minimum changes or unchanged so the heat resulting from compression is minimized.[7,11] The process has a current efficiency of 38% but aims to achieve 80% efficiency.[11]

This chapter presents the working principles of different CAES technologies, technology development, typical technical characteristics, technological and economic challenges, and issues associated with the future development of CAES. The chapter is a research outcome of the Engineering and Physical Sciences Research Council EPSRC UK funded research project "Integrated Market-fit and Affordable Grid-scale Energy Storage" (EP/K002228/1). The overview of CAES technology development was initially summarized in the report prepared for the European Energy Research Alliance in 2014.[1]

2 Description of a CAES System Operation

The common components of a CAES system must include compressors, expanders, and an air storage reservoir. The rest of the system components depend on the system structure and operation principles. The Huntorf and McIntosh CAES plants both have gas turbines (since the process involves burning gas), a main shaft connecting all primary components including the compressors, one reversible electric machine and the turbines; they are engaged for different operation modes through clutches. Figure 3 shows a

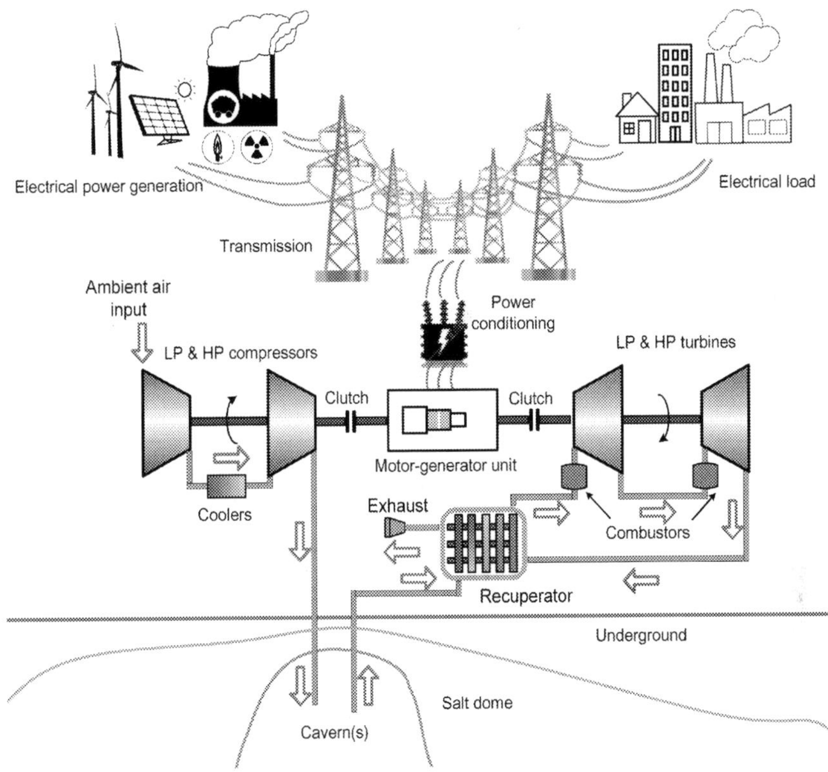

Fig. 3. Schematic diagram of a CAES system.

schematic diagram for a typical large-scale CAES plant which is composed of six major sections:

- A reversible electric motor-generator unit combined with clutch mechanisms to provide alternate engagement as a motor in the compression mode or as a generator in the expansion mode;
- A multi-stage compressor unit operating with intercoolers and aftercoolers to take away the heat generated during air compression;
- One or more cavities for storing a large amount of compressed air;
- A turbine train, containing, in general, both high-pressure and low-pressure turbines, for air expansion;
- A power conditioning system to ensure that the electricity generation matches the requirements;

- Auxiliaries used for control and energy saving, such as fuel storage and heat recuperator units.

The compression mode of a typical CAES plant is activated at the time of low power demand. The surplus electricity is used to run a chain of compressors to inject the pressurized air into the storage reservoir with a typically pressure of around 4.0–8.0 MPa at the temperature of the surrounding formation. Such a compression process can use intercoolers and after-coolers to reduce the working temperature of the injected air, thus improving the compression efficiency and minimizing thermal stress on the storage volume's walls.[4] When the power generation cannot meet the load demand, the expansion mode will be triggered. The stored high pressure compressed air is released from the storage reservoir, heated, and then fed into a high-pressure turbine.[4,5,13,14] In general, the combustion process with the mixed compressed air and fuel (typically natural gas) occurs in the combustor of the high-pressure turbine. Then the gas from the outlet of the high-pressure turbine mixed with additional fuel combusts in the combustor of the low-pressure gas turbine. Both the high-pressure and low-pressure turbines are connected to an electrical generator to generate electricity as shown in Fig. 3. The exhaust heat of the overall system can be recycled before being released into the atmosphere; for instance, a heat recuperator unit is used to recover heat energy from the exhaust at the McIntosh CAES plant. Also, the airflow from the reservoir to the turbine must be high enough to meet the system operation requirements. The low temperature resulting from the incomplete or even the absence of combustion in the expansion mode would pose a significant risk for turbine blades.[4]

The advance of the gas turbine technology has led to improved designs for large-scale CAES applications. For instance, CAES with Air Injection (CAES-AI) technology was patented by Energy Storage and Power Corporation (ESPC). The technology is to directly inject the stored and preheated air into the compressor discharge plenum and into a gas turbine thus increasing the gas turbine's power output. From the simulation and validation tests conducted by ESPC, a total power of ~137 MW in the CAES-AI system can be achieved via a combination of the combustion turbine power of ~112 MW and the additional CAES power of 25 MW generated due to the air injection.[15] Figure 4 shows the schematic layout of this technology. The proposed design has the benefit of eliminating switchover time limitations by decoupling the compression and turbo expander train, thus improving system efficiency.

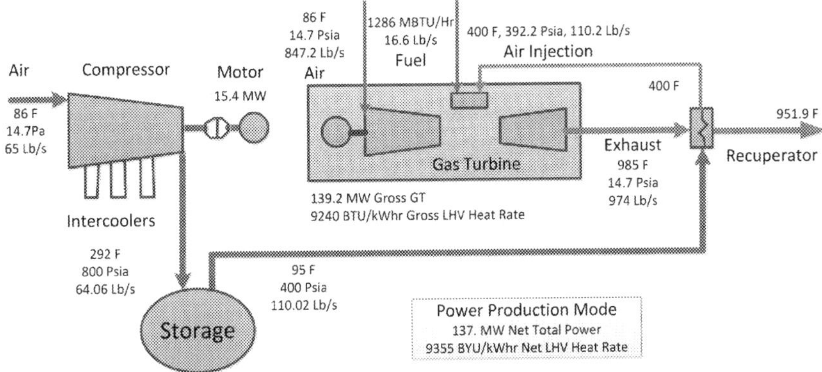

Fig. 4. Schematic layouts for CAES with air injection concept.[15,16]

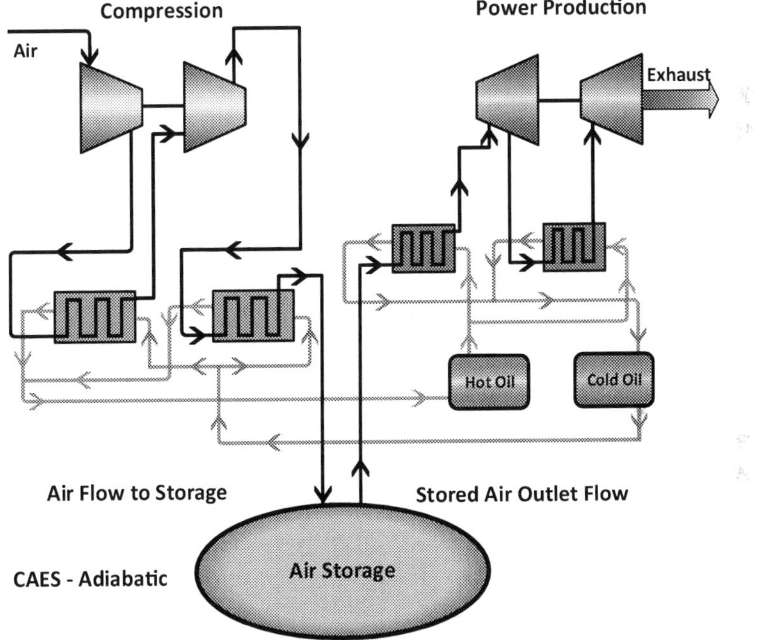

Fig. 5. Schematic layout of an AA-CAES plant.[16]

AA-CAES is a well-accepted CAES technology and an example to demonstrate this concept is illustrated in Fig. 5. In AA CAES, the air is adiabatically compressed and pumped into a storage reservoir. The heat generated during the compression process is stored using a thermal storage mechanism. During expansion, the compressed air is released and heated by passing through the thermal storage unit, so there is no combustion process involved. The key component of such an AA-CAES system is the heat exchanger. Cooling airflow through the compressors and heating the input airflow to each turbine are all achieved via the heat exchangers. The heat exchangers absorb heat from the high-temperature compressed air and store the thermal energy for reheating the air before expansion. However, the heat exchangers may lead to an increase of the overall system cost.[17] Theoretically, the overall round-trip efficiency of AA-CAES is higher than that of the conventional CAES plants.[4,16] AA-CAES technology is currently under development and the main challenges are associated with heat energy recovery including heat exchange and storage technologies.

R&D in small-scale CAES technology has been active in recent years.[18,20] With a small-scale CAES system (less than 10 MW), the storage reservoir can be an over-ground vessel with suitable dimensions. The high-pressure compressed air can be generated by onsite compression facilities and can also be provided through a pressured air product supplier. Liquefied air energy storage (LAES) can be considered as one of the small-scale CAES systems currently in demonstration.[9] Liquefied air has a high expansion ratio between its liquid state (lower than $-196°$C) and its gaseous state, i.e. liquefied air can be expanded to about 700 times in volume when re-gasified[19] so it can be bulk stored above ground in low-pressure tanks. Considering ES material temperature, liquid air storage can be classified into cryogenic ES which employs a cryogen (such as liquid nitrogen or air) to achieve the electrical and thermal energy conversion/storage. Figure 6 shows a schematic layout of a liquid air storage plant.[19] LAES uses an electrical machine to drive an air liquefier and then stores the resultant liquid air in an insulated tank at atmospheric pressure. When electrical energy is required, the liquid air is released and pumped to a high pressure in its liquid state, then vaporized and heated to the ambient temperature; finally the resultant high pressure gaseous air is used to drive a combination of a turbine and an alternator to generate electricity.[19]

3 Overview of CAES Plants in Operation

3.1 *Huntorf CAES Power Plant*

The Huntorf CAES power plant was built in Germany in 1978 and is the first commercialized large-scale CAES facility in the world.[14] The plant was designed to provide black-start power to nuclear power units located near

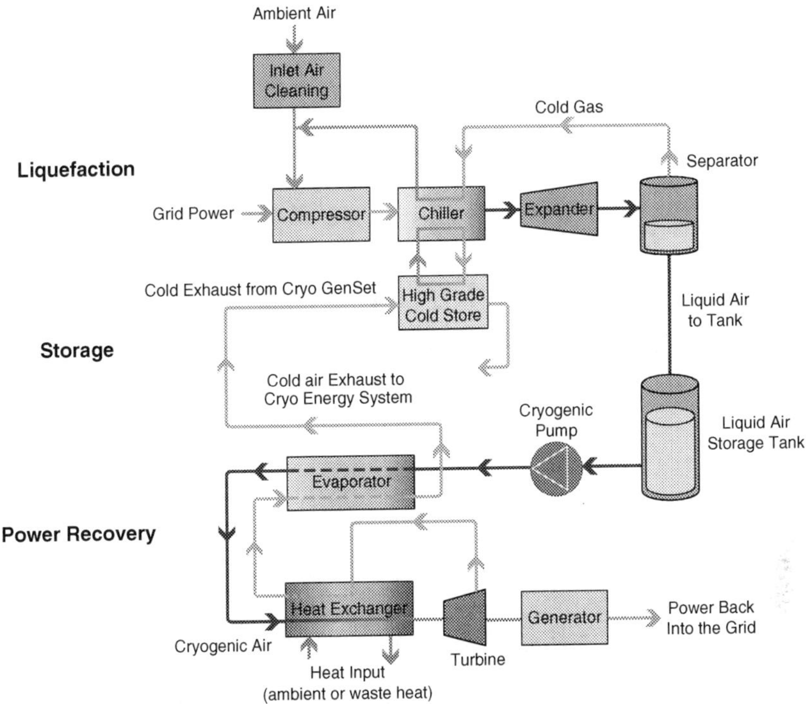

Fig. 6. Schematic layout of a LAES plant (Picture courtesy of Highview).[19]

the North Sea and also served to level and reduce the price of peak power demand. After the Huntorf CAES plant started operation, its mandate was updated to include the support of other facilities, such as back-up for the local power system, filling the energy gap due to the slow response of coal-fired power plants, and buffering against the intermittence of wind energy production in Northern Germany.[4]

The Huntorf CAES plant employs two salt dome caverns to store compressed air. The caverns have a total volume of approximately 310,000 m^3, are located over 600 m below the ground, and are operated at a pressure range between 4.8 and 6.6 MPa.[4,5,14] The CAES compression unit operates with a maximum air output pressure of 10 MPa.[4,14] The plant was operating on a daily cycle with 8 h of compressed air charging and 2 hours of expansion at the rated power of 290 MW.[5] The unit was upgraded to 320 MW in recent years and is currently used for tertiary frequency control with around 100 starts per year. It is reported that the plant consistently shows excellent performance with 90% availability and 99% starting

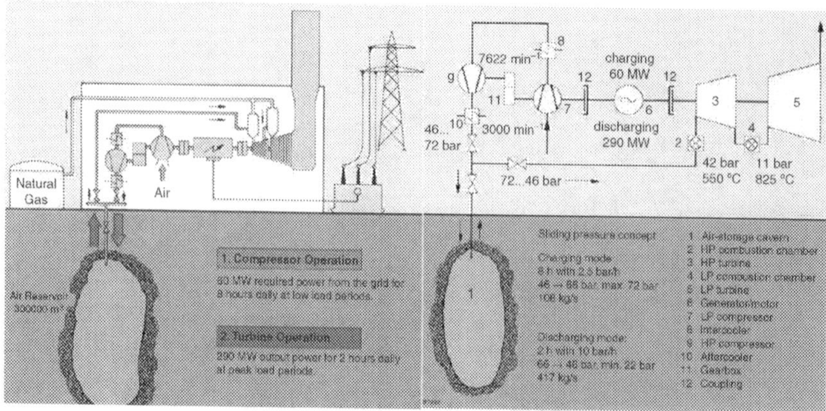

Fig. 7. Schematic layout of the CAES plant at Huntorf, Germany.[14]

Fig. 8. A view of the power house at the Huntorf CAES plant (Curtsey to "UNIPER Kraftwerke GmbH").

reliability.[14] The round-trip (cycle) efficiency of this plant is about 42%.[5] Figure 7 illustrates the schematic diagram for the Huntorf CAES plant and Figure 8 shows a view of the power house at the plant. From these figures, it can be seen that there is a main shaft which is used to connect the compressors, motor/generator, and turbines with clutches; the compression and expansion are both composed of low-pressure and high-pressure stages separately. A 20-stage axial compressor for low pressure and a 6-stage centrifugal compressor for high pressure are used at Huntorf. The compression module is capable of drawing 108 kg/s, while the expansion module is capable of processing 417 kg/s of air.[4,5,14]

3.2 *McIntosh CAES Power Plant*

The second commercialized large-scale CAES facility started operation in McIntosh, Alabama, US in 1991.[5] The project was implemented by Dresser-Rand, and the CAES plant was built by Alabama Electric Cooperative.[4] A schematic diagram to illustrate the structural layout of the McIntosh plant is shown in Fig. 9.

The 110 MW McIntosh CAES plant was designed to be operated for up to 26 h continuously at its full power. The storage capacity of the plant is about 560,000 m^3, utilizing a single salt dome cavern, about 450 m under the ground, to store the compressed air in the range of 4.5–7.4 MPa.[4,5,13] From Fig. 9, it can be seen that the structure of McIntosh CAES plant is similar to that of the Huntorf CAES plant. The major improvement is that the McIntosh plant employs a heat recuperator to reuse part of the thermal energy from the exhaust of the gas turbine. This reduces the fuel consumption by 22–25% and improves the cycle efficiency from ~42% to ~54%, in

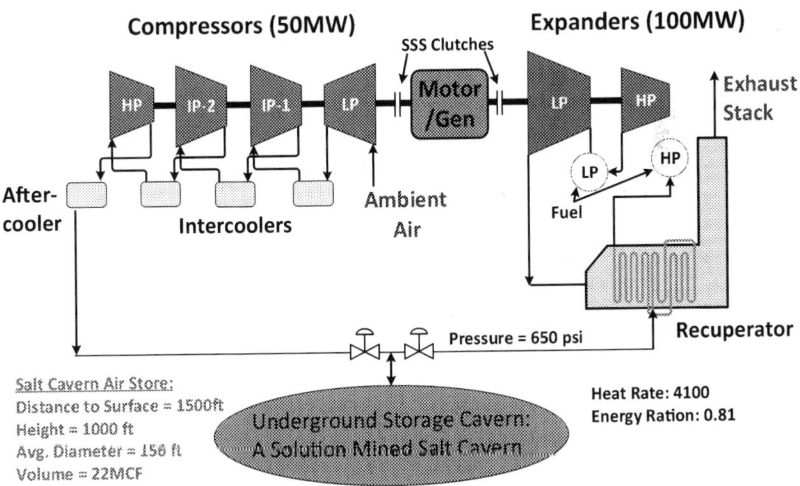

Fig. 9. Schematic layout of CAES plant at McIntosh, Alabama, USA.[15]

comparison with the Huntorf CAES plant.[5,13] Under normal operating conditions, the compressed air storage cavern is partially recharged during the weekday nights and is fully recharged at weekends. During the operation from 1998 to 2008, the plant maintained an average starting reliability of between 91.2% and 92.1%, and an average running reliability of 96.8% and 99.5% for its generation and compression sections, respectively.[4]

4 Current Research and Development of CAES Technologies

4.1 *ADELE — Adiabatic CAES Project*

RWE Power, General Electric, Züblin, and DLR started the world's first large-scale AA-CAES demonstration project: ADELE in Germany. The feasibility study that the project's partners have carried out for the ADELE development program started in 2010.[6] A number of technological challenges were identified in implementing an AA-CAES system such as the

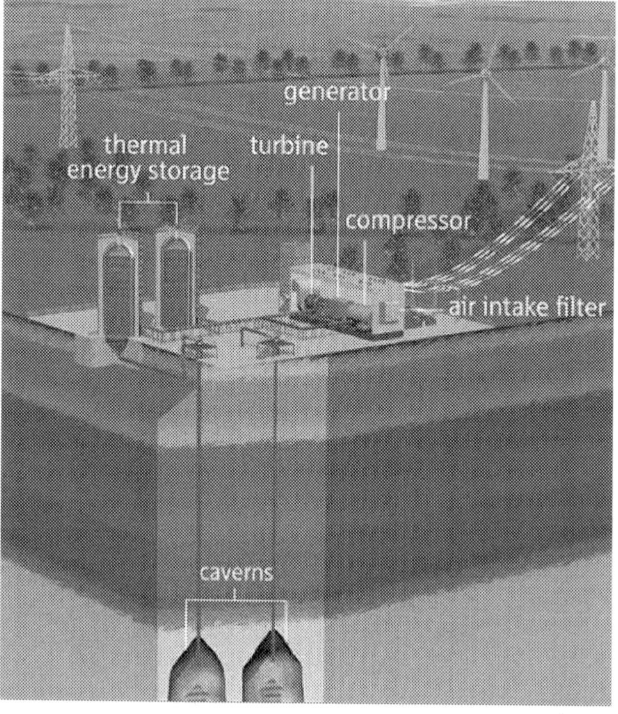

Fig. 10. Schematic diagram of the ADELE AA-CAES project.[6]

system illustrated in Fig. 10. For instance, without using intercooling, the air temperature inside the compressor can exceed 650 K with up to 10 MPa pressure,[21] which may damage the compressor and cavern. Thus, the AA-CAES requires the design of a high pressure and high temperature compressor, taking into account material selection, thermal expansion and thermal stresses, sealing concepts, and thermal limitations for bearings and lubrication.[21,22]

For the compression stage, the ADELE plant is designed to place the resulting heat in an interim heat-storage mechanism and to inject the compressed air into subterranean caverns. The compressed air stored in the caverns can be used to generate power through a turbine — while recovering the heat.[6] The site for this demonstration plant is located in Stassfurt, Saxony-Anhalt, Germany, with a planned storage capacity of 360 MWh and power output of 90 MW. The goal is to achieve 70% cycle efficiency.[6,21,22] The ADELE development phase had obtained a funding of €2 million by 2013, and is supported by Germany's Ministry of Economics and Technology (BMWi) with funds from the COORETEC programme.[6] More details of the world's first large-scale AA-CAES demonstration project can be found on the RWE website. However, the project is currently on hold as it lacks economic viability.

4.2 *Iowa Stored Energy Park Project — Discontinued*

The Iowa Stored Energy Park project was planned by the Iowa Association of Municipal Utilities, and is depicted in Fig. 11. The project intended to build a 270 MW CAES plant coupled with 75–100 MW of wind generation capacity, and to be operational by 2015.[23,24] The plant was designed to take surplus electrical energy generated by the wind farm and use it to compress air into a deep underground aquifer in the northwest of Des Moines. When power demand exceeds the power generation from the wind farm, the stored compressed air can be released through turbines which drive electrical generators to fill the gap between generation and demand.

However, the Iowa Stored Energy Park project had to be terminated in 2011.[23,24] After years of study, the investors concluded that the porous sandstone aquifers in Iowa are not suitable for CAES; the air stored in such aquifers cannot provide air flow fast enough to satisfy the requirements to form an effective CAES site. The porous sandstone geologic formations have successfully been used for underground natural gas storage for decades, but the tests for CAES storage failed the initial onsite test study.[23]

Fig. 11. Schematic diagram of CAES plant with wind farm at Iowa, US.[24]

Although the project was discontinued, it highlighted some promising economic findings and important lessons for other planned energy storage projects. Some of the legacies of the project include: research concerning suitable CAES reservoirs at challenging geological locations; detailed research into the difficult matter of comparable fund estimation and the long-term economics of a CAES plant; insight into the expertize gaps for engineers from different professional fields.[23,24] From a financial point of view, the budget estimation of the Iowa project was inaccurate. The project required the raising of investment capital, but unfortunately this could not be achieved.[25] A detailed study for this discontinued project can be found in the "Lessons from Iowa" Summary Report from the US Department of Energy (DOE).[24]

4.3 *Other Proposed CAES Demonstration Projects*

The Norton ES project by FirstEnergy Generation Corp (FGCO) was announced in 2009. As an initial action, FGCO declared that it had purchased a 92-acre site with a cost of approximately $35 million to develop a compressed air electricity generating plant.[11] The project planned to convert a 600 acre underground idle limestone mine in Norton, Ohio into the storage reservoir for storing compressed air, which can operate within the pressure range from 55 to 110 bar.[4,26.] With around 9.6 million cubic meters of storage, it is planned to build the plant in several phases, from about 270 MW to a total capacity of 2700 MW.[5,26] In July 2013, it is reported that FirstEnergy Corp has delayed building the proposed CAES project due to the current market conditions including low electricity prices and insufficient demand.[27]

A demonstration project in Dong Sheng, Inner Mongolia, PR China, named "advanced large-scale CAES system", with 15–20 MW generation capacity, has recently been proposed; its corresponding initial test system (1.5 MW, Lang Fang, He Bei) was built and has achieved a cycle efficiency of 53%.[28,29] The Institute is now working on a project to build a demonstration plant to scale up the 1.5 MW plant to over 10 MW in Guizhou Province, China.[29]

Ridge Energy Storage & Grid Services L.P. announced a plan to build a 4×135 MW system in Matagorda County, Texas, based on the McIntosh Dresser-Rand design.[14] The proposed system aims to utilize a readily built brine cavern, which has the same geological condition as the ones used at the Huntorf and McIntosh sites. Thus, it is assumed to work well for CAES.[11] In 2007, Luminant and Shell-Wind Energy proposed wind farm projects in Texas and the companies also intended to evaluate the potential for incorporating CAES facilities in conjunction[30] with the wind farms. The demonstration plant was planned to study the possibility of using wind power combined with CAES as base load power generation. After a long wait, in 2013, the project began, aiming for 317 MW of underground CAES.[31] The project is fully permitted and construction-ready. When complete, the plant is expected to provide power for over 300,000 homes and generate millions of dollars in tax revenue each year. The construction is to start in the first quarter 2017 and is expected to be completed within three years.[32]

The US-based LightSail Energy Ltd. is now developing an CAES system by using a reversible electric motor/generator unit and a reversible reciprocating piston machine.[7,20] In the designed system, the heat from compression is captured by the water spray and then stored; during expansion, the stored hot water is sprayed into the compressed air flow. So the air temperature change can be maintained minimum to close to an I-CAES process. The company claimed that high thermodynamic efficiencies can be achieved through the initial tests,[7] without sacrificing performance. LightSail Energy Ltd. was established in 2009 in Berkeley, CA, USA and attracted an initial investment of over $70m.[33] The company has demonstrated the innovative technology and needs market uptake to provide its business opportunity and growth.

Gaelectric partnered with Dresser-Rand is to build a CAES plant to generate 330 MW of power for up to 6 h in the area of Antrim, Northern Ireland, UK.[34] The project is called "CAES Larne" and plans to create two salt caverns for compressed air storage at a depth of greater than 1,400 m underground. The salt source was confirmed in November 2013 and the project has been through seven stages of public consultation. The planning application and Environment Impact Assessment (EIA) were submitted to the government Strategic Planning Division in December 2015. If the CAES Larne is built, it will be capable of providing a range of services and support to power network operation and management.

4.4 *On-going Research Work in CAES*

The improvement of a CAES system relies on various technologies from compressors, turbines/expanders, electrical machines, electric power conditioning, materials, control, thermal storage, and heat exchanger, etc. Research activities can be found in all aspects of the challenging technological areas associated with CAES systems. In addition, the research on geologic formations of caverns is also an active research area which lays the groundwork for seeking suitable locations for building appropriate storage reservoirs.

The research team at the University of Warwick, UK, is currently working on development of a system for hybrid connection of a wind turbine with compressor or expander for compressed air energy conversion through a power splitting device in the form of a planetary gear box.[35–37] Figure 12 presents a simplified block diagram of such a hybrid wind turbine system with CAES integration. From the figure, it can be seen that the turbine drive train transmission makes the system structure different from the currently used electricity–air–electricity system structure for CAES integration.[37] The main benefit of the hybrid wind turbine structure is that mechanical power can be added to the system to help generate electricity during the period of low wind speed through a direct engagement of the expander/compressor unit. In the period of excessive wind power generation, that is, at high wind speed with low electricity demand, the

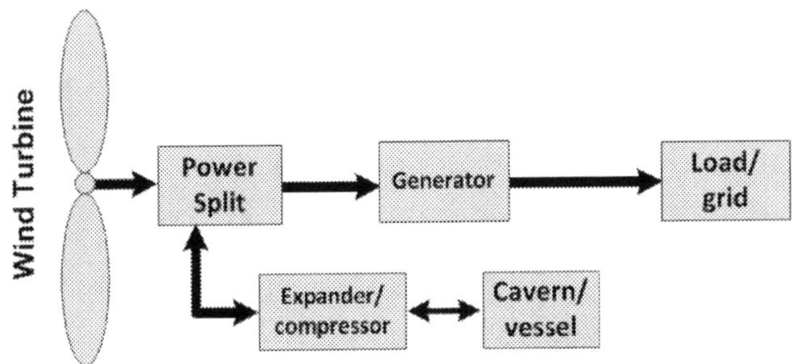

Fig. 12. Schematic diagram of the hybrid wind turbine system with CAES.[37]

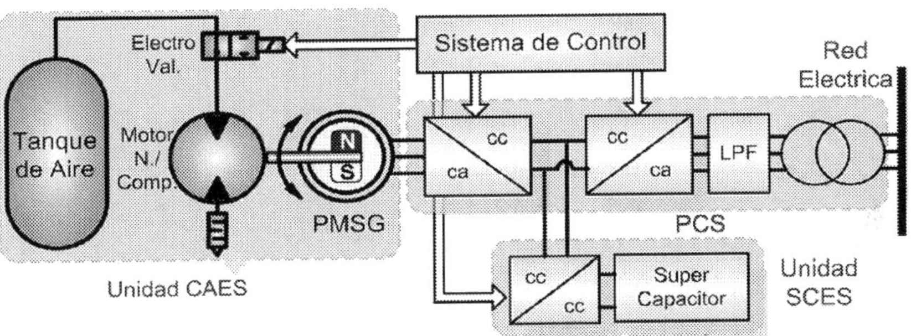

Fig. 13. Structure of the hybrid EES system with CAES and supercapacitors.[38]

excess power can be extracted to drive the compressor to charge the storage reservoir. During a period of low wind speed but high load demand, the compressed air is released to drive the wind turbine generator via an expander. This structure may reduce the cost of the whole system and provides damping to the turbine's drive train during wind gusts, thus prolonging the turbine's lifetime.

The mathematical model for the whole hybrid system is described in the recent publications.[35–37] The simulation study and experimental tests have demonstrated that the proposed hybrid wind turbine system is feasible and brings various benefits to the wind turbine operation, such as improved turbine efficiency.

Martinez *et al.* present a dynamic model and the control design of a hybrid ES system based on CAES and supercapacitors[38] (see Fig. 13). The designed system converts excessive energy from the power supply to stored compressed air energy using a compressor. The energy delivered to the power system is controlled through the intermittent operation of

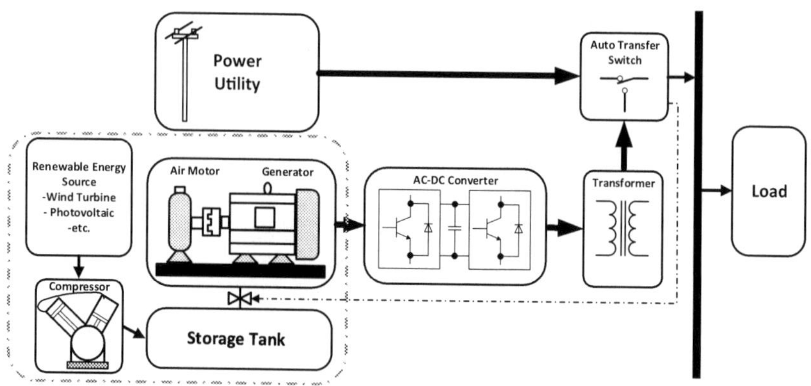

Fig. 14. Simplified diagram of renewable energy integration with EES.[39]

the pneumatic converter. In order to smooth the desired output power of the system, a supercapacitors bank is utilized. Power electronics and their control play a significant role in the integration of the whole electrical ES system with the power network. The dynamic performance of the proposed systems is evaluated by a simulation study using the SimPower Toolbox of MATLAB/Simulink.

A system for providing UPS from intermittent renewable power generation based on CAES was reported, as shown in Fig. 14.[39] The electricity generated from the renewable energy is used to drive the air compressor to produce high-pressure compressed air to be stored in a vessel. From the report, the work focused on the thermodynamic study and analysis of the performance of the proposed system. The air flow under variations of wind speed and pressure ratio conditions was investigated and the system efficiency under varied expansion pressure conditions was studied.[39] This published study suggested that the proposed system can be used for back-up power and peak shaping and other energy management applications.

5 Suitable Geology of Underground Caverns for CAES

There are three main types of geological structures suitable or possibly suitable for forming CAES storage reservoirs: salt, hard rock and porous rock.[4,40,41] The research suggests that many areas in the world have the suitable geological conditions required for CAES. An example is that over 75% of the area of the US has one or more of these three types of geologic conditions.[40,41] In practice, more detailed investigation and evaluation are needed to verify the suitability and feasibility of a specific location for compressed air storage.

5.1 *Salt Cavern*

The salt dome is the most favorable geological structure for solution-mining cavities. This type of cavity can be directly used for compressed air storage. The knowledge acquired from the storage of high-pressure hydrocarbon products, such as liquefied petroleum gas and natural gas, provide the solid evidence of suitability for compressed air storage using the cavities because they operate at a similar range of gas pressure. Both of the two existing commercialized CAES plants (the Huntorf plant in Germany and McIntosh plant in the US) adopt the cavities mined into salt domes as their compressed air storage reservoirs.

Salt caverns for compressed air storage can be built through the solution-mining technique which is a relatively low cost and reliable approach. The initial capital cost of such a storage system is about 2–10$/kWh.[4,42,43] Solution mining is the technique of using water or other liquids to dissolve and then extract salt from a salt stratum, thereby forming a large cavity in the salt with the appropriate size and shape of the cavern. Underground salt deposits exist in two forms: salt domes and salt beds. Salt domes are more suitable for mining to create caverns with large volumes for compressed air storage. Figure 15 shows the potential locations for CAES

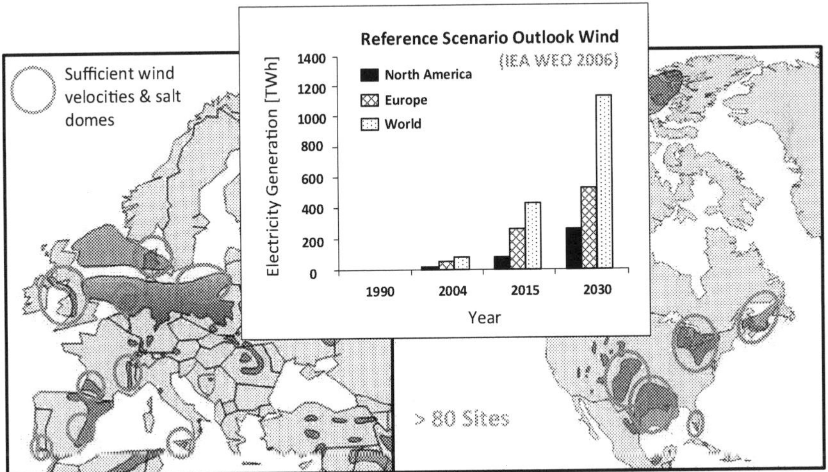

Fig. 15. Potential locations for CAES in the EU and the US (correlation map of High Wind Potential and Salt Domes).[4,44]

in the EU countries and the US, which are chosen to have a close proximity to locations with high wind potential and salt domes.

5.2 *Hard Rock*

Hard rock formations are another geological option for forming underground CAES reservoirs. However, the cost of mining a new reservoir is relatively high, typically 30 \$/kWh.[11,42] The depth suitable for hard rock CAES storage caverns is within the range of 300–1500 m underground.[40,45] Several proposed CAES projects plan to use existing mines (hard rock formations), with the aim of reducing the cost of constructing the reservoirs. The Norton ES project intends to use an underground cavern that was formerly operated as a limestone mine in Norton, Ohio.[4,26] A 2 MW field test program used a concrete-lined tunnel in the former Sunagaawa Coal Mine in Japan.[42] The Electric Power Research Institute (EPRI) and the Luxembourg utility Societe Electrique de l'Our SA employed an excavated hard-rock cavern with water compensation for a CAES feasibility study project.[46] Although a hard rock cavern can be used for large-scale CAES plants, it has restrictions due to its relatively high initial capital cost of construction when compared to the cost of manufacturing salt caverns.

5.3 *Porous Rock*

Porous rock formations may also be an option for storing compressed air underground in a large volume. The significant advantage of using porous reservoirs is its very low cost in cavern construction; it is reported that the estimated cost is approximately 0.10–0.11 \$/kWh.[4,42]

Several feasibility study projects on using porous rock for CAES were attempted. Enel, the largest power company in Italy, operated a 25 MW porous rock-based CAES research plant in Sesta, using porous rock storage that previously held carbon dioxide near a geothermal region.[42] Although the initial air cyclic test was successful, additional tests did not go ahead due to a disturbed geothermal event. Strata Power, EPRI, Nicor, and US DOE, tested the porous sandstone caverns in Pittsfield, Illinois to determine the feasibility of using the porous rock formations for storing compressed air. The test results indicated that compressed air can be stored and cycled successfully. However, the period for air storage was limited as the stored compressed air would react with local pyrites in the sandstone.[47] The Iowa Stored Energy Park project described in the previous section aimed to use porous sandstone aquifers in Iowa. Unfortunately, the geological structures cannot allow the sufficient and efficient compressed air flow required by CAES plants, although it is suitable for compressed air storage.[23,24]

6 CAES Technological Characteristics

This section will introduce the technological characteristics of CAES which are essential for operation of a CAES plant and are important in designing and building a CAES power plant.

- *Rated capacity (or rated energy capacity)*: The total quantity of available energy from the storage system after it is fully charged. The SI unit of storage capacity is Joule (J) i.e. Watt-hour (Wh). Megawatt-hour (MWh) is usually used instead of Watt-hour. The existing CAES plants demonstrate a rated capacity of up to 2860 MWh[4]; the rated capacity of small-scale CAES is dependent on applications which can vary from a few kWh to over 100 MWh.[4,5,9,20]
- *Specific energy*: The energy per kilogram of the storage medium, measured in Joule per kilogram (J/kg) or Watt-hour per kilogram (Wh/kg). The specific energy of the existing CAES-related plants is from 30 to 60 Wh/kg (large-scale CAES); 140 Wh/kg at 300 bar (stored compressed air in high-pressure cylinders); 214 Wh/kg (LAES).[4,48]
- *Specific power (or power-to-weight ratio)*: The amount of obtained power per kilogram of the storage medium, in the unit of the Watt per kilogram (W/kg). There is no acceptable available data of the specific power of CAES from the literature survey and study.
- *Energy density*: the amount of energy stored per unit volume in a given system or region of space, in Watt-hour per cubic meter (Wh/m^3). In EES applications, the energy density is different for different applications and is related to the pressure of air storage. For the same volume, the higher the energy density of the chosen storage medium, the higher the amount of energy that can be stored or transported. The energy density of CAES is in the range of 2–6 Wh/L depending on the air pressure.[51,52]
- *Power rating*: The power rating indicates the maximum rate at which the system can discharge energy, expressed in units of kilowatt (kW) or megawatt (MW). The power ratings of a CAES plant can range from a few kW to around 1000 MW (large-scale CAES).[8,51,52]
- *Part-load operation*: CAES has a high part-load operation range, that is, it can operate properly with a variable load below its rated power.[4] The CAES plant output can be controlled by adjusting the airflow rate with inlet temperatures kept constant at the multi-expansion stage, which

leads to better heat utilization and higher efficiency during part-load operation.[53]

- *Discharge time duration at rated power*: The maximum time required for discharge at the rated power without recharging.[51] The discharge time duration, as a characteristic of system adequacy, is determined by the depth of discharge and operational conditions of the system. For current CAES plants in operation, the discharge time duration is up to 26 hours (large-scale CAES), which is dependent on the storage volume and restriction of operational range.[4]
- *Response time*: The rate of power increase or decrease with time. The McIntosh plant can increase or decrease the power output at around 18 MW per minute, which is about 60% higher than a comparable typical gas turbine facility.[21] The proposed Matagorda Plant is designed to be able to bring its 4 × 135 MW power train modules to full power in 14 minutes (or 7 minutes for an emergency start).[11]
- *Self-discharge rate*: The rate of stored energy released while the energy storage system is on standby or unused. The self-discharge rates for both CAES and LAES are very small as the seals of the storage can be very good.[4,8]
- *Heat rate*: the fuel consumed per kWh of output for a CAES plant involved in fossil fuel burning.[11] Heat rates for CAES systems without a heat recovery system are typically 5500–6000 kJ/kWh Lower Heating Value (LHV) and heat rates with a recuperator are typically 4200–4500kJ/ kWh LHV.[4,53] The Huntorf plant, with a rated output of 290 MW over 3 hours and an overall efficiency of 42%, has a heat rate of 5870 kJ/kWh LHV; the McIntosh plant recuperates the turbine exhaust heat, thus improving the overall efficiency to 53%, with a heat rate of 4330 kJ/kWh LHV.[4,54]
- *Recharge rate*: The rate at which power can be pushed for storage. The recharge rate for CAES can be described as the quantity of compressed air per unit of time that replenishes a reservoir or a cavern.
- *Lifetime*: The service time of a unit or a system. It varies with technology and intensity of use. PHS has the longest lifetime in EES systems, approximately up to 50 years; CAES and LAES have lifetimes of around 20 to 40 years.[4,51,55]
- *Cycling times* (*or cycle life*): The total number of cycles of completed common charge and discharge cycles during its life time.[51] Normally, cycling time varies with the different rates of discharge depth. The large-scale CAES plants have cycling times of 8,000–12,000 and

the small-scale compressed air battery system can have about 30,000 stop/start operations which has been verified by a system test.[8,56]

- *CAES reservoir operation methods*: According to the geological conditions, the operation methods for the compressed air reservoir of a large-scale CAES system mainly consists of two approaches described as follows[4,11,53,57]:

  (1) Constant volume: The storage volume is fixed and the air pressure inside the reservoir varies over an appropriate range, which leads to two design options to control the reservoir output: (a) the output pressure of the reservoir is varied and the high-pressure turbine inlet pressure is controlled to follow the reservoir output pressure; (b) the output pressure of the reservoir is varied but the high-pressure turbine inlet pressure is controlled at a constant level. Both Huntorf and McIntosh CAES plants adopt (b); the former operates at the air pressure of 46 bar while the cavern operates between 48 and 66 bar; the latter operates at 45 bar while the cavern operates between 45 and 74 bar.
  (2) Constant pressure: It aims to keep the storage reservoir at a constant pressure throughout operation by using a water compensation system with an aboveground reservoir as shown in Fig. 16.

- *Energy transfer process of CAES plants*: For a traditional CAES plant, the energy transmission and conversion is schematically shown in Fig. 17. During the compression stage, the surplus electrical power from the grid or from power generation drives the compressor. Thus, the electrical power is converted into mechanical power, and in turn to the compressed air internal energy. During the expansion stage, the turbine train can restore the energy carried by the compressed air and utilize the chemical power obtained from the combustion process to generate kinetic energy to drive the mechanical shaft which is connected to an electric generator for electricity generation to the grid. For such an energy conversion and transmission process, energy losses are inevitable, including heat losses, combustion losses, mechanical losses due to friction and vibration, etc.
- *Cycle (round-trip) efficiency*[a]: This is a combination of the charging and discharging efficiencies, which is the ratio between the useful energy output and the total input energy.[55,56] In many cases, the system cycle efficiency is obtained through the charging process efficiency multiplied

[a]Cycle efficiency and round-trip efficiency are interchangeable terms.

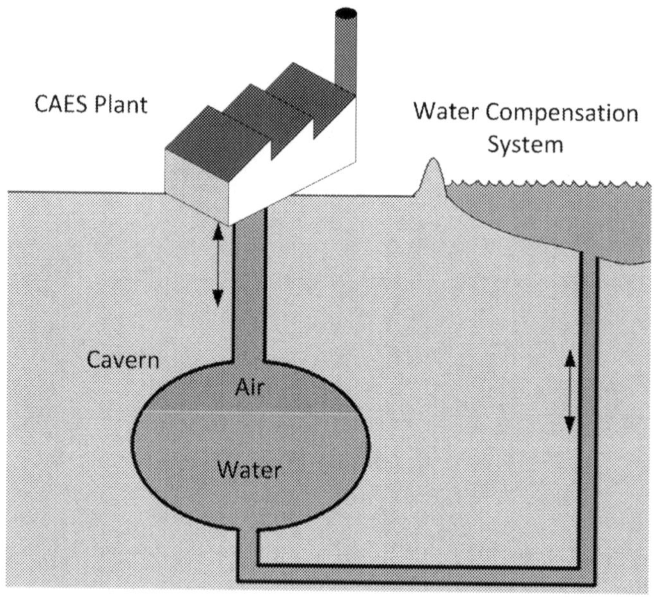

Fig. 16. Schematic diagram of CAES plant with constant-pressure compensation system.

by the discharging process efficiency. For conventional CAES plants, the input electrical energy is used to drive the compressors and the fuel energy released in the combustion process is used to increase the air internal energy prior to expansion. Thus, it is necessary to consider the energy input from both sources, that is: η_{RT} = (*energy output*)/(*total energy input from electricity and fuel*), where η_{RT}, represents the round-trip efficiency.[4,5,11] The detailed calculation method is described in a recent publication.[59] The efficiencies of current CAES systems are summarized below:

(1) The discharge efficiency (i.e. the energy efficiency from the compressed air energy to electric energy) of conventional large-scale CAES is about 70–79%.[60]
(2) Considering the two existing commercial CAES plants, the Huntorf plant has an round-trip efficiency of 42%; the round-trip efficiency of the McIntosh plant, using heat recuperation, is about 53~54%.[4,5,14,15,20]
(3) The round-trip efficiency of AA-CAES is expected to be over 70%, which is similar to PHS.[6,16,29]

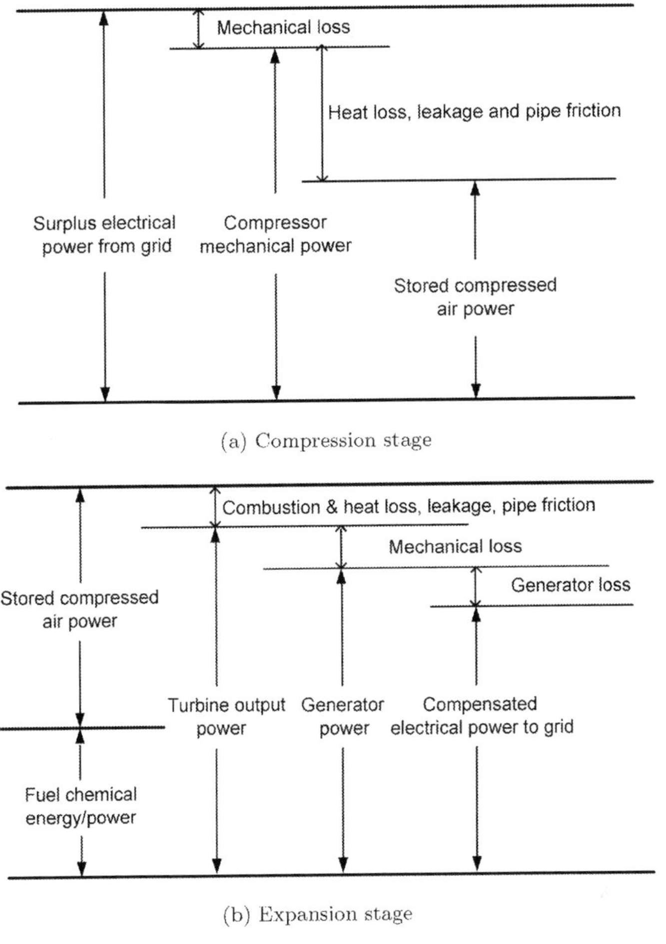

Fig. 17. Energy transmission and conversion process of a typical CAES plant.[1]

- *Operation switching time*: For the two existing large-scale CAES plants, the turbine normally drives the machinery train to start rotating and increases its speed until synchronization is achieved; it is the same when shutting down the turbine.[4] The turbine needs to be engaged for operation in both the compression and expansion modes. At the Huntorf CAES plant, the switch from one operating mode to another requires a minimum of 20 minutes.[11] To reduce the switching time, one possible solution is to redesign the overall system structure by separating the compression and turbo expander components rather than linking them through a common shaft via the clutch mechanisms. This has been considered in more advanced CAES plant design.

7 Analysis of Suitable CAES Application Areas

The power rating of a CAES system can range between a few kWs up to 1000's of MWs. So CAES can be used in many different energy sectors and various applications. Small-scale CAES can replace traditional chemical batteries in many suitable applications. With the moderate response time and good partial load operation, CAES can work well with intermittent power generation from renewable energy for load shifting, back-up power, smoothing power outputs, etc. In recent years, the research on CAES technologies and its integration with grid operation has rapidly increased. A number of examples of this are provided in the first few sections. Table 1 summarizes and predicts the industry and the power grid applications of CAES in the near future.

8 Research Challenges and Needs

Similar to the PHS, the major barrier for building the CAES plant with underground caverns is the availability of appropriate geographical storage locations for electricity generation and grid integration. Traditional large-scale CAES plants require combusting fossil fuels, which will lead to CO_2 emissions and environmental pollutions.

As described in earlier sections, research is ongoing worldwide to address all the barriers encountered. To further the development of CAES technology, the following research challenges need to be addressed.

(1) Formation of salt caverns

Salt caverns for compressed air storage can be developed through solution-mining techniques which can provide a low cost and reliable methodology (see Sec. 5.1). To make the process economic, the process should maximize the size/volume of the caverns for each salt mining well drill. Also, the shapes of caverns will affect the size of its volume and result in different impacts on the cavern structure when the air pressure rapidly changes for charging/discharging. Research is required into site characterization and the optimal forms of storage caverns to enable high efficiency operation. Costs related to the location of the energy source relative to the storage site(s) need to be examined in the techno-economic analysis.

Table 1: Application potentials of CAES related technology.[1]

Application area	Characteristics ([18, 50, 61∼65])	Suitable or potential CAES related technology
Power quality	<1MW, response time (milliseconds, <1/4 cycle), discharge duration (milliseconds to seconds)	Hybrid systems with small-scale CAES and battery or supercapacitor or other EES technologies with fast response
Energy management	Large-scale (>100 MW), medium/small-scale (<100 MW), response time (minutes), discharge duration (up to days)	Large-scale energy management (large-scale CAES); Small-scale energy management (small-scale CAES, LAES)
Renewable back-up power	100 kW–40 MW, response time (seconds to minutes), discharge duration (up to days)	Multi-scale CAES, hybrid systems with CAES and capacitor or others with fast response may need, possible LAES
Emergency back-up power	Up to 1 MW, response time (milliseconds to minutes), discharge duration (up to ∼24 hours)	Possible small-scale CAES, hybrid systems with small-scale CAES and other technologies with fast response
Time shifting	1MW–100 MW and even more, response time (minutes), discharge duration (3–12 hours)	Multi-scale CAES and LAES
Peak shaving	100 kW–100 MW and even more, response time (minutes), discharge duration (hour level, <10 hours)	Multi-scale CAES and LAES
Load levelling	Up to several hundreds of MW, response time (minutes), discharge duration (up to ∼12 hours and even more)	Multi-scale CAES, possible LAES
Seasonal ES	Energy management, 30–500 MW, discharge duration (weeks), response time (minutes)	Possible large-scale CAES and LAES
Black-start	Up to 40 MW, response time (∼minutes), discharge duration (seconds to hours)	Multi-scale CAES, possible LAES
Spinning reserve	Up to MW level, response time (normally up to a few seconds), discharge duration (30 minutes to a few hours)	Possible the hybrid system with small-scale CAES and capacitor or other EES technologies with fast response

(Continued)

Table 1: (*Continued*)

Application area	Characteristics ([18, 50, 61∼65])	Suitable or potential CAES related technology
UPS	Up to 5MW, response time (normally up to seconds), discharge duration (10 minutes to 2 hours)	hybrid system with small-scale CAES and supercapacitor or other EES technologies with fast response, e.g. Pnu power
Standing reserve	Around 1–100 MW, response time (<10 minutes), storage time at rated capacity (∼1–5 hours)	Promising multi-scale CAES and LAES
Transmission upgrade deferral	10–100+ MW, response time (minutes), storage time at rated capacity (1–6 hours)	Promising multi-scale CAES and LAES

(2) Improvement of round-trip efficiency

One of major barriers for acceptance of CAES is its low round-trip efficiency. The Huntorf plant has a cycle efficiency of 42% and the cycle efficiency of the McIntosh plant is about 53∼54%. The low efficiencies of CAES result from the heat losses in the compression and expansion modes, air leakage throughout the whole CAES system, and internal energy losses due to the air compressibility. A-CAES combined with adequate thermal energy storage will reduce heat losses and is expected to achieve 70% cycle efficiency. To improve CAES round-trip efficiency, the following researches are required: (i) suitable thermal storage procedure and facility design to maximize the utilization of thermal energy stored, such as high-pressure thermal storage; (ii) the individual components or devices working at their optimal status does not mean the whole system is in its optimal status due to complicated coupling effects. Therefore, whole system-efficiency analysis, system optimal design, optimal coordinate control, and operation optimization are needed. The key technology challenge is the efficient utilization of the thermal energy released from the compression process.

(3) Innovation in turbo machinery design and manufacturing

The UK company, Flowbattery is a good example of the importance of technology innovation. By adopting an innovative scroll air expander technology, the company developed its product, the Compressed Air Battery which achieves a high air-to-electricity conversation efficiency. Technical innovations and technology breakthrough are essential, especially for high-pressure compressor and turbine technologies, such as developing improved sealing methods for compression and expansion machinery to suppress internal leakage and discovering approaches to minimize losses associated with secondary flows in compressors and turbines. Research on innovative technologies is the key to the realization of the full potential of CAES.

(4) Innovation in small scale CAES systems (<10 MW)

In general, the larger the scale of CAES is, the more efficient the system will be. However, this is changing. Small scale CAES units have started to demonstrate their value in recent years when they are used to replace chemical batteries in some applications. Also, small scale CAES can be used in distributed ES to serve distributed power generation. In addition to research efforts made on large-scale CAES, research on the technologies used in small-scale CAES will also be rewarding.

(5) Integrated technologies

CAES on its own has relatively low round-trip efficiency. Integrated utilization of energy through the whole process can increase the round-trip efficiency as the energy losses can be recovered via the integrated process. Integration should focus on (i) integration of CAES with wind/solar power generation which will store energy locally and maximize the renewable power generation; (ii) integration of CAES with heating and cooling systems in which the heat from compression and the heat losses from expansion can be used directly, for example, a compressed air battery can be installed in a data center and the cool air from the exhaust circulated around the center for cooling; (iii) integration with thermal storage to maximize its compression efficiency; (iv) integration with other ES technologies to operate at its most efficient operation range.

9 Summary

This chapter introduces the concept of CAES systems and describes their basic operation, capabilities, and potential. The chapter also provides an overview of the current state-of-the-art CAES technology development, generated as an outcome from the EPSRC funded research project — "Integrated Market-fit Affordable Grid-scale Energy Storage". The areas of CAES technology that require further research for CAES plants to be widely deployed and their full potential achieved are also discussed.

CAES plants store energy in the form of compressed air (or sometimes in liquid air form) in underground geological chambers (such as salt or hard rock caverns) or above ground in suitable storage tanks. The energy stored in the compressed air is then converted to another form to carry out useful work. The application of a CAES system depends on its size and locality. However, typical uses include: storing surplus energy generated by renewable sources such as wind farms; providing an energy source to meet peak demand; providing black-start power for nuclear power stations; smoothing power outputs and providing an alternate to chemical energy storage for back-up power sources.

CAES plants are the only ES technology which have a power rating comparable to pumped-hydro storage plants with the power rating range of CAES plants being between a few KW to around 1000MW,[8,51,52] meaning they can store and deliver large amounts of energy. However, CAES plants have low energy and power densities (typically around 2–6 Wh/L pressure and 0.5–2 W/L, respectively,[49,50]) meaning that they require a large storage volume. CAES plants also have a relatively low round-trip efficiency (around 42%) due mostly to heat and mechanical losses in the compression and expansion phases. This efficiency can be improved if thermal ES methods are utilized to make use of the heat lost due to compression and the exhaust in the expansion phase; a round-trip efficiency of ~54%[4,5,14,15,20] is already achieved and 70%[6,15,29] is expected in the future using these methods.

The future research priorities concerning CAES systems should focus on: finding suitable locations and researching improved methods for forming salt caverns to store the compressed air; improving the relatively low round-trip efficiency by reducing air leakage and utilizing/reducing heat losses; optimizing the machinery used in high pressure systems to reduce leakage and other performance inhibitors; developing small scale CAES systems to support distributed power systems and to provide an alternative to chemical ES and integrating other technologies such as thermal storage to improve the overall performance.

Acknowledgments

The authors would like to thank the funding support from Engineering and Physical Science Research Council, UK (EP/K002228/1, EP/L019469/1) and the support from China 973 Research Programme (2015CB251301) to enable the collaboration with Chinese Academy of Science to understand the 1.5 MW and 10 MW CAES demonstration plants.

References

1. X. Luo and J. Wang, Overview of the current development on compressed air energy stroage, EERA Report. Available at: http://integratedenergystorage.org/. Accessed on March 4, 2016.
2. J. Wojcik, Compressed Air Energy Storage animation (IMAGES). Available at: https://www.youtube.com/watch?v=MNVG7x9ZKBc. Accessed on March 4, 2016.
3. B. J. Davidson *et al.*, Large-scale electrical energy storage, *Physical Science, Measurement and Instrumentation, Management and Education Reviews, IEE Proceedings A* **127**(6) (1980), pp. 345–385.
4. S. Succar and R. H. Williams, Compressed air energy storage: Theory, resources, and applications for wind power. Technical Report, Energy Systems Analysis Group, Princeton Environmental Institute, Princeton University (2008).
5. H. Chen, T. N. Cong, W. Yang, C. Tan, Y. Li and Y. Ding, Progress in electrical energy storage system: A critical review, *Prog. Nat. Sci.* **19**(3) (2009), pp. 291–312.
6. RWE power, ADELE — Adiabatic compressed-air energy storage (CAES) for electricity supply, ADELE Brochure. Available at: https://www.rwe.com/web/cms/mediablob/en/391748/data/364260/1/rwe-power-ag/innovations/Brochure-ADELE.pdf. Accessed on March 3, 2016.
7. LightSail Energy Ltd., We store energy in compressed air. Available at: http://www.lightsail.com/. Accessed on March 8, 2016.
8. Compressed air battery systems. Available at: http://www.flowbattery.co.uk/. Accessed on October 8, 2015.
9. E. Gent, Liquid air energy storage could become £1bn industry, *Engineering and Technology Magazine*, May 9, 2013.
10. Birmingham Centre for Cryogenic Energy Storage. Available at: http://www.birmingham.ac.uk/research/activity/energy/research/centre-energy-storage/cryogenic-energy-storage/index.aspx. Accessed on March 4, 2016.
11. D. Wolf and M. Budt, LTA-CAES — A low-temperature approach to Adiabatic Compressed Air Energy Storage, *Appl. Energ.* **125** (2014), pp. 158–164.
12. M. Budt, D. Wolf, R. Span and J. Yan, A review on compressed air energy storage: Basic principles, past milestones and recent developments, *Appl. Energ.* **170** (2016), pp. 250–268.
13. S. Samir, *Large Energy Storage Systems Handbook*, ed. Jonah G. Levine, (CRC Press, Boca Raton, 2011), pp. 112–152.
14. J. A. McDowall, High power batteries for utilities — the world's most powerful battery and other developments, in *Proc. of 2004 Power Engineering Society General Meeting*, (June 6–10, 2004), pp. 2034–2037.
15. M. Nakhamkin, M. Chiruvolu and C. Daniel, Available Compressed Air Energy Storage (CAES) plant concepts, in *the Proc. of Power-Gen Conference*, December (2007).
16. ConvenEnergy Storage & Power LLC, CAES — advanced generation, "Adiabatic CAES concept. Available at: http://www.espcinc.com/. Accessed on February 28, 2016.

17. F. Díaz-González, A. Sumper, O. Gomis-Bellmunt and R. Villafáfila-Robles, A review of energy storage technologies for wind power applications, *Renew. Sust. Energ. Rev.* **16**(4) (2012), pp. 2154–2171.
18. X. Luo, J. Wang, M. Dooner, J. Clarke and C. Krupke, Overview of current development in compressed air energy storage technology, *Energy Procedia*, **62** (2014), pp. 603–611.
19. Highview power storage — liquid fluid storage. Available at: http://www.engineering.com/ElectronicsDesign/ElectronicsDesignArticles/ArticleID/5780/Cryogenic-Energy-Storage.aspx. Accessed on October 12, 2015.
20. LightSail Energy Ltd., Innovation in energy storage — isothermal compressed air energy storage. Available at: http://www.lightsail.com/wp-content/uploads/2014/06/Innovation-in-Energy-Storage.pdf. Accessed on December 8, 2016.
21. University of California, 2020 strategic analysis of energy storage in California. Available http://www.energy.ca.gov/2011publications/CEC-500-2011-047/CEC-500-2011-047.pdf. Accessed on October 8, 2012.
22. M. Finkenrath, S. Pazzi, M. D'Ercole *et al.*, Status and technical challenges of advanced Compressed Air Energy Storage (CAES) technology, International Workshop on Environment and Alternative Energy, Munich, Germany (2009).
23. H. Dan, Scrapped Iowa project leaves energy storage lessons. Available at: http://www.midwestenergynews.com/2012/01/19/scrapped-iowa-project-lea leaves-energy-storage-lessons/ (January 19, 2011). Accessed on February 17, 2016.
24. R. H. Schulte, N. Critelli *et al.*, Lessons from Iowa: Development of a 270 megawatt compressed air energy storage project in Midwest independent system operator: A study for the DOE energy storage systems program, Sandia National Laboratories. Available at: http://www.sandia.gov/ess/publications/120388.pdf (January 2012). Accessed on February 23, 2016.
25. A. Jeannine, Iowa compressed air storage project fails, but leaves some lessons, hopes for future bulk energy storage, *Public Power Daily* (February 6, 2012). Available at: http://www.publicpower.org/Media/daily/ArticleDetail.cfm?ItemNumber=33842. Accessed on February 3, 2016.
26. S. Linden, Review of CAES systems development and current innovations that could bring commercialization to fruition, in *the Proc. of Electrical Energy Storage Applications & Technology Conference*, San Francisco (2007).
27. J. Funk, FirstEnergy postpones project to generate electricity with compressed air (July 5, 2013). Available at: http://www.cleveland.com/business/index.ssf/2013/07/firstenergy_postpones_ project.html. Accessed on March 8, 2016.
28. H. Lund and G. Salgi, The role of compressed air energy storage (CAES) in future sustainable energy systems, *Energ. Convers. Manage.* **50**(5) (2009), pp. 1172–1179.
29. The Institute of Engineering Thermophysics Chinese Academy of Sciences. Available at: http://www.etp.ac.cn/jgsz/kybm/kybm6/. Accessed on November 7, 2015.

30. Comanche peak nuclear power plant, units 3 & 4 COL Application, part 3 — environmental report. Available at: http://pbadupws.nrc.gov/docs/ML1118/ML11186A372.pdf. Accessed on October 4, 2015.
31. J. S. John, Texas to Host 317 MW of Compressed Air Energy Storage (July 9, 2013). Available at: http://www.greentechmedia.com/articles/read/texas-calls-for-317mw-of-compressed-air-energy-storage2. Accessed on March 5, 2016.
32. Bethel Energy Center. Available at: http://www.apexcaes.com/project. Accessed on December 8, 2016.
33. B. Haislip, Energy-Storage Startup LightSail Plots Long-Term Game Plan, *The Wall Street Journal*, February 22, 2016.
34. LARNE, CAES power plant in Antrim Northern Ireland. Available at: http:// www.project-caeslarne.co.uk/project-updates/. Accessed on February 20, 2016.
35. H. Sun, X. Luo, and J. Wang, Management and control strategy study for a new hybrid wind turbine system, in *the Proc. of the 50th IEEE Conference on Decision and Control and European Control Conference*, USA, December (2011).
36. H. Sun, X. Luo and J. Wang, Hybrid wind turbine, Feasibility study of a hybrid wind turbine system — integration with Compressed Air Energy Storage, *Appl. Energ* **137** (2015), pp. 617–628.
37. C. Krupke, J. Wang, J. Clarke and X. Luo, Dynamic modelling of a hybrid wind turbine in connection with compressed air energy storage through a power split transmission device, *IEEE/ASME Int. Conf. on Advanced Intelligent Mechatronics*, Busan, Korea (July 7–11, 2015).
38. M. Martinez, M. G. Molina, P. F. Frack and P. E. Mercado, Dynamic modeling, simulation and control of hybrid energy storage system based on compressed air and supercapacitors, *IEEE Lat. Am. Trans.* **11**(1) (2013), pp. 466–472.
39. V. Vongmanee, The renewable energy applications for uninterruptible power supply based on compressed air energy storage system, in *the Proc. of IEEE Symposium on Industrial Electronics & Applications* **2** (2009), pp. 827–830.
40. K. Allen, "CAES: The underground portion, *IEEE T. Power Ap. Syst.* **104 (1985)**, pp. 809–812.
41. B. Mehta, CAES geology, *EPRI Journal* **17** (1992), pp. 38–41.
42. EPRI-DOE, *Handbook of Energy Storage for Transmission and Distribution Applications* (EPRI, Department of Energy (DOE), Palo Alto, CA, Washington DC, 2003).
43. EPRI, Preliminary design study of compressed air energy storage in a salt dome, EPRI Report EM-2210 (April 1982).
44. B. Calaminus, *Innovative Adiabatic Compressed Air Energy Storage System of EnBW in Lower Saxony*, in *Proc. of 2nd International Renewable Energy Storage Conference* (IRES II) Bonn, Germany (2007).

45. L. P. E. James, Compressed air energy storage, course text, Decatur Professional Development, LLC. Available at: http://www.pdhengineer.com/courses/m/M-3051.pdf. Accessed on October 8, 2012.
46. EPRI, Compressed-air energy storage using hard-rock geology: Test facility and results (Vol. 1–2) (EPRI, Palo Alto, 2005).
47. Handbook of energy storage for transmission or distribution applications, EPRI 1007189. Available at: http://www.w2agz.com/Library/EPRI Sources & Reports/. Accessed on March 5, 2016.
48. Liquid air energy network forms in UK: Focus on transportation and energy storage (May 9, 2013). Available at: http://www.greencarcongress.com/2013/05/laen-20130509.html. Accessed on February 27, 2016.
49. Electrical energy storage: White paper, Energy Storage team, IEC Market Strategy Board, 2011. Available at: http://www.iec.ch/whitepaper/pdf/iecWP-energystorage-LR-en.pdf. Accessed on February 15, 2016.
50. T. Lombardo, Cryogenic energy storage, May 2013. Available at: http://www.engineering.com/ElectronicsDesign/ElectronicsDesignArticles/ArticleID/5780/Cryogenic-Energy-Storage.aspx. Accessed on February 28, 2016.
51. F. A. Farret and M. G. Simões, *Integration of Alternative Sources of Energy* (John Wiley & Sons, Inc., Hoboken, 2006), pp. 262–300.
52. P. Taylor, R. Bolton. D. Stone *et al.*, Pathways for energy storage in the UK report, March 2012. Available at: http://www.lowcarbonfutures.org/pathways-energy-storage-uk. Accessed on March 11, 2016.
53. I. Hadjipaschalis, A. Poullikkas and V. Efthimiou, Overview of current and future energy storage technologies for electric power applications, *Renew. Sust. Energ. Rev.* **13**(6–7) (2009), pp. 1513–1522.
54. R. Schainker and M. Nakhamkin, Compressed Air Energy Storage (CAES): Overview, Performance and Cost Data for 25MW to 220MW Plants, *IEEE Trans. Power Appar. Syst.* **104**(4) (1985), pp. 790–795.
55. K. W. Li, *Applied Thermodynamics: Availability Method and Energy Conversion* (Taylor & Francis, London, 1995), pp. 3–54.
56. X. Luo, J. Wang, H. Sun, J. W. Derby and S. J. Mangan, Study of a new strategy for pneumatic actuator system energy efficiency improvement via the scroll expander technology, *IEEE/ASME T. Mech.* **18** (2013), pp. 1508–1518.
57. J. Eyer, J. Iannucci and P. C. Butler, Estimating electricity storage power rating and discharge duration for utility transmission and distribution deferral: A study for the DOE energy storage program, SANDIA Technical Report (2005).
58. Electricity storage: technology brief, IEA-ETSAP and IRENA Technology Policy Brief, April 2012. Available at: https://www.irena.org/DocumentDownloads/Publications/IRENA-ETSAP%20Tech%20Brief%20E18%20Electricity-Storage.pdf. Accessed on February 20, 2016.
59. X. Luo, J. Wang, C. Krupke, Y. Wang, Y. Sheng, J. Li, L. Xu, D. Wang, S. Miao, H. Chen, Modelling study, efficiency analysis and optimisation of large-scale Adiabatic Compressed Air Energy Storage systems with low-temperature thermal storage,*Appl. Energy* **162** (2016), pp. 589–600.

60. S.M. Shoenung, Characteristics and technologies for long- vs. short-term energy storage: A study by the DOE energy storage systems program. Technical Report, United States Department of Energy (2001).
61. The Energy Research Partnership, The future role for energy storage in the UK main report, June 2011. Available at: http://erpuk.org/wp-content/uploads/2014/10/52990-ERP-Energy-Storage-Report-v3.pdf. Accessed on February 28, 2016.
62. M. Beaudin, H. Zareipour, A. Schellenberglabe, and W. Rosehart, Energy storage for mitigating the variability of renewable electricity sources: An updated review, *Energy Sustain. Dev.* **14**(4) (2010), pp. 302–314.
63. Electricity Storage Association (ESA), Energy storage technologies, Available at: http://www.electricitystorage.org/technology/storage_technologies/technology_comparison. Accessed on February 22, 2016.
64. Nationalgrid. Available at: http://www2.nationalgrid.com/uk/services/balancing-services/. Accessed on March 7, 2016.

Chapter 4

Batteries for Energy Storage

Paul A. Connor

School of Chemistry, University of St Andrews,
St Andrews Scotland
pac5@st-andrews.ac.uk

This chapter discusses the use of batteries for storage of electrical energy. It covers the general principals of how batteries work, with more specific details on the common types of rechargeable batteries available: *viz.* Lead acid, NiCd, NiMH, NaS, and Li-ion. Some of the practical aspects of connecting the cells together into energy storage systems are discussed, as well as the management systems need to operate the battery systems safely and efficiently. The chapter finishes with a brief section on a selection of applications that exist for battery energy storage, with a focus on electric vehicles (EVs).

1 Introduction: What is a Battery?

If you need to store electricity locally, batteries are the storage method of choice. Batteries are critical for the storage of electrical energy in many applications including transportation and communications. If you ask someone to describe a battery, they would probably talk about a small package with some form of electrical contacts on the outside that can store and supply electricity. They might even say that batteries can be recharged when they stop producing power. (And then go on to complain that the batteries in their modern gadget of choice do not last long enough.) There are few other energy storage technologies that are as widely known to the general public. If shown a wide range of batteries such as a lead-acid car battery, an AA flashlight battery, a small hearing aid cell, or a laptop battery pack, most people would recognize them all as different kinds of batteries.

However, when talking about energy storage, the size and shape of the battery varies substantially, and a rigorous definition of a battery is difficult. The simplest definition is that batteries are electrochemically-based electrical storage devices that are self-contained. This is clear for lead-acid batteries or laptop battery packs but this definition also includes flow cells and some fuel cell storage systems, and excludes some hearing aid cells, zinc–air, and lithium–air cells. The slightly broader description used in this chapter is that a battery is an electrochemically-based electrical storage device that stores the energy inside the cell in at least one of the electrodes, with no moving parts. This definition excludes flow cells and fuel cell storage systems as they have moving parts or store all their energy outside of the cell (and are covered elsewhere in this book), and does include zinc–air and Li–air cells.

In common English usage, a battery is a device for supplying electrical power but it literally means a large grouping of things, e.g. a gun battery, a battery of tests. Sometimes the word "cell" is used as well, as in "This battery is a button cell". In engineering terms, however, a cell and a battery are not the same. For any serious energy storage system, there will be multicell batteries, which are in turn joined to make battery modules, with the modules making up the system, containing hundreds or thousands of cells. In this chapter, the term "cell" is used to describe the basic repeat unit, and "battery" for a group of cells. The term "battery" is also used to describe the chemistry of the systems, e.g. Li-ion battery or lead-acid battery.

All cells have the same basic internal structure as shown in Fig. 1, a positive electrode, separated from a negative electrode by an ionically conducting electrolyte, all encased to either keep a liquid electrolyte in or

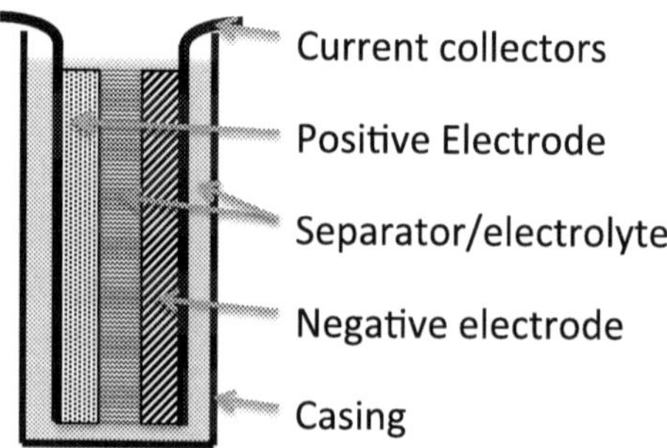

Fig. 1. A schematic of a generic cell with the labeled parts. This is for a flooded cell with large amounts of electrolyte; some cells have the electrolyte just between the electrodes.

air out. Two current collectors must pass through the casing to get the electricity in and out of the cell. The cells may be rolled or folded in a variety of ways to pack more electrode into a smaller volume.

1.1 *What Can You Use Batteries for?*

The uses of rechargeable batteries are wide and varied. They can be used wherever there is a source of power to charge them and an electrical demand to meet, with some separation between charging and demand. This can be either separation in space, for mobile batteries, e.g. mobile phones or electric vehicles (EVs), or separation in time, for stationary batteries, e.g. solar storage.

Separation in space, or using the energy stored in moveable batteries when the use of wires to supply the power is unavailable or unwieldy, is the most common use of battery systems. This ability to move the energy in a portable system such as a phone, laptop, or electric drill is one of their main selling points. The other main type of mobile batteries are for traction where the batteries are used to move vehicles (and the batteries) such as in EVs and forklifts.

Stationary systems are used to give separation in time, their only job is to store energy until it is needed at a later time. There are many cases where there is power available at one time, but the demand is at a different time in the same location. The power may be produced by a source such as solar electricity (from photovoltaic (PV) cells), which is only available during the day, typically when people are at work, and there is low demand for electricity. At night the demand increases so to meet the demand, the electricity must be stored till it is needed. This is especially true for off-grid systems where all the power must be generated on site. Almost all renewable sources of electricity are intermittent: solar, wind, or tidal, and so require storage to cover peak demand. The scalability and fast response of batteries make them ideal for these systems where there may be a need to change from storing to supplying electricity very quickly.

The storage of solar electricity during the day to use at night shows both the need for storage systems and the scalability of battery systems. The now ubiquitous solar path lights are small stand-alone devices that collect and store electricity during the day in a small battery, and then turn on at night until the battery is exhausted. Slightly larger-scale systems are used in boats, caravans, RVs, etc., where small solar panels (50–100 W) charge standard 12 V batteries to run lights, TVs, etc. On a larger scale is the Tesla PowerWall,[1] which is a 100 kg box which mounts on a wall and connects to an array of solar cells. This collects power from the cells during the day, and can supply enough power to run the house completely for several hours. This all sits on the direct current (DC) bus from the solar panels, and then passes through an inverter to produce alternating current (AC) for the house. With some extra electrical systems, this can also be used as an uninterruptable power supply (UPS) to provide electricity in the house during a power cut.

Time shifting can also be used on larger grid connected systems to supply a number of advantages to the grid. This works by charging the batteries when there is low demand for electricity, which is therefore cheap, and then selling it back when the demand (and price) is high. This can be used on constrained systems where there is a power bottleneck, and so not enough power can be transferred at peak times. A battery system can sit just past the bottleneck, and can store electricity from low demand times and help smooth out the peaks. Batteries are also used on small grids (for example, on islands[2,3]) to both time shift renewable generation and to supply power when switching between different sources. Large battery systems can also sit on the grid and be used to supply ancillary services such as voltage and frequency control, which has a significant effect on grid stability.

2 Why Use Batteries for Storage?

Batteries have many desirable properties for electrical energy storage. They are simple electrochemical systems, which basically store electrons directly. They have no (internal) moving parts, and so have little chance of mechanical failure and also little wear under normal use. Battery systems may require fans or pumps for ventilation and cooling, but these are external and easily maintained. The low failure rate of individual cells and the redundancy built into most battery systems means large numbers of batteries can be used together with little down time.

The lack of moving parts means batteries are not limited to the efficiencies of the Carnot cycle. This gives batteries a high energy efficiency, typically from 70% to 90% for electricity in to electricity out.[4] This efficiency is much higher than comparable storage systems like flywheels or hydrogen storage, with this high efficiency available over timescales from seconds to weeks depending on cell chemistry. High efficiency means less of the (expensive) input energy is lost, giving the battery system lower energy running costs. The high efficiency also reduces the waste heat output of the system, making thermal management simpler. The electron efficiency (number of electrons in to electrons out) of almost 100% means that hundreds or even thousands of cycles are possible.

The other main advantage of batteries for energy storage is their scalability. The size, and hence capacity, of an individual cell or battery is limited by manufacturing methods. However, the high reliability of batteries means that the joining of cells together to form batteries, and joining of batteries into larger groupings (often called modules), can be used to increase the voltage, power, or capacity of a system. Adding cells in series increases the voltage, and in parallel increases the current, with both increasing the total energy stored. This gives scalability to match most demands, from a simple phone battery (typically just one 3.7 V cell), through to six cell, 12 V lead-acid batteries strong enough to start an internal combustion engine (ICE), to hundreds of volts which can propel an EV,[5] on to systems offering multi-MW[2] grid scale storage.

2.1 *Why You Shouldn't Use Batteries for Storage?*

The main issue with batteries for storage is their scalability. While it is easy to see that if you want twice the capacity, you must use twice the number of cells. However, this will take up twice the space, cost at least twice as much and easily double the complications of managing the system. This is not the same for most competing systems, which often use gas/liquid storage systems, where to double the capacity you just need a bigger tank, which does not typically cost twice as much, nor does it add any extra complexity to the system. This means that a battery system's cost scales as the number of cells used, whether for voltage, current or capacity, whereas most other systems scale at less than capacity, once over a certain size. This favors battery systems for smaller capacity applications and other storage systems for very large capacity storage, unless the battery price is very low.

The other option to increase capacity is to make larger cells, but there is a practical limit to most cell designs, which limits their maximum size and capacity. Cells can be purchased with over 300 Ah capacity[6] in several different types, but this is much smaller than the capacity required to run a car, or help balance the grid. Single cells rarely supply enough voltage to provide reasonable power demand and need to be stacked in series to raise the battery voltage to a usable level.

The other issue is that capacity is linked to power. The more cells used to provide higher power also gives higher capacity and *vice versa*. In most other storage systems, power and capacity are decoupled, such as in a flow cell or fuel cell systems where the storage tank size gives the capacity and the active cell area limits the maximum power, and these can be varied separately. For flywheels, the mass and rotation speed of the flywheel control the energy stored, and the motor/generator limits the maximum power, which again can be independently adjusted. With batteries it may be that the number of cells, and hence cost, needed to achieve the desired voltage or current is higher than the needed capacity. While this gives extra life, it does increase the cost and size unnecessarily. The use of smaller capacity cells may help this problem. For small systems this is not an issue but may affect costs as the systems are scaled up.

3 How a Battery Works

All batteries operate on the same simple principle. When charged, the positive electrode holds electrons in a material at high energy, while the negative electrode has sites to receive the electrons and hold them at low energy. On discharge, the electrons fall from the high energy material down to the low energy locations. The electronically insulating electrolyte forces the electrons to travel through the external circuit where they can do work (this is summarized in Fig. 2). When charging the battery, the electrons are pushed back up using an external energy source, from the negative electrode to the positive one (this is summarized in Fig. 3). The electron flow is the battery's current, and the energy difference between the electrodes gives the energy of electrons, which is called the voltage. The more electrons you can store, the longer the cell can produce current, and so it has a higher capacity. The higher the voltage the more energy is stored per electron. For various

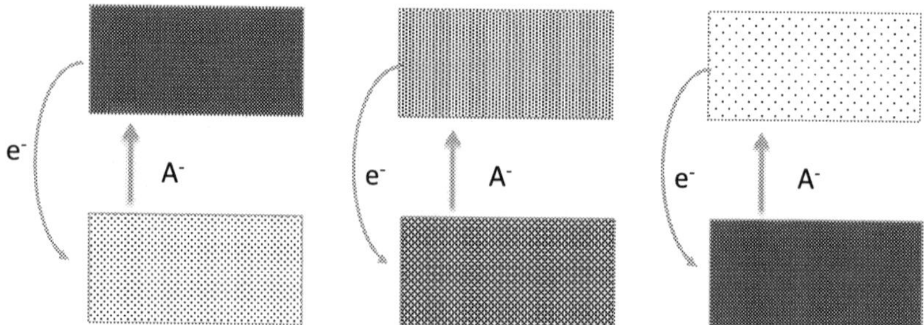

Fig. 2. An illustration of the process of discharging a battery, with electrons dropping from the high energy electrode to the lower one via the external circuit, with negative ions moving the other way through the electrolyte to balance the charge. Shading indicates the number of electrons in the electrode.

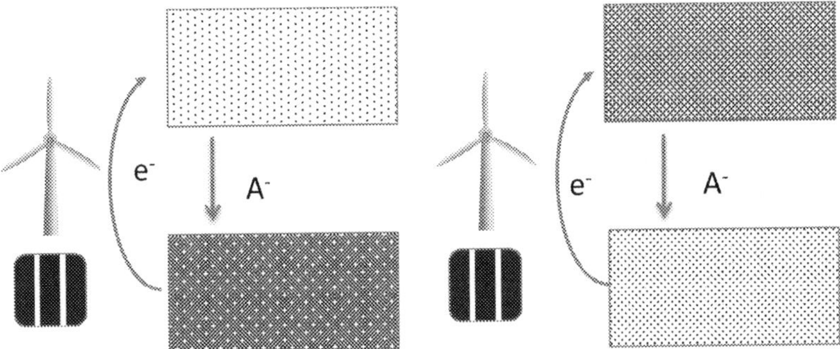

Fig. 3. An illustration of the process of charging a battery, with electrons being pushed from the low energy electrode to the higher one via a source of energy in the external circuit, with negative ions moving the other way through the electrolyte to balance the charge. Shading indicates the number of electrons in the electrode.

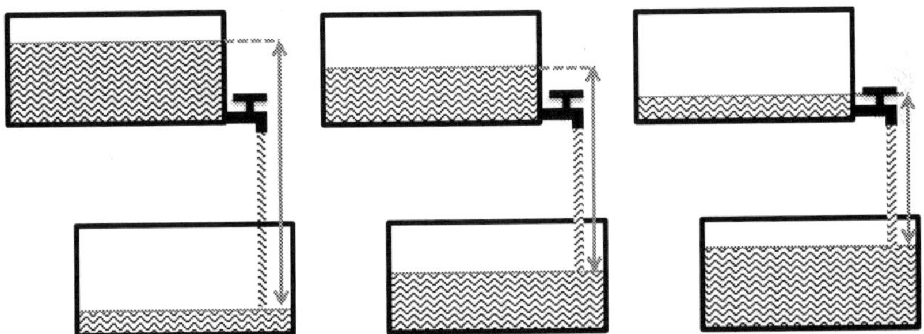

Fig. 4. Discharging of hydro storage system as an analogy to a battery, where the arrows indicate the head of water, or the energy possible to extract from the system, analogous to the cell voltage.

reasons, more energy is needed to move the electrons up to the positive electrode, than can be obtained on discharge.

The simplest analogy for this is a pumped-hydro storage system, with a high (potential energy) reservoir and a low (potential energy) reservoir. On discharge, the water falls down and gives up its energy (to a spinning generator for hydro generation). The amount of energy stored depends on the difference in height of the two reservoirs (equivalent to cell voltage) and the size of the reservoirs (cell capacity); with the size of the pipe and generator limiting how much power can be extracted (see Fig. 4).

As the water level in the upper reservoir drops and the lower one rises the potential energy of the water drops (shown by arrows in Fig. 4).

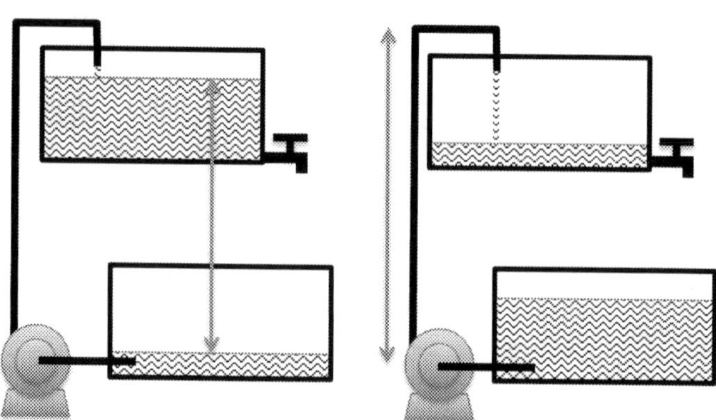

Fig. 5. Hydro storage analogue of charging a battery. The arrows indicate the maximum head and the height the water needs to be pumped to get into the top reservoir. It will always be slightly higher than the head that can be extracted.

To charge the system, the water needs to be pumped back up to the top reservoir (see Fig. 5). This involves pumping the water up and over the top of the reservoir, and having it drop down to the water level in the top reservoir. This drop is lost energy that cannot be recovered from the discharge process.

3.1 *Losses*

A pumped-hydro system shows similar losses to a battery system with the height of the water outlet above the higher reservoir being analogous to the battery voltage that is higher on charge than on discharge and the frictional losses in the pipes of a hydro storage system are similar to the resistive losses in a cell. In both systems, if you require more energy, you need to move the reservoirs further apart (higher cell voltage), and for more current you need bigger pipes and generator (lower cell resistance). There are extra losses in a hydro system due to the inefficiencies of the energy conversions from gravitational to kinetic to electro-magnetic energies to make the electrons move (current) in a wire, whereas the battery uses electrons directly.

The losses in a battery are due to effects of both the current flowing through the cell (kinetic effects) as well as the extra energy needed to move the electrons back up to the positive electrode (thermodynamic losses). It always takes at least a little extra energy to move the electrons back into the positive electrode than you get out. This is intrinsic to the active materials used in the electrodes and cannot be influenced greatly by any engineering, and is largely independent of the size of the current. These losses are unavoidable but may be reduced by the use of an alternative active material with smaller losses.

The other losses in a cell arise from the voltage drop due to the current flowing through the components of the cell. All the components of the cell will have some electrical resistance. When a current flows through a resistance there is a voltage produced across the resistance, which can be calculated from Ohm's law.

$$V = IR, \tag{1}$$

or for a cell which has an open circuit voltage (OCV) of V_{OC}, the actual voltage across the terminals is given by;

$$V_{\mathrm{cell}} = V_{\mathrm{OC}} - IR_{\mathrm{cell}}. \tag{2}$$

This loss is called *IR* or Ohmic loss. This voltage always acts against the external voltage across the terminals and so lowers the voltage on discharge, giving less energy out of the battery than is stored in the electrons, and raising the voltage required to charge the battery. Any metallic components like the current collectors and some electrodes should have a low resistance, whereas the electrolyte and any non-metallic electrodes have higher resistances. These resistances are controllable to some extent by engineering, by increasing the size of the current collectors, increasing the conductivity and/or reducing the thickness of the electrolyte, adding a more conductive phase to the electrodes or increasing the area of the cell. Reducing cell resistance both minimizes the losses in the system, and increases the maximum current available from the cell. However, these may add to the weight, and cost, of the cell.

So far we have only discussed electrons. If one could simply use electrons alone, huge amounts of energy could be stored. However, just moving electrons means that a very, very large electrostatic charge would build up on each electrode during charge or discharge. Any large charge concentrations have a very large energy cost, which means systems always stay basically neutral with respect to charge. This is what limits the energy stored in supercapacitors. In any real battery system, ions are required to move through the electrolyte from one electrode to the other to counteract the negative charge imparted by the moving electrons. This could be positive ions (e.g. Li^+ or Na^+) moving in the same direction as the electrons, or negative ions (e.g. OH^-, SO_4^{2-}) moving in the opposite direction. These must be incorporated in the electrodes in some form of chemical interaction (electronic and chemical process — electrochemistry) or the system is simply an electrolytic capacitor, with good cycle ability, but much, much less energy can be stored. There is no simple equivalent of this in the pumped-hydro model.

So, overall the electrons are removed from one electrode and added to the other electrode, with ions moving to ensure that there is no extra charge on either electrode. What matters is how well the electrode material can repeatedly move electrons and ions in and out of each electrode.

3.2 *A Basic Refresher of Electrochemistry and Electrochemical Terms*

This combination of electrons moving and chemical species changing are called electrochemical reactions and use a specific set of terms to describe and understand batteries.

The **oxidation state** is a number that quantifies the number of extra, or fewer, electrons in a system compared to a reference state. To help separate the charge on a species from its oxidation state, oxidation states are labeled by roman numerals, e.g. 5+ is V and 2− is –II as in $P^{V}O^{-II3-}_{4}$ which has an overall 3− charge. Some species can have multiple stable oxidation states and so can undergo electrochemistry when they change. A species that has an easily accessible set of oxidation states is called a **couple** e.g. Li(m) and Li^{+} (Oxidation states 0 and I), Pb(m) and $PbSO_4$ (0 and II). Loss of electrons is called **oxidation**, and causes the oxidation state of the system (or at least a component) to become more positive. Gaining electrons is called **reduction** and causes the oxidation state of the system (or at least a component) to become more negative. The loss or gain of electrons, and some ions as well, at any electrode, can be written as a **half cell equation**. Note that while half cells are useful ways of describing the process happening at one electrode, the half cells can never exist by themselves, a complete cell is required to obtain a voltage and current. An example is the cadmium half cell

$$\mathrm{Cd}^{0} + 2\mathrm{OH}^{-} \Leftrightarrow \mathrm{Cd}^{\mathrm{II}}(\mathrm{OH})_2 + 2\mathrm{e}^{-}, \tag{3}$$

where the oxidation state of the Cd changes from 0 to II, with the release of two electrons. The extra charge on the cadmium atom is compensated by the OH^{-} ions. Going from left to right is the reaction for a cell under discharge for this system, which is an oxidation. And conversely going from right to left is reduction, and is the charging reaction for this system.

Electrochemists call the electrode where oxidation occurs the **anode**, and the electrode where reduction occurs the **cathode**. This means that the cathode and anode flip when the current changes direction when swapping between charging and discharging cells. However, the positive electrode in a cell is always positive and the negative electrode is always negative. This chapter uses the terms positive and negative electrode as they are more consistent between charge and discharge, but cells are regularly referred to as having a cathode (positive) and anode (negative), which assumes that discharge is the normal state of a battery.

3.3 *Materials Requirements*

All batteries are made from individual cells, which in turn are made from the same four basic components, as shown in Fig. 1. There is a positive electrode, a negative electrode, each with a current collector, and the electrolyte. The electrodes are the active parts, which have an electrochemical couple to store the electrons. The current collectors are usually metals for their good conductivity, and are either foils with the electrodes coated onto the surface, or may be the electrode itself or the case of the cell, if these are metals. The last part is the electrolyte that blocks the electrons moving directly from one electrode to the other (i.e. is an electronic insulator) but conducts ions (charged chemical species). Typically the electrolyte is a liquid containing suitable ions, which are contained in an inert porous film (separator), which acts as a mechanical separator to keep the electrodes apart. The liquid must be stable at each electrode under both charge and discharge conditions. Water-based electrolytes are popular as they are cheap, have high conductivity and cannot burn. However, water is liable to be broken up by electrolysis into oxygen and hydrogen gases at higher voltages. Hydrogen is a flammable gas, which is explosive even at low concentrations, and raised levels of oxygen can make normally safe materials burn. Therefore, these water-based cells are limited to about 1.2 V as a maximum cell voltage and require gas removal systems and electrolyte management to replace any lost water.

A good electrode must meet several criteria, some of which are electrochemical and are set by the material used, and some of which can be engineered. They must hold the electrons, be conductive to transfer electrons in and out, and be stable over the life of the battery. Often a single material cannot achieve all the desired properties, and so a composite is used to make a good electrode. The part of the electrode that stores the charge (usually called active material) must have some form of redox couple that can reversibly hold electrons at high energy/voltage for the positive electrode, or low energy/voltage for the negative electrode, which is selected by the nature of the species storing the electron. The ultimate speed of the electrochemistry is also largely controlled by the species chosen as the active material. The electrodes must also conduct electricity to move the electrons in and out. Often the active material is not conductive enough and an extra conductive phase needs to be added to give the desired conductivity. This of course adds to the weight and volume of the cell. This is why metallic electrodes are popular, as they need neither current collectors nor an additional conductive phase. The electrode must also have enough surface area to interact with the ions in the electrolyte, and so electrodes are typically made from high surface area powders. For maximum current out of a cell the conductivity of all the components must be high, there must be a large surface area to get the current through and the electrochemistry must happen sufficiently quickly. For long cycle life the reversible storing of the charge must be very, very reversible, and the electrolyte stable at the electrodes.

3.4 *The Key Parameters of Batteries*

There are several key functional parameters for any battery regardless of the internal chemistry. The most commonly reported numbers are the capacity and the battery voltage, which is typically measured under load. These numbers are only a guide, because at almost all times the battery does not have the exact voltage or capacity reported. The battery voltage depends on the number of cells and the voltage of each cell, and the load (current) the battery is under. For instance, a 12 V lead-acid battery has six 2 V cells, but a single cell Li-ion battery will have a nominal voltage of about 3.7 V. The "12 V" lead-acid battery terminal voltage can vary from 14 V to below 11 V depending on the state of charge (SOC) and magnitude of current flowing. Likewise the 3.7 V Li-ion cell's voltage may vary from 3.4 to over 4 V.

An easier value to measure is the OCV (V_{OC}, also called the open circuit potential) which is measured with the cell not connected to any load, or at least with no current flowing. The OCV can be used as an indication of the level of charge remaining in the battery, for some cells at least. The cell voltage when under load is always lower than V_{OC}, due to internal resistances in the cells (see Fig. 1)).

The other number given on most cells is the capacity, which is the average amount of charge that the battery can reversibly store, under some arbitrary conditions, typical those sensible or most favorable for the cell's performance. For practical measurement reasons, the capacity is quoted in ampere hours (Ah or Ahr) or mAh (also mAh, milliampere hours) and not the standard coulomb (C). This capacity gives an approximate indication of the amount of time the cell can produce a fixed current. A capacity of 2 Ah for a battery, means that the battery should be able to provide 2A for 1 h, or 200 mA for 10 h or 4 A for 1/2 h. However, in reality the higher currents typically give lower than expected capacities due to internal losses in the cells. The capacities printed on cells are normally collected under favorable, low current discharge conditions, which may not be the current used for the desired application and so may be larger than practically observed. Also, as batteries age they lose capacity, and so will have less than the labeled capacity.

The product of the battery voltage and capacity gives an indication of the energy that can be stored in the battery. Hence a 12 V, 6 Ah battery would contain 72 Wh or 260 kJ, whereas a 3.7 V, 10 Ah cell would be able to store 37 Wh or 133 kJ.

The most useful data is the actual discharge and charge curves for batteries under a real load. An example is shown in Fig. 6 for a NiMH battery. Note that while the nominal voltage for one of these cells is 1.2 V (labeled

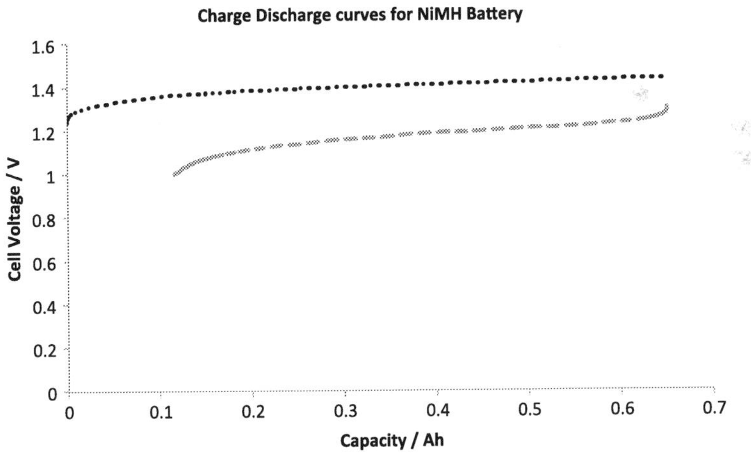

Fig. 6. The charge and discharge curve for a NiMH battery. Note the charging voltage is always higher than the discharge voltage, and for NiMH cells a significant portion of the electrons are lost during charge.

on the figure) the voltage is below this value for much of the discharge and a voltage of about 1.4 V is needed for charging. Note due to side reactions in the cell, charge capacity is larger than discharge capacity. This data is usually collected by a battery cycler, which controls and records the voltage and current, where the charge parameters can be controlled. This allows the actual performance of the batteries to be measured under (simulated) conditions that they should experience in real operations, e.g. constant charge and discharge for a managed system, intermittent charge but regular discharge for storing solar power, regular charge and variable discharge for an electric car with variable driving conditions, or the extremely quick pulse discharges for digital transmission systems such as cell phones.

The voltage profile gives much more useful design information. It gives the difference in voltage between charge and discharge and gives capacity under real conditions. It also shows the change of voltage of the cell with capacity, as the voltage of a cell always drops as the cell is discharging. This can be used as an indicator of how much charge can be extracted from the battery when discharged to any arbitrary cut-off voltage.

If the actual voltage window of the desired application is used, which may not be the entire voltage window available, (see Figs. 6 and 8) this gives the practical capacity of the cell.

This cycling may be repeated hundreds or thousands of times to show how the capacity changes with health of the cell. The cycle life is defined as the number of cycles it takes for the capacity to drop below an arbitrary percentage of the initial capacity, usually somewhere between 50% and 80% depending on the application. Usually these curves are collected at several different C rates to show how voltage and capacity change with current.

3.4.1 *C-rate*

The current used to charge and discharge affects the performance and capacity of the battery. Absolute current is not helpful as a large lead acid battery can easily produce tens or even hundreds of amps, whereas a small coin cell may only produce mA. C-rate is used to normalize the size of the current to the purported capacity of the battery.

The C-rate is the current that would take 1 h to discharge the cell based on the nominal capacity of a new battery. So the 1C (or just C) current will discharge the cell in 1 h, 2C will take 1/2 h to discharge, 1/2C (or C/2) current will discharge in 2 h. Since all real batteries have losses when a current flows, as the current increases (i.e. a higher C-rate) the losses also

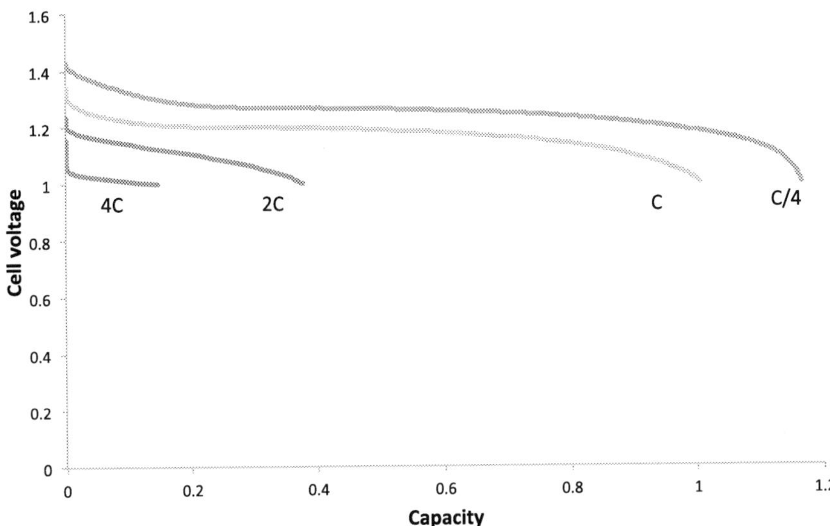

Fig. 7. A plot of capacity of a small NiMH cell vs. C-rate down to a cutoff voltage of 1 V.

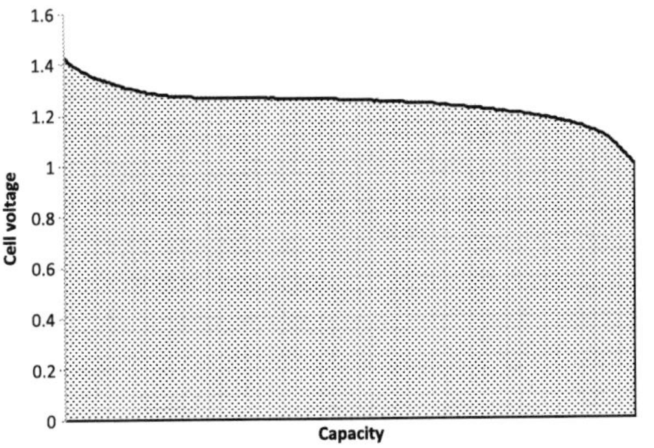

Fig. 8. A typical discharge curve of a cell. The area under the curve gives the total energy available from the cell.

increase and reach the voltage cut off sooner, and so the total capacity will be lower. Therefore, it is normally advisable to measure the capacity at the C-rate at which a particular application demands. Figure 7 shows the loss of capacity with increasing C-rate.

C-rate capability is also a very useful metric in choosing a battery system for a given storage application. Cells with good current capability, are needed for short-term cycling, such as cells powering electric drones, which may only have a few minutes flying time, or a UPS which may only need to run for a few minutes till larger, slower replacements can come online. The systems will need to be able to produce good capacity at a C-rate on the order of 10C to 40C. Cells that run a pure EV may need a C-rate of about 1C or C/2. A laptop, phone, or solar storage system typically cycle on the order of once per day, say 10 h or C/10. As with all engineering solutions there is typically a trade-off between C-rate capability (power) and capacity, and of course cost.

3.4.2 *Densities*

While Ah-based capacities for cells are the prime measure for a battery, there are more useful metrics depending on the desired use.

The discharge curve can provide much useful information. The power at any time is the instantaneous voltage multiplied by the current. The practical capacity can be read off from where the lower voltage limit is reached. If the application is power limited this will set the lower limit or there may be separate system based on a lower voltage cutoff. The total energy in the cell can be calculated by integrating the area under the discharge curve which gives the Wh (or kWh) of the battery, which is the total electrical energy in the system, (1 Wh = 3.6 MJ, 1 kWh = 3.6 GJ). To compare various types or designs of battery these total energy numbers are converted to energy densities by dividing by the mass or the volume of the battery (including the case, etc.). This gives Wh/kg which is useful for transportable cells such as cell phones and EVs. Wh/L or Wh/m^3 gives a guide to the space needed to store energy. This is summarized in Fig. 9 which shows the energy density of the various chemistries commercially available. The higher up and the further to the right in the plot, the smaller and lighter the battery. Cells that need to be carried in a small phone must be both small and light, whereas for stationary applications, weight is less of an issue, but volume or foot print may be important.

The other important metric from this measurement is to divide the price of the cell by the energy in kWh to give cost per kWh. This is the important information in larger stationary systems where the cost of the capacity is the key.

For small, portable systems such as phones and laptops the cost of the battery is small compared to the system cost, and energy density is the most important metric. For large stationary installations size and weight are less critical and cost is the overriding factor. This is good as cost is

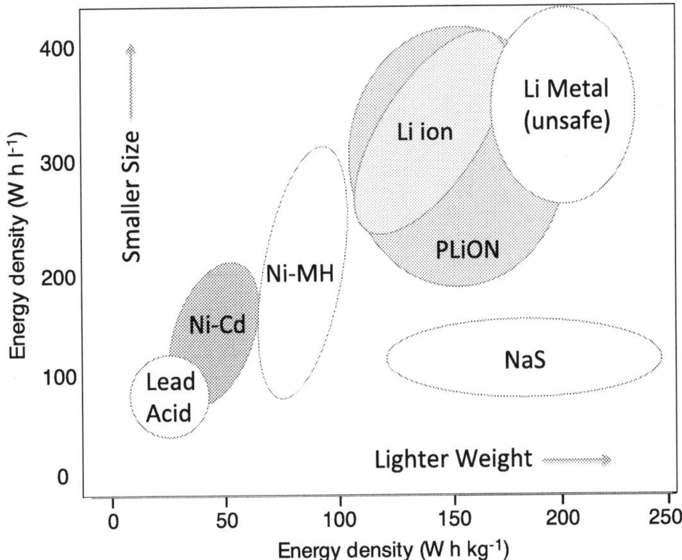

Fig. 9. Comparison of energy density of various cell chemistries. Used with permission from J. M. Tarascon and M. Armand[7] but modified by the author.

usually inversely related to the size and weight. For EVs the energy density is still key, but the battery now makes up a significant cost of the system and price per kWh also becomes critical. This gives a hard engineering balance to achieve batteries small enough to fit in the vehicle, but still at a competitive price.

The charge curve can be analyzed the same way to give the total energy needed to charge the battery. The ratio of the charging energy to discharge energy gives the energy efficiency of the cell but will be lower in a real system due to other losses outside the cells, and the possibility of the application not using the full voltage of the cells, and so not using all the available energy in the cells.

The ratio of input capacity to output capacity gives the electron efficiency of the cell. This shows how much capacity is lost to irreversible storage reactions in the cell. Both lead-acid and NiMH cells always split some water on charge, and so their round trip electron efficiency is lower, however the discharge to discharge efficiency is high. The electron efficiency must be well over 99% to achieve a sensible number of cycles as shown in Fig. 10. At 99% electron efficiency, after 100 cycles the capacity will already be only about 36%. At 99.9% electron efficiency, after 100 cycles there is a

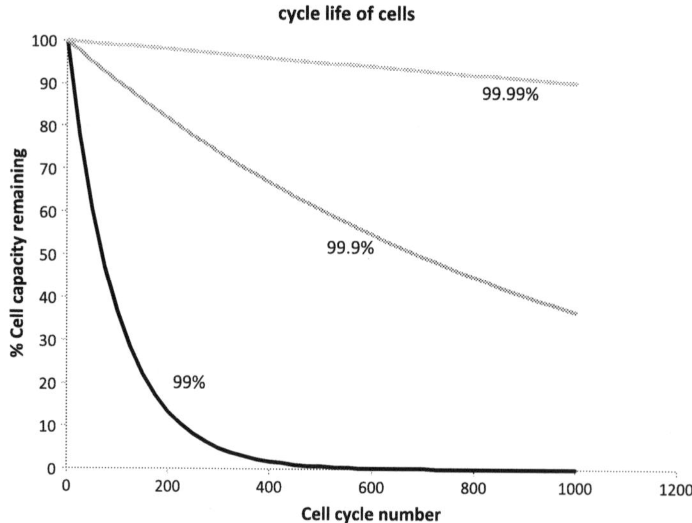

Fig. 10. The loss of capacity for a cell with 99%, 99.9%, and 99.99% of electron efficiency. Approaching 99.99% is needed for sufficient cycling.

useable capacity of 90% remaining, and after 1,000 cycles it again decreases to 36%. Assuming a 70% residual cut-off capacity for the end of cell life 99% efficient cells would only last 35 cycles, 99.9% for 356 cycles, but 99.99% cells could last over 3,500 cycles.

3.5 *State of Charge (SOC) and State of Health (SOH)*

SOC is the level of charge in the cell, varying between 0% and 100% for a fully charged cell. As the capacity of all cells degrades with time, temperature and cycling the 100% charged capacity drops with cell age. Hundred percent SOC represents the maximum capacity of the cell at its present age and condition. This loss of capacity compared to a new cell is recorded as the SOH of the cell. This is a more arbitrary number, which starts at 100% for a new cell, and gets smaller as the cell ages. How SOH is measured, how it corresponds to capacity, and at what SOH a cell fails at, is manufacturer and cell dependent. SOH will typically come from empirical measurements on real cells, often under accelerated testing regimes to reduce testing time and speed cell capacity loss. However as long as the capacity at any SOH is known it can be used to track the life of a cell.

4 Battery Types

4.1 *Lead-Acid Battery*

This is the oldest rechargeable battery still commercially available, and was invented by Gaston Planté in 1859.[8] It is interesting to note that this was after Grove's gas cell (1839)[9] which is the precursor to fuel cells. The original Planté cell used two lead (Pb) sheets separated by a cloth soaked in a sulfuric acid electrolyte. The active material in the positive electrode is PbO_2, which converts to $PbSO_4$ on discharge. The negative electrode is based on Pb metal, which also converts to $PbSO_4$ on discharge. Both these reactions use up the SO_4^{2-} from the aqueous electrolyte, so the electrolyte concentration changes from concentrated H_2SO_4 solution when charged, to a much more dilute solution on full discharge. This gives a separate measure of the SOC of the battery, by measuring the specific gravity of the solution. This also means that there must be enough SO_4^{2-} in the electrolyte to react with both electrodes, either needing a high concentration and/or a significant volume. This also means that the available current is less from a fully discharged battery due to the low conductivity of the cells.

The basic electrode half equations are:

$$\text{Positive electrode } \mathrm{Pb^{IV}O_2 + SO_4^{2-} + 4H^+} + 2e^- \Longleftrightarrow \mathrm{Pb^{II}SO_4 + 2H_2O}$$

$$\text{Negative electrode } \mathrm{Pb^0}(m) + \mathrm{SO_4^{2-}} \Longleftrightarrow \mathrm{Pb^{II}SO_4} + 2e^-. \qquad (4)$$

The voltage difference between the couples is nominally 2 V, usually grouped to form 6, 12, or 24 V batteries. Planté's original cells were based on two pure lead sheets rolled up together and covered in sulfuric acid, and then charged and discharged repeatedly to form the working PbO_2 electrodes, in what is called a "formation" reaction. Modern cells use a paste of PbO_2 on a metal (Pb or Pb alloy) support, which also acts as the current collector. These only need one "formation" cycle to form the working electrodes, and so make the cells faster and cheaper to produce. By engineering the structure of the metal supports and of the PbO_2 paste, the electrodes can be formed into different shapes to give either higher current, longer cycle life, or the ability to cycle more deeply, depending on the required application of the cells.

Far and away the biggest use for lead-acid batteries is in the automotive industry, where, due to their ability to supply large currents for short periods of time, they are ideal for powering the starter motor for the main internal combustion engine. These batteries also give continuous power to the lights and ignition systems and so are called starting, lighting, ignition (SLI) batteries.

While having good volumetric energy density and high current capabilities, the weight of the lead plates, the large amount of water-based electrolyte and the casing to keep the acid secure, make the cells very heavy and so they have a poor mass energy density. This makes them less useful for other portable storage applications. Stationary applications are less affected by weight and so lead-acid cells have long been used for emergency backup power supplies. Their well-engineered robustness and high availability makes them popular as a low cost option for small storage systems.

Lead-acid batteries have some system integration issues. The use of highly concentrated sulfuric acid as the electrolyte makes moving and handling these batteries somewhat dangerous. The cells are heavy and the acid can leak out if the battery is tipped. Advances with the use of gelled electrolyte have reduced the problem of leakage but with an attendant loss of maximum current. The use of large amounts of toxic materials containing lead along with the strong acid means that the batteries need to be recycled properly, as mandated by law in most countries.[10] While the recycling systems are in place, the required disposal at the end of life of the battery may make these cells less desirable for large installations.

Another system problem with lead-acid batteries is their high cell voltage of 2 V which is higher than the voltage required to split water and means that the cells release H_2 on charge, and possibly in the case of quick discharge too. When a large number of cells are placed together this necessitates good ventilation to ensure the hydrogen produced never reaches flammable or explosive levels. This release of hydrogen uses up the water from the electrolyte and so the cells need regular maintenance to replace the lost water. Large cells are typically open, to aid monitoring and facilitate the regular electrolyte top up. Modern low-maintenance batteries are sealed and include a catalyst to recombine the H_2 and O_2 back to water to keep the electrolyte level constant. These cells need to be charged more carefully to ensure the recombination catalyst can keep up with the rate of hydrogen production. The cells are fitted with pressure release valves in case of battery over pressure.

They are still used for three main reasons: every car/truck has a lead acid battery for SLI purposes, so millions are produced annually. SLI batteries need to supply a large current for engine starting. They also handle the abuse in an engine bay well, particularly in coping with extremes in temperature.

Lead-acid cells have had the longest development time of any rechargeable cell, and so are well engineered, reproducible, are intrinsically safe in use (not made from flammable materials) and long lasting. Lead-acid batteries are cheap per unit, and are readily available, and so are the standard battery used for home storage. The low capacity per unit makes them less likely to be used for large installations where size and cost issues become significant. If a cell costs twice as much, but has three times the capacity it still pays to use the better cell for large installations.

4.2 *NiCd/NiMH*

NiCd batteries were invented by Waldemar Jungner in 1899,[11] 40 years after Plante's lead-acid battery. They use a water-based alkaline electrolyte with a nickel oxy-hydroxide positive electrode and a cadmium-based negative electrode, with a cell voltage of about 1.2 V. The basic electrode reactions are:

$$\begin{aligned}&\text{Positive electrode } \mathrm{Ni^{III}O(OH) + H_2O} + e^- \Longleftrightarrow \mathrm{Ni^{II}(OH)_2 + OH^-}\\&\text{Negative electrode } \mathrm{Cd^0 + 2OH^-} \Longleftrightarrow \mathrm{Cd^{II}(OH)_2} + 2e^-.\end{aligned} \tag{5}$$

While the OH^- is involved in the electrode reactions, unlike the sulfate in the lead-acid case, it is both used and reformed at the same rate, and so is constant throughout cycling giving more consistent performance but is therefore not an indicator of the state of charge.

The chemistry is very reversible, able to produce 1,000s of cycles and still having good capacity. The cells do suffer from side reactions and so their efficiency is lower than other cells, and they do self-discharge, losing capacity while sitting over days and weeks. The cells are best used for short-time scale storage.

The cells are typically made by rolling up a nickel foil electrode and a cadmium foil electrode with a porous insulating separator to form the electrolyte as shown in Fig. 11. These may be flat or with holes (pocket plate) to increase surface area and current. There is also a non-porous layer to isolate the positive and negative electrode when rolled. This is placed into a cylindrical casing which is then filled with electrolyte. The large surface area in a small volume gives a low-cell resistance and the ability to produce high currents. Like the lead-acid cell, (and any cells with a water-based electrolyte) H_2 can be produced at high rates of discharge or on overcharge. Therefore, the cells are fitted with recombination catalysts

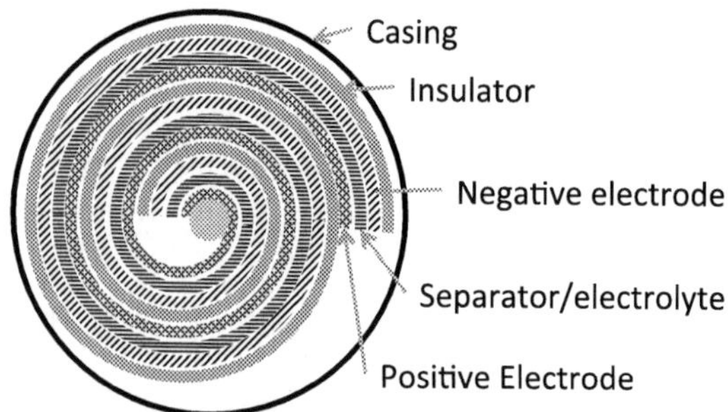

Fig. 11. The geometry of a rolled cell. Cell needs an extra insulating layer to separate positive and negative electrodes as they are rolled (roll not to scale).

and over pressure release valves. The sealed cells also allow the cells to be used in any orientation with no leakage.

Some larger-scale NiCd cells are unsealed, such as those sold by Saft Batteries and Storage Battery Systems, much like a lead-acid battery, requiring regular maintenance to replace the water lost from the electrolyte, but saving the weight and volume of the pressure vessels and allowing for much bigger area cells.

Cadmium is classed as a toxic material in most countries now,[10] and cannot be used in new battery systems. This encouraged the replacement of the cadmium negative electrode with a metal hydride-based one, helped by the increase of capacity in the metal hydride electrode.

The metal hydrides emerged from hydrogen storage research in the 1970s,[12,13] but were not put into practical cells till the 1980s. The metal hydride electrode utilizes the same basic reaction as a fuel cell, with the water being split to form H_2 (and OH^-). In this case, it is protons (H^+) from the water-based electrolyte being converted into H_2 which is stored in a metal as a hydride, i.e. as a solid. The reaction at the negative electrode for metal hydride cells is

$$\mathrm{H}_2^0 + 2\mathrm{OH}^- \Longleftrightarrow \mathrm{H}_2^{\mathrm{I}}O + 2e^-. \tag{6}$$

This is used with the same NiOOH positive electrode and KOH electrolyte as the NiCd cells. Storing the H_2 as a gas gives a very low energy density, unless very high pressures are used (see fuel cell storage, Chapter 6), but the metal hydrides can store the hydrogen at much higher densities, and low pressures. The initial cells were based on TiNi materials, which are still used, but most small cells now use LaNi-based electrodes.[14] These electrodes give higher capacities than the cadmium electrodes, with much the same performance. The main disadvantage is that the charging process for NiMH cells is different from that of the NiCd cells and the same charging methods cannot be used. NiMH cells do have a significant self-discharge rate of several percent per day and so are not useful for long-term storage. Also like the NiCd cells, the NiMH cells have significant side reactions on charge, and so show lower overall energy efficiency.

(a) Memory effect

One of the well-known issues with NiCd-based cells is the memory effect. If the cells are only partially discharged repeatedly, they "remember" their level of discharge, and lose the extra, unused capacity. The only real memory effect was observed for cells in satellites with repetitive cycling profiles.[15] The NiCd (and NiMH) cells are often observed to have these effects but they are in fact issues with the charging regime of the cells, and can be easily avoided with a sensible charge–discharge regime.[15]

4.3 *NaS Battery*

These batteries are unlike the other batteries covered in this chapter, in that they run at high temperatures (200–300°C), and also have liquid electrodes with a solid electrolyte, the reverse of the previous cells. This is also the first practical new rechargeable cell developed for almost 80 years after the NiCd cell, coming out of research by Ford[16] in the 1960s and 1970s to make better EV batteries. It was also the first battery that could be said to be designed from a theoretical approach, instead of the empirical "lets put this in a cell and see if works" approach of the earlier cells.

These cells work at elevated temperatures to ensure that the electrodes (Na and NaS_n, sodium polysulfide) are molten, which gives them their ionic conductivity and good contact to the solid electrolyte. The high temperature is also needed to increase the conductivity of the solid electrolyte to a useable level. However, this elevated temperature makes these cells more complicated to run and manage thermally. The thin ceramic electrolyte and the requirement to maintain the temperature high enough means they are now typically only used for stationary applications.

The positive electrode material is a melt of sulfur and sodium polysulfide (NaS_n) contained in a metal box, which also acts as a current collector. The metal needs very good corrosion protection from the highly corrosive polysulfide mix. The phase diagram of NaS (Fig. 12), shows the various forms of NaS that occur as the Na content varies, from pure S through Na_2S_5, Na_2S_4 to Na_2S_2. To ensure that no solid polysulfide precipitates out,

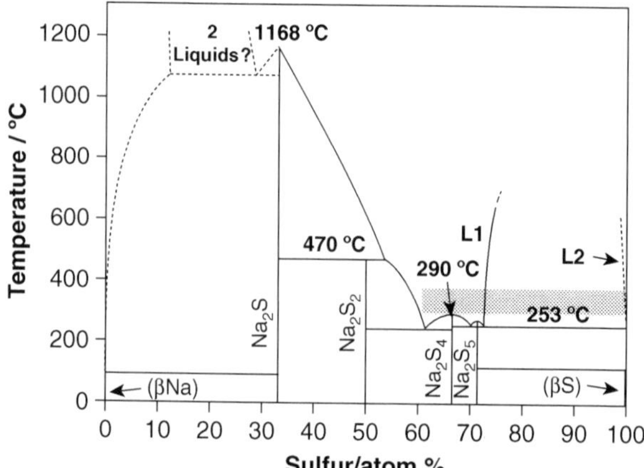

Fig. 12. Phase diagram of the Na-S system. Gray box shows the operating region of NaS battery. Slightly modified from original © P. Adelhelm, P. Hartmann, C. L. Bender, M. Busche, C. Eufinger and J. Janek From Beilstein *J. Nanotechnol.*, 2015, 6, 1016–1055 Creative commons licence (CC BY).[19]

under normal operating conditions (300–400°C) the maximum Na content is limited to about 40% (i.e. 60% atomic S), where solid Na_2S_2 starts to form. Often the cells capacity is lowered to only form Na_2S_5 to give a single discharge plateau. Note the volume of the polysulfide changes with SOC and so needs expansion volume.

The negative electrode is molten sodium metal, which is separated from the molten sulphide by a thin layer of β''-alumina,[a] which is a very good Na^+ ion conductor. β''-alumina is a slight modification of β-alumina, a sodium aluminum oxide. The β-alumina family have been known as refractory material for a long time[17] (where the presence of the sodium was overlooked, hence the name), but it was only in the late 1960s and 1970s when the fast sodium ion conduction in these materials was discovered and utilized.[18] At elevated temperatures, as used in NaS cells, the conductivity of the ceramic can even surpass that of Na^+ ions in solution. The solid electrolyte is very stable and so the cell voltage can be higher, with no

[a] β-alumina: Crystals made from the same elements can come in different crystal forms. This are usually labeled using Greek letters, e.g. α-quartz, β-quartz. β-alumina was the second alumina structure described, but was modified artificially to give better conductivity, to form β'-alumina and then β''-alumina.

risk of electrolyte decomposition or out gassing as with an aqueous-based electrolyte.

$$\text{Negative electrode } \mathrm{Na}^0 \Longleftrightarrow \mathrm{Na}^{\mathrm{I}+} + e^-$$
$$\text{Positive electrode } \mathrm{xNa}^+ + \mathrm{S}^0 + e^- \Longleftrightarrow \mathrm{Na_x}S^{-\mathrm{Ix}}$$
$$\mathrm{Na_xS} \text{ is usually written as } \mathrm{NaS_n}. \tag{7}$$

The main operating issue with NaS batteries is their high temperature, which requires careful control and management to not lose too much efficiency. There needs to be multiple cells per hot module, to minimize the heat loss through a smaller surface area–volume ratio.

All the previous battery chemistries have had water-based electrolytes and non-flammable electrodes, and so would not burn, even on extreme overcharge. The molten sodium, and to a lesser extent the polysulfide, are highly flammable on coming in contact with air and even more reactive with water. This means that any leaks in the containment vessel can lead to a fire, which is hard to extinguish by conventional means. There was a significant fire in the NaS batteries at the Tsukuba Plant of Mitsubishi Materials Corporation in September 2011.[20] No one was injured, and no other buildings were affected, but it did raise safety issues with these batteries. These safety issues arose from the arrangement of cells into modules. This problem has now been solved, and there have been no subsequent fires.[20]

Any high energy density storage system is prone to fires, with over 200,000's fires in gasoline-powered cars per year in the US alone.[21] Lithium-ion batteries often burn as well, with several high profile EV fires[22] and the problems with the Samsung Note 7 leading to its complete recall.[23] The total number of Note 7 that burned is small, with only 92 in the US,[24] out of several million sold. This is about a 0.01% failure rate which is very low. However, the mode of failure was very destructive. It is a wonder that, with the number of battery-powered devices in the world, there are not more fires. The need to use highly reactive materials to obtain the energy densities required for reasonable energy storage, either chemical or electrical, means that fires are always a possibility.

4.4 *Lithium-ion Batteries*

Lithium-ion batteries are the most recent of the existing commercial battery chemistries, coming out of research from the 1970s to 1980s.[25] Lithium is both the lightest metal and the metal with the lowest reduction potential

(−3 V vs. the standard hydrogen electrode) and so should make high capacity and high voltage (energy) batteries.

These batteries do have some issues in that the very low reduction potential means very few solvents are stable in the presence of Li metal, and so only a small group of organic solvents can be used as electrolytes. Lithium, like its fellow metals sodium and potassium, reacts violently with water, and so must be kept away from water and air. Li is so reducing it can even react with the usually inert nitrogen from the air. The high-cell voltage (typically 3.6 V) means that they produce almost three times the energy compared with a similar sized NiMH cell (1.2 V).

The initial lithium rechargeable batteries used pure Li metal as the negative electrode, which produced the highest voltages and energy density, but they were prone to forming dendrites on cycling. These dendrites would eventually penetrate the electrolyte/separator and short circuit the cells, causing overheating, fires, and explosions.[26] Li-metal batteries are now only available as high power single use (primary) batteries.

In 1991, Sony commercialized the use of carbon as a negative electrode, which could reversibly store Li^+ ions at low potentials similar to those of Li metal, but with no dendrite formation. This was coupled with $LiCoO_2$ as a high voltage (energy) positive electrode to produce a 3.7 V cell with good cycling properties. These batteries had no free Li metal, with only Li^+ ions present, hence the name. Both of these materials are layered materials that can reversibly insert the small Li^+ ions between the layers, in a process called intercalation (see Fig. 13). This insertion then moves the layers further apart depending on the amount of Li inserted. Both $LiCoO_2$ and graphite have only a small movement of the layers and so are very

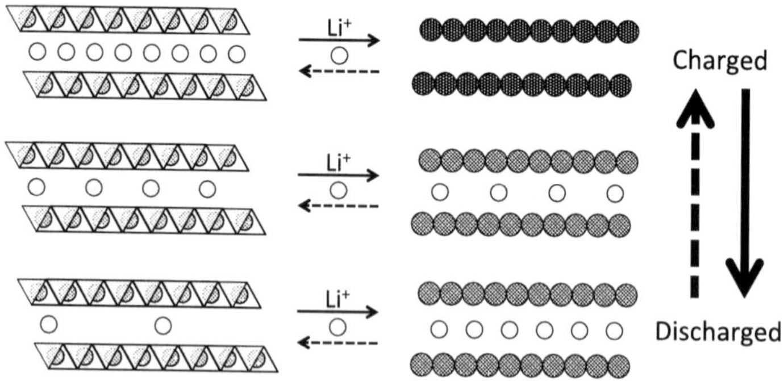

Fig. 13. A diagram of how Li ions move between electrodes in a battery. The electrons move in the same direction in the external circuit.

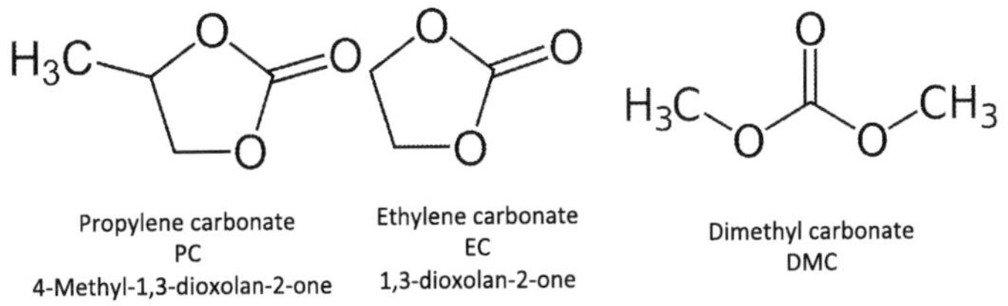

Propylene carbonate
PC
4-Methyl-1,3-dioxolan-2-one

Ethylene carbonate
EC
1,3-dioxolan-2-one

Dimethyl carbonate
DMC

Fig. 14. Structure of carbonate electrolytes.

reversible, and cells can cycle many hundreds of times.

$$\begin{aligned}&\text{Negative electrode } C_6^0 + x\text{Li}^+ + xe^- \Longleftrightarrow C_6^{-x}\text{Li}_x\\&\text{Positive electrode } \text{LiCo}^{\text{III}}\text{O}_2 \Longleftrightarrow \text{Li}_{1-x}\text{Co}^{\text{III}}_{1-x}\text{Co}^{\text{IV}}_x\text{O}_2 + x\text{Li}^+ + xe^-\\&\text{where } 0 < x < 1(\text{max}).\end{aligned} \tag{8}$$

The electrolyte for these cells needs to both conduct Li^+ ions well and be stable over the voltage range of the electrodes. The electrolyte varies with manufacturer but is typically a mixture of various ester compounds, known as carbonates, see Fig. 14 for some examples, coupled with a Li^+ salt such as $LiClO_4$ or $LiPF_6$. There are often other proprietary additives to improve ion conduction, shelf life, fire retardants, etc.[27] Like other NiMH or lead acid cells, the electrodes are held apart by a porous separator and flooded with electrolyte. The initial cells were made in the 18650 form, where the cell was rolled up and inserted into a cylindrical casing (18 mm diameter, 65 mm long) and then topped up with electrolyte, as per Fig. 11.

The extreme reduction potential of the Li/C negative electrode is such that it decomposes the electrolyte on the surface of the electrodes. This forms a layer on the surface called the solid-electrolyte interphase (SEI) which is made of decomposed electrolyte at the electrode surface and may lead to lowered capacity by blocking the surface, or by using up the free electrolyte. This is a physical phenomenon and can only be mitigated, not stopped. It is used in some primary Li cells to enhance shelf life.

Unlike Lead-acid and Ni batteries which can cope well with low levels of overcharge by using the electrons to electrolyse water, then using a catalyst to reform the water, there are no relatively benign side reactions for Li-ion cells. This means that any overcharge or overdischarge starts damaging the electrodes and electrolyte quickly. Overcharge is the most potentially damaging, as it leads to heating and overpressure in the cells which can causes them to rupture and perhaps burn. To combat this, all Li-ion cells need some form of electronic control circuitry to monitor the conditions including the temperature of the cells and to ensure the cells are charged and discharged safely. This control circuit is called a battery management system (BMS), and is vital for the successful and safe use of all Li-ion cells. The protection circuits can be in the battery itself, or in the charging circuits outside the cells, such as in a laptop or cell phone. More will be discussed about these BMS in Sec. 5.3

The demand of users for higher power, higher capacity, longer lasting and cheaper cells is ongoing. There are engineering methods to improve cell performance, by decreasing cell resistance, or packing more active material into the cells, however there is a limit to what can be achieved by this approach. As there must always be a BMS system between the battery and the load/charger there is more flexibility on the cell reactions than for other battery types. This allows the replacement of the $LiCoO_2$ materials with other better performing materials. This had led to Li-ion batteries becoming a family of high voltage cells that utilize the reversible insertion of Li^+ ions into two dissimilar materials. This is different to other systems where for example NiCd and NiMH cells are treated differently, even though there is only one electrode changed. Most development has been on the positive electrode materials, as they have lower performance (capacity, mAh/g) than the existing negative electrodes. The negative electrode is still typically some form of carbon, but now often with higher surface areas to enhance Li-ion capacity. The form of the electrolyte has also become more varied.

4.4.1 *Positive electrode (Cathode) materials*

$LiCoO_2$-based cells are still in production as their capacity and cycle ability are well known. However, the use of relatively expensive and toxic cobalt is an issue. Replacing the Co oxide with a cheaper metal oxide both lowers cost and toxicity, but needs to keep the same capacity (or hopefully even increasing it). Positive electrode materials are typically crystalline oxide-based materials which have both some form of redox couple, which sets their voltage, and locations to insert the Li ions. To minimize the mass of the electrode, the positive electrode materials must be made from the oxides of the first row transition metals. This limitation coupled with practical voltage and supply reasons leads to either Fe or Mn oxide materials being the best candidates, perhaps with small amounts of other metals to modify structure or performance. There are few stable layered metal oxide compounds analogous to $LiCoO_2$ and so other structures, such as spinel and olivine, which have empty sites for Li insertion are used. These structures do not have the same easy access for the Li^+ ions as the layered materials. $LiMnO_2$ and several LiFe oxides are also used at present in a variety of cells. These all have slightly different cell charging voltages but this is controlled by the BMS systems and all that must be known is the charge and discharge voltage of the system.

One of the most promising new materials is $LiFePO_4$, a lithium iron phosphate with olivine structure. This is not a layered structure so cannot intercalate, but has a 3D arrangement of sites to hold Li ions. Cells made from this material are often called LFP cells, named after the symbols for the three constituents. Both iron and phosphate are non-toxic and both are more abundant and cheaper than the cobalt oxides, with equivalent capacity.[28] Their main issue in use is that they are not a very good electrical conductor, and so need extra electronic conduction added to the material. This is usually achieved by coating the particles with a conductor, typically carbon.[28] The nominal cell voltage is 3.2 V, lower than cobalt cells, but with better capacity and cycleability.

Other positive electrodes, based on Ni, Mn, or Fe metals are also used in commercial batteries. The voltage varies with the mix of metal atoms used, with the capacity, power, and price varying between materials. The capacity is broadly similar for all the materials, with voltages varying as shown in Fig. 15.

$$\text{Positive electrode: } \mathrm{LiFe^{II}PO_4} \Longleftrightarrow \mathrm{Li_{1-x}Fe^{II}_{1-x}Fe^{III}_{x}PO_4} + x\mathrm{Li}^+ + xe^- \tag{9}$$

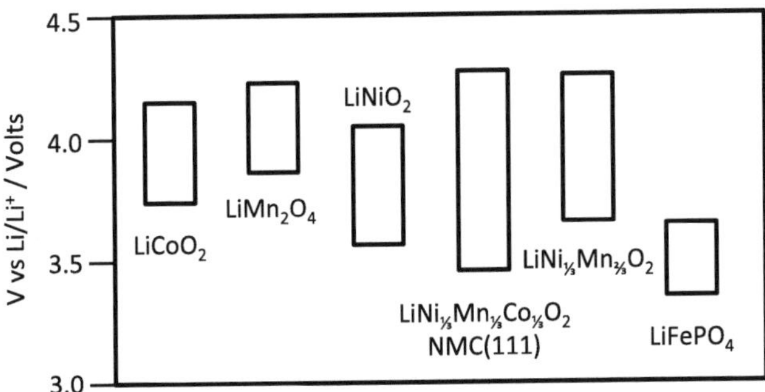

Fig. 15. Examples of positive electrode materials with their voltage ranges vs. Li/Li^+ reference. Higher voltage means more power per electron. Battery voltage is about 0.6 V lower than the positive electrode voltage due to the use of a carbon negative electrode instead of a lithium metal one.

4.4.2 *Negative electrodes (Anode) materials*

For cost and practical capacity reasons, the negative electrodes are still based on various forms of carbon, from different organic starting materials. These operate at a voltage above that of Li metal, and so give a cell with less energy than a Li metal-based cell.

A major issue with carbon, and even more so for Li inserted carbons, is that they are flammable, and can burn if the cell overheats. On overcharge, once the carbon electrode is full of lithium, the voltage of the negative electrode can drop down to where there is Li metal formation and the surface of the carbon becomes coated in Li metal, called Li plating. This forms a finely divided, highly reactive Li layer on the surface. As the heat increases, the cell may rupture and the gases released can ignite, which will likely set fire to the rest of the cell. This can release a lot of energy, as the reduced C, the electrolyte and the cell casing are all flammable.

One method to make the cells safer, less reactive to electrolyte, and safer from plating is to use higher voltage negative electrodes, e.g. Ti-based materials which are about 1–1.5 V above Li metal. This means the cells will have a lower voltage, which means less energy density, but hopefully longer life. These would be safer, particularly in EV applications, where cells can be ruptured in an accident. This is less of an issue in stationary applications where the batteries are less likely to be damaged.

4.4.3 *Electrolytes*

The resistance of the electrolyte often limits cell performance, and there are safety issues with the liquids. However, the electrolytes with the best performance are still the liquid carbonates with Li salts, which have the highest Li^+ ion conductivities at room temperature. The conductivities in these organic solvents are still much smaller than those in water-based electrolytes and so the electrolyte thickness needs to be much smaller than in water-based cells to get the same low resistance. Liquid cells still need a separator membrane to physically hold the electrodes apart while pressure is needed to minimize the distance between them. This pressure can be achieved by rolling tightly such as in the common 18650 cylindrical format cells, which use strong casings to contain both the cell pressure and the electrolyte. This casing adds mass to the cell, lowering the energy density and the round cells do not pack very well into a battery pack, thus also lowering the volume density (Fig. 16). This type of cell is used in Tesla's

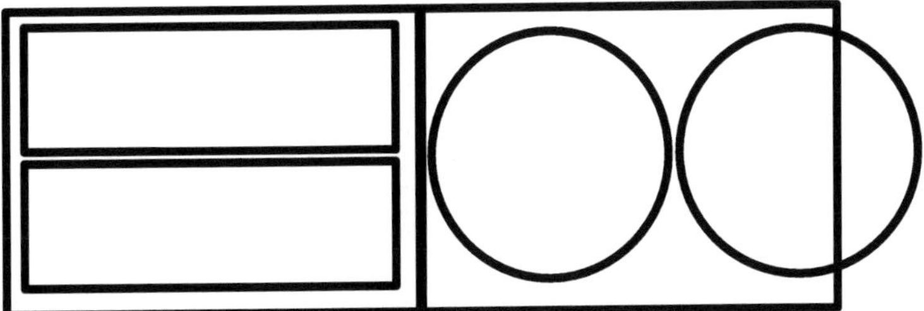

Fig. 16. Packing of cells. The circles and rectangles both have the same area, which equates to volume and capacity of cell, but the rectangular cells pack into a smaller battery than the circles.

battery packs, as they are cost-effective and robust, and can provide high currents.

The first improvement to the electrolyte was the use of gels. Gelation has been used in lead-acid batteries to minimize cell leakage for some time and has been used successfully in Li-ion cells as well. The liquid electrolyte is trapped in a porous polymer membrane, which can act as separator and the source of the electrolyte. This removes the need for the separator and added liquid electrolyte and so makes the cell smaller and lighter. The conduction mechanism is still through the liquid phase. However, the liquid is trapped in pores in the polymer. Any side reactions with the liquid electrolyte will still occur.

One significant version of this is the plastic Li-ion battery (pLiON) where the positive electrode, negative electrode, and separator are based on the same porous polymer and can be laminated together. This forms a simple monolithic cell assembly that can readily be filled with electrolyte to form a working cell. This does not need any external pressure to hold the cell together, removing the need for the rolling or a strong casing, allowing the cells to be simply sealed in a plastic pouch to keep the air out and the electrolyte in. This reduces the cell weight considerably, at the expense of some robustness of the cell. These cells in the sealed plastic pouches are known as Lipoly (LiPo) cells. These have the highest energy density as they have no unnecessary casing and being basically rectangular, can pack well into a small space (see Sec. 4.5 and Fig. 17). The name polymer cell should really be reserved for the cells with a solid Li-ion conducting polymer electrolyte and so contain no liquid at all. These layers can be made very thin to lower resistance and make the cells smaller whilst also removing any side reactions by replacing the liquid with a much more stable, long polymer chain. This increased stability could possibly allow the use of Li

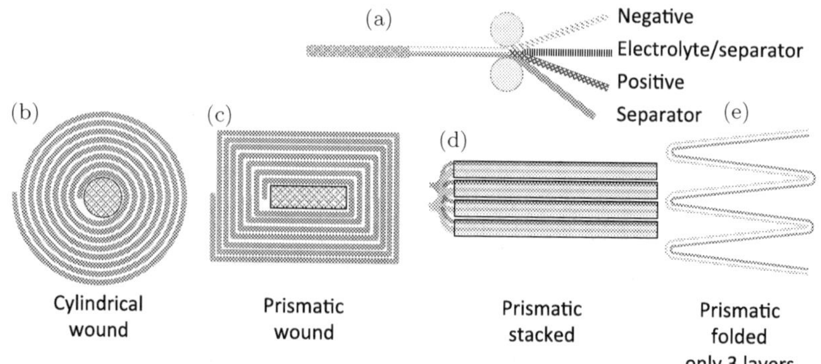

Fig. 17. Various packing geometries of Li-ion cells. The various components of the cells (both electrodes and at least one separator) are laminated together and then rolled or folded to form the cell. The wound cells come in a cell casing, whereas the stacked and folded cells also come in polymer encased "coffee bag" cells.

metal electrodes again, significantly increasing cell capacity. There are not yet any cells commercially available based on these conductive polymers.

4.4.4 *Current collector*

All Li-ion cells need a current collector to extract the current from the electrodes. These are usually light Al foils for the positive electrode and more expensive, but non-reactive Cu for the negative electrode. The electrodes are simply coated onto the metal foils, cut, and either rolled up or laminated and sealed into the cells. The thinner the current collectors, the lighter and smaller the cell, and the higher the energy density.

4.4.5 *Packaging*

All Li-ion batteries need to be sealed to keep water and oxygen away from the highly reactive negative electrode. This is normally done on the individual cell level. Also the electrode layers are often very thin, and so do not have large capacity. Therefore, a large area needs to be packed into each cell. This is done by either cutting out positive and negative electrode sections and stacking them up and joining all those with the same type of current collector to make one cell which is then sealed. The other method is to laminate the electrodes together with a separator on each side, so they can be rolled or folded up (see Fig. 17(a)).

The original and standard cell is the cylindrical 18650 form factor with the cell rolled up and place inside a typically metal casing (Fig. 17(b)).

These cells are still produced in very large numbers due to their low cost and good current and heat management. This cell design will have a further boost from Tesla's Gigafactory development. The round casing does not pack very well into a larger system, and the casing can be heavy, affecting the overall mass and volume energy density.

The other older battery casing is the prismatic cell which packs together better to form batteries (see Fig. 16). Again, to maximize the cell area and capacity, many cell layers are packed in the casing. To achieve this, the cell is rolled about a rectangular former, or folded or cut and stacked into a square package (see Fig. 17(c)).

The more recent cell design is the pouch cell, where the cell parts are folded or stacked and placed in a plastic bag, which is sealed tightly about the cell. Occasionally, these cells are called coffee bag cells as the casing is similar to vacuum packed coffee bags.

4.4.6 *New technologies*

A new technology that is gaining prominence is the Li-air battery.[29] This cell promises much higher energy density as it does not require the active positive electrode material but instead uses the freely available "air" just as in an ICE which uses air to combust the gasoline. The Li-air cell also uses metallic Li as the negative electrode which has a much higher capacity than carbon. The positive electrode only uses the oxygen from the air, and so this is really a Li-O_2 cell. Some of the other components of the air, especially CO_2, can poison the cell so there needs to be a way of extracting just the pure O_2 from the air. There is much work to be done on this chemistry to move it from the laboratory and into a commercial system.

Li-S cells, as an analogy to the NaS cells, are also gaining research interest,[30] as they have higher energy density and can operate at ambient temperatures.

5 Practical Aspects

5.1 *Recycling*

One of the main uses for battery storage systems is to lower carbon emissions by storing electricity and then using it elsewhere such as in an EV, and so replacing fossil fuel based energy. The other main use is to time shift electricity, generated by renewable energy sources, to when it is needed, again to reduce carbon emissions. This is less convincing if the system itself has high carbon emissions during production or is not recyclable.[31]

Lead-acid batteries are completely recyclable, containing just lead-based electrodes, sulfuric acid and plastics, all of which are separated and recycled.[31] With the toxicity of lead there is typically mandatory recycling of lead-acid cells, with an over 95% recycle rate in the US[31] and similar in the EU. With the low smelting and melting point of lead, it is energy and cost-effective to recycle the cells.

NiCd/NiMH cells are also recycled, as most of the cell construction consists of metals (Ni, steel, etc.) which are much less energy intensive to recycle than to smelt fresh ore.[32] The NiCd cells are recycled to minimize the release of Cd more than to re-use the Ni *per se*.

Li-ion cells however are typically not recycled. There is often not a large amount of metal in the cells, just plastics, oxides and carbon, the current collector foils and the electrolyte. With the large variety of positive electrode materials, and the high cost of re-purifying oxides, there is no recycling of the positive electrodes. At present, the supply of Li salts is cheap enough so that these are not recycled either. There will need to be some form of recycling for these cells as usage and demand increases and Li resources become more limited. Large battery packs may be reused for other applications, such as using older EV battery packs as stationary cells, when they have lost some but not all of their capacity.

5.2 *Safety Issues*

Whilst the actual choice of cell chemistry will affect the placement and arrangement of battery modules, it does not affect the electrical interface of the storage system, as there is almost always a computer-controlled management system, a BMS, that is between the electrical input and the cells themselves. The BMS also ensures that the cells are operating under safe conditions and also to help balance the capacity of the cells. When batteries are used for large-scale energy storage, the system is largely independent of the battery chemistry chosen as any desired DC power can be produced by the correct combination of any type of cells. There will be inverters and rectifiers to facilitate the interconversion of AC to DC. The choice of chemistry will typically be set by price and available space, and the nature of any extra safety or maintenance designs needed for the system.

What the choice of cell chemistry does change is the safety aspects of the systems if there is a failure of cells or management systems. Water-based, electrolyte-based chemistries such as lead-acid and NiMH cannot react with water or air, and so are basically non-flammable. Their issue is with their electrolytes, which are corrosive, which is a problem if they leak. There is also the decomposition of the water in the electrolyte into H_2 and O_2 gases during normal use but especially on overcharge. In open cells, these gases are released and so there needs to be some method to extract them safely without risk of combustion. The water lost from the electrolyte would then need to be replaced as part of the maintenance of the cells. Automatic systems are available for this. Sealed systems use catalysts to recombine the gases, which stops the release of the gases and keeps the water in the cell. These catalysts can only cope with so much gas production, so with excessive overcharge the gas production can overwhelm the catalyst with a large increase of internal pressure, which could cause cells to vent or rupture. This could damage the cells and will release large amounts of flammable (even explosive) gas.

The non-water-based cells (Li ion and NaS) do not have any gas formation under normal conditions, but are made from materials which are not stable when exposed to air and release a lot of energy when burned. The Na and Li metals and the reduced LiC_6 are pyrophoric in the presence of air and water. This means any loss of containment can lead to a fire at the negative electrode, which is fed by the organic liquid electrolyte or polysulfide positive electrode, depending on the cell. The energy density of charged cells approaches that of TNT (see Table 1), but is much less than the energy available from the cell components during combustion. This is the problem of needing high energy, unstable materials to store significant amounts of energy.

Li and Na metals, the carbon negative electrodes and the carbonate electrolytes have an energy density about the same as gasoline, and so produce a large amount of heat when burned. For individual cells, such as in a phone or a laptop, failure will lead to destruction of the cells and can

Table 1: Energy densities of selected materials.

Material	Energy density MJ/kg
Li-ion cell	0.5–0.8
Combustion	
Carbon	33
TNT	4.184
Li metal	43
Na metal	13
Gasoline	46
Ethylene carbonate	12

also destroy the device. There are a large number of images of this on the Internet. In a large energy storage facility there may be thousands of cells, and so precautions must be taken to contain any damage to the cell or possible single battery or module level. This was the problem in the NaS battery fire[20] where the failure of one cell affected other cells in the module, causing them to also combust.

Li batteries do fail and burn and do cause headlines,[22,23,26] but considering that billions of cells are produced each year, the catastrophic failure is actually fairly rare. The failure is almost always on charge where energy is being pumped into the cells or on puncture, with air getting into the cells. When they are working properly and absorbing the charge there is no problem, but when the charging current continues past the point of full charge (called overcharge) problems may occur. The extra electrons must go somewhere once the electrodes are full, leading to unwanted side reactions. These side reactions, even at a low level, can degrade cells, and so cause accelerated aging of the cells and loss of capacity. In a vicious cycle, this loss of capacity means cells are more likely to the overcharged, and so continue to lose capacity and suffer more overcharge. This continues until the cell fails.

High overcharge currents and/or high levels of overcharge can lead directly to cell failure. For aqueous electrolytes water splitting producing hydrogen and eventually losing the electrolyte is the usual problem. In Li cells, overcharge can lead to Li plating and/or electrolyte decomposition, which is coupled with the release of extra heat. This heating can melt the Li making it more reactive, and also evaporate the electrolyte raising the internal pressure of the cell till it ruptures and the air causes ignition. This seems to have been the issue with the Samsung Note 7,[24] with the cell failing on charge, and starting fires in houses, cars, etc. Many pictures and videos are available on the Internet that show this phenomenon.

While all cells should start off with very similar properties, with cycling, the batteries can age and lose capacity at different rates. This means that to ensure that the cells do not charge past the fully charged state, time cannot only be used as a reliable guide for a complete system of cells.

5.3 *BMS*

The actual choice of cell chemistry does not affect the electrical interface of the storage system, as there is almost always a computer-controlled management system between the electrical input and the cells themselves. The input and output are simply voltage (usually DC) and the current which can be set by almost any combination of cells. The choice of chemistry only changes the cost of the system and the nature of any extra safety designs needed for the system.

The BMS controls the flow of current into or out of the system, as well as ensuring the cells are operating in a manner that is safe and gives optimal capacity. The BMS also helps balance the capacity of the cells.

BMSs are needed whenever cells are joined together to produce significant power and capacity. The management system is needed to ensure that all the cells are operating within their optimal conditions. The BMS needs to make sure that cells are neither overcharged, nor overdischarged, are neither too hot nor too cold, and that the cell is still performing adequately. To accomplish this, the BMS needs to monitor cell temperature and voltage, and to control and record the charge transferred into and out of the cell. The BMS is the gateway between the external current flow and the internal battery charging, and it is the BMS that limits the current into and out of the battery. Lead-acid and NiMH batteries can manage a small amount of overcharge and so can often use a very simple BMS whereas Li-ion batteries always need a sophisticated and detailed BMS as they are easily damaged with overcharge.

The BMS can be mounted externally in "smart chargers" where the cells are charged in one system and discharged in another, such as portable power packs. The management electronics can also be external to the batteries, such as in laptops and phones, but with the battery tightly integrated into the device where the BMS sits with connections to sensors in the battery packs. This is why there are usually more than two connections to a battery in a laptop or phone. In any large battery system the BMS is usually integrated in the battery or battery modules, where it can be tied to the cells intimately.

Overcharging and overheating is bad for all battery chemistries causing them to lose capacity. A failing cell in a battery will have less capacity, and so become overcharged without protection. The overcharge then causes more damage and lowers the capacity even more. This leads to complete failure of the cell very quickly.

The BMS connects the battery system to the outside world, and so controls the amount of power into the system. For instance when the cells become too hot the BMS will decrease the current, and as the cells reach a fully charged or discharged state, the BMS will likewise reduce the current.

The BMS may also be connected to a cooling system to keep the cells at the correct temperature.

The BMS must control the flow of current into, or out of, the cells. This is usually done with some form of solid state switch, e.g. transistor or MOSFET, depending on the power needing to be switched, controlled from the BMS microprocessor. There are several choices as to how they are connected to the cells, with different levels of complexity and efficiency, and therefore cost. The implementation will be dependent on the final use of the system. Details of these can be found in BMS and power electronic sources.

5.4 *Battery Models*

To know both the SOC and SOH of a cell, the BMS needs a model of the cell that it can compare with the actual cell.

The DC electrical model of all batteries is simply a voltage source in series with a resistor, as shown in Fig. 18. The voltage of the source is the OCV of the cell, and the resistance is the internal cell resistance (measured from an IV curve, current interruption or other AC impedance measurement). The internal resistance is mainly due to the electrical resistance of the electrodes and electrolyte, with some due to the chemical processes at the electrodes. This model predicts that the terminal voltage of the battery should be linearly dependent on the current flowing. This holds for reasonable levels of current, but at high currents the electrodes cannot keep up and the resistance increases and becomes non-linear. For small changes in capacity (i.e. low currents and short times) this is a good model of all battery types.

However, both the V_{OC} and cell resistance are both dependent on the SOC and SOH of the cell. The V_{OC} comes from the difference in energy of the electrons in the system and so drops as the cell discharges. The internal cell resistance can also be affected by the level of discharge of the cell. In a lead-acid cell, the SO_4^{2-} ions in the electrolyte are used up on discharge,

Fig. 18. DC electrical model of a battery.

increasing the resistance. In some Li–ion cells the positive electrodes become more conductive on reduction, and so lower the resistance of the cell.

The internal cell resistance can be measured directly by varying the current or by using AC methods, where a small varying voltage is applied to the cell and the current is monitored at the same frequency, which gives the AC resistance of the cell. Different choices of frequency can measure the various components of the cell like the electrolyte or the electrodes.

This means that the model needs to track the SOC of the cell to be an accurate model. It also needs to track the SOH of each cell to know how much the capacity drops with age. Both SOC and SOH, as well as temperature, etc., need to be known and tracked for every cell in the battery/module.[33] To achieve this requires a reasonable amount of processing power and a lot of sensors and controllers, which need a large number of wires per cell and so complicates the building of the modules.

5.5 *Cell Balancing*

Another important job of the BMS is to balance the cells in a battery. This is to protect any weaker cells in the battery, as when connecting cells together in series and parallel, any cells with lower capacity than the others can drop the capacity and performance of the whole battery. Also, without protection the weaker cells will degrade faster leading to premature failure of the whole battery.

The worst arrangement for cells is in series as any weaker cells drop the series voltage and will be overly discharged. If, for instance, one cell has only 80% of the capacity of the others, as the cells discharge all is fine at the start. But as the 80% discharge is reached on the bulk of the cells, the weak cell is now completely discharged. If the discharge is kept running, the weak cell will be overdischarged, inducing side reactions and probably further damaging the cell. On charging again, the weak cell will be fully charged when the others reach about 80%, and will suffer from overcharge, again lowering its capacity. The only apparent observation on the outside of the battery is a lower-than-expected battery voltage.

When cells are connected in parallel, the voltage of all the cells must be the same so that any bad cells will still have a good voltage. There is also no way to measure the individual cell voltage when cells are connected in parallel. With enough parallel cells the current lost from one bad cell is lost among the others. However, if the cell drops in resistance it may draw current from the better cells and so start a self-discharge of the system.

5.6 *Cell Grouping Strategies*

Cells can be connected in both parallel and series or most commonly both to achieve the desired voltage, current, or capacity. Normally one or two of these will be limiting. There are however several ways to arrange the cells which can have an effect on the cost and life of the battery system. For instance, if you have an application that needs 12 Ah, at a minimum of 10 V and the cells available are 1 Ah 3.7 V Li-ion cells. Twelve cells are needed to supply the capacity, and at least three cells need to be wired in series to give the required voltage with space to discharge below the nominal 3.7 V. There are two basic ways to connect the cells; four stacks of three cells in series (3S4P) or four cells in parallel connected to two others in series (4P3S) (see Fig. 19).

Both arrangements provide the same current, voltage and capacity, but the means of failure and control are different. If the indicated cell fails to open circuit, the current from the entire string is lost from (a) and so while the voltage stays high, 1/4 of the total capacity is lost. In (b) the other three cells in the parallel connection still produce current, and will be worked harder to make up for the loss, and while the initial loss is only 1/16, this will lead to earlier failure of the rest of the group. If however the cells fail to short circuit, (a) is the better arrangement as the other two cells of the

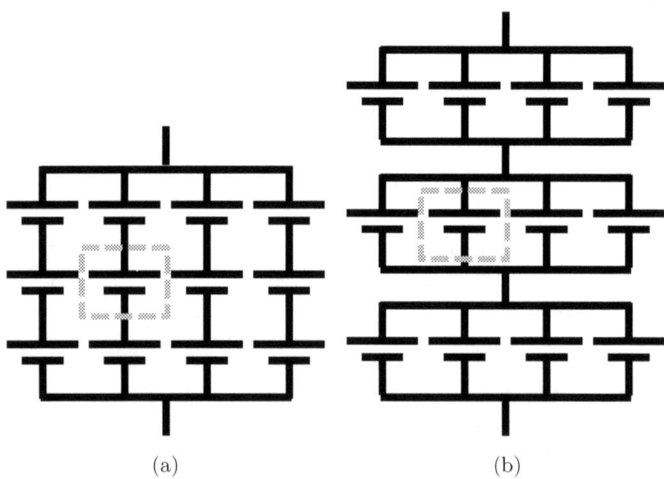

Fig. 19. Possible arrangement of 12 cells to make a battery; (a) is three cells in series each set connected as four sets in parallel (3S4P), whereas (b) is four cells connected in parallel with each parallel set connected as three sets in series (4P3S). Example cells are shown in dashed boxes.

string continue to work, and hopefully the other three strings can support the voltage. Overall capacity to the 10 V cut-off voltage will be lower. In the case of (b) this would be fatal for the battery pack as the shorted cell removes the voltage from the entire parallel set. The pack would then only give 7.4 V and not the required 10 V. To make things worse the shorted cell will take current from the other cells with the likelihood of overcharge and overheating.

There are also issues in monitoring the cells. In the 3S4P cells, the currents and voltages can be measured and/or controlled for every cell in the pack whereas all the cells in each parallel collection on the 4P3S arrangement must have the same voltage, and so it is not clear what current flows into which cells and so the voltage and current control can only be done on the level of parallel connections. However, this reduces the number of channels needed to monitor the cells, reducing wiring complications and cost.

A better solution to the above case is to use three 4 Ah batteries in series and monitor them all. There is a limit in the maximum size of a cell so for large-scale systems there will always be the need for series and parallel connectivity.

6 End Uses for Battery Storage

Batteries compete with all other storage systems, and will only be used if there are compelling technical or commercial reasons for using them. The most likely continued use is in small portable high-powered devices such as phones, laptops, etc. where there is nothing that can provide the small size and reliability that batteries can deliver. Similarly batteries will still be used for small-scale storage in the domestic market as well, powering things like portable power tools, remote solar lights, and so on.

The biggest rise in battery use will be in the tens of kWh range, which is big enough to run an EV or act as an electricity store for a domestic solar installation. It is about this size when other technologies, as covered elsewhere in this book, will be able to compete. Fuel cells, coupled with an electrolyser, could store electricity for a house or a car, and flow batteries become practical for domestic storage as well. These systems require more engineering and maintenance than the simple battery systems and so need economies of scale to be affordable.

The next level up is grid-scale levels approaching the MW and MWh level. The scaling cost for cells means that for these bigger capacities the batteries start to become more expensive than some of their competitors. This does not mean they will not be used, as due to their fast response times and general robustness they can add value to a product, over capacity alone.

Hybridization will also be a growing market for battery technology, where a smaller battery system is coupled to a larger energy system to help overcome transients or to do peak shaving. This is exemplified by hybrid cars where there is both a battery system and an ICE. The battery gives the short-term acceleration and stop/start motion, plus deceleration through regenerative braking, and the ICE gives longer running time and "base load" power. Some hybrid systems use both the ICE and batteries to give extra power for acceleration such as in F1 cars from 2009 on,[34] or the BMW i8.

6.1 *Transportation Systems*

Batteries have been historically used in vehicles where the low emission/quietness has significant added value. A large number of the very early motor cars were electric[35] when batteries and electric motors were more developed and dependable than the newer, more complicated ICE, with over 30,000 EVs on the road at the start of the 20th century.[36] However, the greater power and range of the ICE pushed these out of the market.

Later usages include the milk floats, electrically driven milk delivery vehicles in the UK, which replaced horse drawn vehicles for early morning door-to-door milk delivery. Their quietness and low speed made them ideal for this job. Their fixed route and return to depot each day removed any range problems. Electric fork lifts have also been used where the quietness and lack of exhaust gases were required. The extra mass of batteries even helped out as the counter weight required in lift trucks.

Another interesting use for batteries is in diesel-electric submarines, which have electric propulsion motors and can be run from generators or batteries. The diesel generators give good range and power but are both noisy and needed oxygen/air to operate. The large banks of batteries could be used to power them while submerged and while trying to avoid detection. The batteries are typically lead-acid systems due to age of the submarines and the robustness of the cells. The batteries were charged from the diesel generators when on the surface. This is the same type of system in modern hybrid cars.

The main issue with using batteries for transportation is that battery systems still do not contain the energy density of fossil fuels such as gasoline

Table 2: Comparison of raw energy density between fossil fuels and batteries.

	Wh/kg	Wh/L
Gasoline	13,000	9,400
Present battery[37]	150+	200
2020 battery[37]	250	400

or diesel, and so cannot give the same range. See Table 2 for the raw energy densities (not counting mass/volume of motor and transmission). As batteries and electric motors are much more efficient than gasoline engines the disparity is actually less in practice, but still less than 1/4 of an equivalent gasoline powered vehicle. There are no practical battery systems in sight to improve this significantly (except the unproven Li-air cells). The charge rate is also an issue when a battery may take several hours to recharge, when a gasoline vehicle can be "recharged" in a few minutes at a gasoline station.

Without increasing battery capacity, there are two basic engineering solutions to get around the lack of capacity, one is to have a hybrid power source, with another source to complement the battery power such as pedaling on a bicycle, or the diesel engines of the submarine, or gasoline engine in a hybrid car. Hybridization lowers the total capacity needed, but often at the expense of extra complexity or weight in the system. The other method is to ensure that the demand is always less than the capacity by only traveling on a fixed route or staying close to the charging point such as in milk floats or forklifts, which only are used in limited locations and have regular down time to charge up.

The efficiency of electric drive trains (i.e. battery to wheels) can be over 70%[38] compared to the 25–30% for ICE, requiring only 1/3 of the energy to move the same distance. This makes electrical vehicles more cost effective to operate, helping to overcome their initial higher cost. The fewer moving parts of an electric system also results in less expensive maintenance. However, to fully compete with gasoline powered cars, EVs need to at least approximately equal the range of gasoline powered cars. This means the batteries still need to be smaller, lighter, and cheaper, which is hard to achieve.

Heating cars in cold climates is also a problem. ICE powered cars have large amounts of waste heat (which is why they are not very efficient), and so tapping off some heat for keeping the car warm is not a problem. An EV is much more efficient and so has little waste heat, therefore, the heating

Table 3: Table of performance and cost of various vehicles.

Car	Power	Efficiency	Capacity	Range/ miles (EPA)	MSRP USD
Ford Focus S 2017	160 HP (120 kW) 2L ICE	26 city/36 highway/30 combined (EPA)	12.4 gal (47 L) tank	372	17,225
Ford Focus Electric	107 kW (143 hp)	97 (EPA)0	23 kWh	Up to 76	29,170
Tesla S 75D	240 kW (320 HP) electric drive		70 kWh	Up to 259	79,500
Nissan Leaf	80 kW (107 HP) electric drive		30 kWh	Up to 107	29,010

energy must come from the energy stored in the battery packs which further reduces their range.

Two examples of fully electric cars on the market are the Tesla S and the Nissan Leaf (see Table 3 for details of each). The Tesla has two-sized battery systems a 240 kW, 70 kWh with a claimed range of over 340 mile and a larger 310 kW, 90 kWh system with a claimed range of over 400 miles.[39] The smaller Leaf has an 80 kW, 30 kWh system and a claimed range of 107 miles.[5]

Both of these EVs were built from the ground up to be electric powered but other manufacturers are offering electric versions of their usual cars. The Ford Focus line has both ICE and electric cars available, as shown in Table 3. The power is similar for both cars, but the range of the EV car is only 76 miles, with a ticket price twice that of the ICE car. This extra price, coupled with the lower range, makes the EV less attractive to the customer. Various governments are pushing EV take up[40] (to reduce CO_2 emissions) by supplying purchase subsidies, lower road taxes, etc.

The lower running costs of an EV do also help mitigate the initial cost, which is shown by the EPA equivalent of 97 mpg for the electric Focus. However, as shown in Table 4, the cost saving is only 2.1 p per mile and so the extra US$12,000 cost would only be recouped after 570,000 miles! Even with a subsidy on the cost to bring it down to US$2,000 extra it would take

Table 4: Comparative costs for running the Ford Focus ICE or EV cars, based on average US and UK prices for 2016.

Engine type		Cost per unit US	Cost per unit UK
ICE	30 mpg	US$2/Gal gives 6.67 c/mile	GB£4/US Gal gives 13.3 p/mile
EV	76 miles/23 kWh = 3.3 miles/kWh	US$0.15/kWh gives 4.54 c/mile	GB£0.15/kWh gives 4.54 p/mile

95,000 miles to recoup the cost. The story is a little different in the UK, where the more expensive gasoline price, over twice that of the US, makes the differential much better. However, to recoup the £8,000 difference in cost between EV and ICE in the UK would still take 90,000 miles. These numbers do not of course include depreciation which affects both cars, and the rising costs for fuel in the future. The only way to make the EV's more financially sensible at present, is to reduce the cost and weight of the batteries.

One of the driving forces for EVs is government regulation and incentives. In 2013, the US DOE produced a document called "EV Everywhere Grand Challenge Blueprint"[38] which outlined the US government's goals for EVs. The major goals of the Batteries and Energy Storage subprogram are to reduce the production cost of an EV battery by a quarter, whilst also halving the size and weight of an EV battery by 2022. In hard numbers this is to lower the battery cost from about $500/kWh to $125/kWh, whilst also increasing energy density from about 100 Wh/kg to 250 Wh/kg and 200 Wh/L to 400 Wh/L, and increasing the power density from about 400 W/kg to 2,000 W/kg.

Commercial 18650 cells are already available (2016) with costs less than $200/kWh, and energy densities of over 150 Wh/kg, and 240 W/L. The cell casing of these cells lowers power density. Large-scale production of cells, such as in a Tesla Gigafactory, aims to lower this even further.

6.2 *Solar Storage*

The proliferation of cheaper PV panels is increasing the global use of solar energy. In a typical home, the peak solar collection is during the day, with the peak demand in the evening or at night. It would then be sensible to have a storage system local to the PV system. Where the system is not readily connected to the electric grid, 12 V solar panels and charge controllers for lead-acid batteries are readily available, and widely used in mobile vehicles (boats, caravans, RVs). These solar panels charge cheap 12 V batteries during the day, and run low powered (12 V LED) lights and other small electrical devices at night.

In a house the electrical demands are more significant, with fridges, freezers and ovens, etc., which is much more than a small set of 12 V lead-acid batteries can readily supply. Storage systems at the house level are available such as the Tesla Powerwall. The power wall costs approximately US $3,000 for the basic 6.4 kWh unit (plus installation costs).[1] This is a wall mounted battery pack that connects to the DC bus from a solar installation, and returns power through the same inverter as the PV power. With several kW of power it has enough capacity to run a significant sized house for many hours.

The usual system with solar installations at present is to use the grid as a storage system, sending the excess electricity generation to the grid during the day and purchasing electricity back at night. However the cost of buying the 6.4 kWh battery pack, assuming about US$0.15/kWh as the cost of electricity, is about US$0.96 per cycle (6.4 kWh $\times$ US$0.15 = $0.96). It would therefore take over 3,000 cycles to cover the cost of battery alone, which is over 8 years at one full cycle per day. The capacity fade of the pack will increase this payback time, as it will not using the full capacity each night. The payback also suffers if the electricity could be sold instead of storing it, lowering the effective cost of the later purchases. Of course in regions where the electricity price is higher, such as islands and other remote areas, the payback time, will be shorter.

The domestic electrical storage systems have other added value features, which can make them more viable. The storage system does displace CO_2 emissions from thermal generation of electricity, and so reduces that cost. Also if the local grid is not very reliable the batteries can be used as a backup power supply, keeping the electricity on, which does have a considerable value.

So for stationary storage the main driver has to be cost, with volume and weight much less of an issue. For large grid connected systems the margins on the cost of electricity are small and so the cells must either be very cheap, or be supported by other revenue streams.

7 Dreams for the Future

With no obvious new electrochemical storage systems on the horizon, except Li-O_2 cells, the move will be to cheaper, lighter, and safer cells. With the availability and cost of lithium salts, cells based on the much more abundant sodium or magnesium ions would allow many more cells to be made. These less reducing ions makes the cells safer, at the cost of lower voltages and so less energy density. The Mg^{2+} ion also stores two electrons per atom, and so can give higher volume energy density than Li-based cells, but lower mass energy density.

With the huge numbers of cells produced now, which will only increase in the future, there needs to be a global system of reusing and recycling Li-ion cells, to help reduce the environmental footprint of manufacturing the cells.

Ultimately the future may bring cheaper cells which will be available for all to be able to use them, to help store and use the widely available but intermittent renewable forms of energy around the world. Not just in the developed nations but in all places where an inexpensive solar panel or wind turbine coupled with cheap reliable batteries could ensure lighting, communication and education for all.

Other Readings

C. A. Vincent and B. Scrosati, *Modern Batteries: An Introduction to Electrochemical Power Sources* (Butterworth-Heinemann, London, 1997).

R. Dell and D. A. J. Rand, *Understanding Batteries* (Royal Society of Chemistry, London, 2001.

K. C. Kam and M. M. Doeff, Electrode materials for lithium ion batteries, *Material Matters* **7**(4) (2012), pp. 19–25.

Many Online resources e.g. http://batteryuniversity.com

References

1. Available at: www.tesla.com/powerwall. Accessed on September, 2016.
2. Energy transition initiative islands: Energy snapshot Bonaire DOE/GO-102 015-4655, Saft Batteries Press release (June 2015). Available at: http://www.saftbatteries.com/press/press-releases/saft-energy-storage-system-support-caribbean-island-bonaire-power-grid-switch.
3. Saft And Enercon's megawatt-scale energy storage system to help faroe islands stabilize its grid while increasing wind power usage, Saft Batteries Press release (April 13, 2015).
4. B. Dunn, H. Kamath, J.-M. Tarascon, Electrical energy storage for the grid: A battery of choices, *Science* **334**(6058) (2011), pp. 928–935.
5. nissanusa.com, 2016-nissan-leaf-en.pdf (2016).
6. Alibaba.com Various sales (2016).
7. J. M. Tarascon and M. Armand, Issues and challenges facing rechargeable lithium batteries, *Nature* **414** (2001), pp. 359–367.

8. Recherches Sur L'eílectriciteí, Gaston Planteí Aux Bureaux De La Revue La LumieÍre Eílectrique (1883).
9. W. R. Grove, On voltaic series and the combination of gases by platinum, *Phil. Mag. Ser.* **3**(14) (1839), pp. 127–130.
10. Directive 2006/66/EC of the European Parliament.
11. UK patent GB189907892 (A) 1900-03-24 Improvements in electrical batteries. Jungner Ernst Waldemar UK patent GB190007768 (A) — 1901-03-23 An improved negative accumulator electrode. Jungner Ernst Waldemar. US Patent US670024 (A) — PROCESS OF MAKING ACTIVE MATERIAL FOR ACCUMULATOR-PLATES. Jungner Ernst Waldemar.
12. S. R. Ovshinsky, M. A. Fetcenko and J. A. Ross, A nickel metal hydride battery for electric vehicles science, *Science* **260** (1993) pp. 176–181.
13. K. Beccu, Negative electrode of titanium-nickel alloy hydride phases (1974) US Patent 3, 824, 131.
14. F. G. Will, Hermetically sealed secondary battery with lanthanum nickel anode (1975) US Patent 3, 874, 928.
15. M. Zahran and A. Atef, Electrical and thermal properties of nicd battery for low earth orbit satellite's applications, in *Proc. of the 6th WSEAS International Conference on Power Systems* (2006).
16. N. Weber and J. T. Kummer, Sodium–sulfur secondary battery *Annual Power Sources Conference*, **21** (1967), pp. 37–39.
17. W. L. Bragg, C. Gottfried and J. West, Zeitschrift für Kristallographie, *Crystalline Materials* **127** (1968), pp. 94–100.
18. J. T. Kummer, W. Neill, Ford Motor Co., US Patent 3404036 A 2 May 1966.
19. From lithium to sodium: cell chemistry of room temperature sodium–air and sodium–sulfur batteries P. Adelhelm, P. Hartmann, C. L. Bender, M. Busche, C. Eufinger and J. Janek Beilstein J. Nanotechnol. 2015, 6, 1016–1055.
20. Press Release: June 7, 2012 NGK Insulators, LTD. Available at: http://www.ngk.co.jp/english/announce/111031_nas.html. Accessed on August, 2016.
21. U.S. Vehicle Fire Trends And Patterns Marty Ahrens, National Fire Protection Association Fire Analysis and Research Division (June 2010).
22. After 3 Fires, Safety Agency Opens Inquiry Into Tesla Model S, *New York Times*, November 20, 2013, p. B1.
23. "Samsung Junks Popular Phone Plagued by Fire, *New York Times* October 12, 2016, p. A1.
24. US CPSC Recall number: 16-266 Recall date: September 15, 2016.
25. M. S. Whittingham, Electrical energy storage and intercalation chemistry, *Science*, **192** (1976), pp. 1126–1127.
26. The Warnings Are Many on the Risks of Lithium Batteries, *New York Times*, March 3, 2016, p. D2.
27. Sheng Shui Zhang, A review on electrolyte additives for lithium-ion batteries, *J. Power Sources*, **162**(2) (2006), pp. 1379–1394.
28. T. V. S. L. Satyavania, A. Srinivas Kumara and P. S. V. Subba Rao, Methods of synthesis and performance improvement of lithium iron phosphate for high rate Li-ion batteries: A review, *Eng. Sci. Technol.* **19**(1) (2016), pp. 178–188.

29. L. Grande, E. Paillard, J. Hassoun, J.-B. Park, Y.-J. Lee, Y.-K. Sun, S. Passerini and B. Scrosati, The lithium/air battery: Still an emerging system or a practical reality? *Adv. Mater.* **27**(2015): 784–800.
30. L. F. Nazar, M. Cuisinier and Q. Pang, Lithium-sulfur batteries, MRS Bulletin, **39**(5) (2014), pp. 436–442.
31. Sustainable Materials and Technologies 1–2 (2014), 2–7.
32. C. J. Rydh and M. Karlström, Life cycle inventory of recycling portable nickel–cadmium batteries, *Resour. Conserv. Recy.* **34**(4) (2002), pp. 289–309.
33. E. Prada, J. Bernard, V. Sauvant-Moynot, Ni-MH battery ageing: From comprehensive study to electrochemical modelling for state-of charge and state-of-health estimation, *IFAC Proc. Vol.* **42**(26) (2009), pp. 123–131.
34. 2009 Formula One Technical Regulations FIA (2009).
35. How about that? World's first electric car built by Victorian inventor in 1884, *The Telegraph*, April 24, 2009.
36. The EV City Casebook IEA (2012).
37. EV Everywhere G rand Challenge Blueprint US DOE (2013).
38. M. Miller, A. Holmes, B. Conlon and P. Savagian, The GM "Voltec" 4ET50 Multi-Mode Electric Transaxle, *SAE Int. J. Engines* **4**(1) (2011), pp. 1102–1114.
39. Available at: www.tesla.com/models. Accessed on September 2016.
40. Office for Low Emission Vehicles, US IRS Internal Revenue Bulletin: 2009-48 (November 30, 2009).

Chapter 5

Capacitive Energy Storage

Wentian Gu[*], Lu Wei[*,†] and Gleb Yushin[*,‡]

[*]*School of Materials Science and Engineering*
Georgia Institute of Technology
Room 288, 771 Ferst Drive NW
Atlanta, GA 30332-0245, USA
[†]*School of Materials Science and Engineering*
Huazhong University of Science and Technology
Wuhan, Hubei 430074, China
[‡]*yushin@gatech.edu*

Capacitors are electrical devices for electrostatic energy storage. There are several types of capacitors developed and available commercially. Conventional dielectric and electrolytic capacitors store charge on parallel conductive plates with a relatively low surface area, and therefore, deliver limited capacitance. However, they can be operated at high voltages. As an alternative, electrochemical capacitors (ECs) (also called supercapacitors) store charge in electric double layers or at surface reduction–oxidation (Faradaic) sites. Thanks to the large surface area of the electrode and the nanoscale charge separation, electrochemical capacitors provide much higher capacitance, filling in the gap in the energy and power characteristics between batteries and conventional capacitors. However, they offer a lower energy density than batteries and commonly lower power than traditional capacitors. In the past decade, intensive research on ECs brought about the discovery of new electrode materials and in-depth understanding of ion behavior in small pores, as well as the design of new hybrid systems combining Faradaic and capacitive electrodes, which are essential for the enhancement of the performance of ECs.

This chapter presents the classification, construction, performance, advantages, and limitations of capacitors as electrical energy storage devices. The materials for various types of capacitors and their current and future applications are also discussed.

1 Introduction

Capacitors are passive electrical devices which store electrostatic energy in an electric field. The basic form of a capacitor (previously called a "condenser") consists of two parallel conductors separated by a dielectric. The ability to store charge can be characterized by a single quantity, the capacitance, with the unit Farad (F), which is defined as the ratio of the charge on one of the conductors divided by the voltage applied across the dielectric that induces this charge.

There are three major categories of capacitors: dielectric, electrolytic, and ECs (Table 1). ECs can be subdivided into EDLCs, pseudocapacitors, and hybrid capacitors. Dielectric capacitors are fundamental electrical circuit elements that store electrical energy (on the order of micro-Farads, μF) and assist in frequency filtering. Electrolytic capacitors are the next generation of commercial capacitors, which have similar cell construction as batteries. A fundamental evolution happened with the invention of the EDLC, in which the electrical charge is stored at a metal–electrolyte interface. The main component in the electrode construction of commercial EDLCs is activated carbon. The prototype of this construction has been initialized and industrialized for 40 years, yet there was little or no research progress until recently. The revival of interest in EDLCs arises from the increasing demands of long cycle life and very short high power pulses for electrical energy storage in several growing applications, such as digital electronic devices, implantable medical devices, industrial lifts and cranes, offshore wind farms, stop/start operation in vehicle traction as well as the electrical grid. However, EDLCs suffer from low energy density ($\sim$100 times smaller than common Li-ion rechargeable batteries).

To solve this problem, researchers produced novel carbon electrodes with enhanced energy storage capacity and incorporated materials that perform reversible reduction–oxidation (redox) reactions, such as transition

Table 1: Classifications of capacitors.

Classification	Basis of charge or energy storage
(i) Dielectric capacitors	Electrostatic
(ii) Electrolytic capacitors	Electrostatic
(iii) Electrochemical capacitors (ECs)	Electrostatic
(a) Electrochemical double-layer capacitors (EDLCs)	Faradaic charge transfer
(b) Pseudocapacitors	Faradaic and electrostatic
(c) Hybrid capacitors	charge transfers

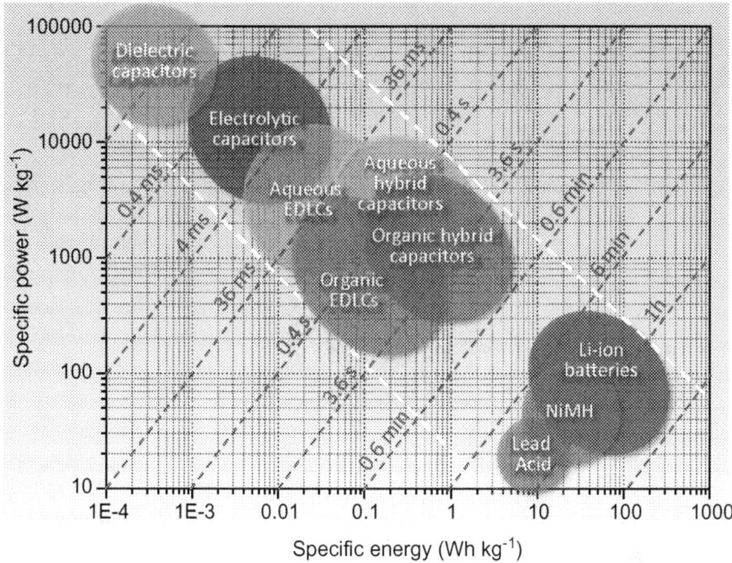

Fig. 1. Schematic illustration of the specific power vs. specific energy for various electrical energy storage devices.

metal oxides and conductive polymers, into the electrode materials. Specifically, the reversible redox processes usually enhance the specific capacitance by 2–100 times compared to plain EDLCs without redox materials, depending on the nature of the electrode and electrolyte system. These ECs fall in the range of pseudocapacitors.[1–3] The latest important alternative approach under serious investigation to produce high energy density is to develop hybrid (asymmetric) capacitors. This approach can overcome the energy density limitation of the EDLCs because it employs a hybrid system of a battery-like (Faradaic) electrode and a capacitor-like (non-Faradic) electrode, producing higher working voltage and capacitance.[4]

The energy and power densities of different storage technologies are shown in the Ragone plot in Fig. 1, with dark dash lines indicating characteristics charge–discharge times (in milliseconds, seconds, minutes, or hours). There are a number of desirable qualities that make capacitors a valuable option as energy storage devices. For example, they have reversible storing and releasing charge capabilities that make them able to withstand a very large number of charge–discharge cycles and are also able to charge discharge more quickly than batteries. The advantages and limitations of capacitors[5] as electrical energy storage devices are listed in Table 2.

Table 2: Advantages and disadvantages of capacitors as electrical energy storage devices.

Advantages

Long cycle life, >100,000 cycles, some systems up to 10^6;
Quick charge–discharge rate;
Good power density;
Simple principle and mode of construction;
High working voltage (for electrolytic capacitors);
Cheap electrode and electrolyte materials (particularly for aqueous embodiments);
Combines state-of-charge indication, $Q = f(V)$;
Can be combined with rechargeable battery for hybrid applications (EVs).

Disadvantages

Poor specific (mass normalized) energy density;
Poor volumetric energy density;
Non-aqueous embodiments require pure, H_2O-free materials, more expensive;
Requires stacking for high potential operation (EVs);
Hence, good matching of cell units is necessary.

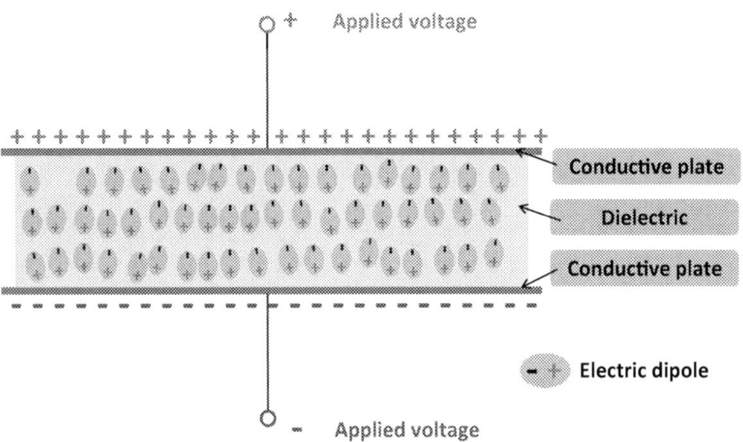

Fig. 2. Schematic of a dielectric capacitor.

2 Dielectric Capacitors

A dielectric capacitor typically consists of two conductors (parallel plates) separated by a dielectric material (dielectric), as shown in Fig. 2. When a potential difference (voltage) is applied across the conductors, a static electric field develops across the dielectric, which induces a flow of electrons from the side of higher potential to the power source, and from the power source to the opposite side. As a result, an electron deficiency develops on one side, which becomes positively charged and an electron surplus develops at the opposite side, which becomes negatively charged. This electron flow continues until the potential difference between the two sides is equal to the applied voltage. Energy is stored in the electrostatic field.

The basic equation for the capacitance of such a device is:

$$C = \varepsilon S/d, \tag{1}$$

where C is the capacitance, ε is the permittivity of the dielectric, S is the surface area of the electrode and d is the thickness of the dielectric.[6]

Dielectric capacitors yield capacitance in the range of 0.1–1 μF with a voltage range of 50–400 V. Various dielectric materials have been adopted for such capacitors, including but not limited to paper, paraffin, polyethylene, insulated mineral oil, polystyrene, ebonite, polyethylene tetraphtharate, sulfur, mica, mylar, steatite porcelain, Al porcelain, plastics (polymers), and glass.[1] Table 3 gives the physical properties of some common dielectrics and the electrochemical performance of their corresponding capacitors.

Dielectric capacitors are mainly used in:

(a) electronic circuits for blocking direct current, while allowing alternating current to pass;
(b) filter networks for smoothing the frequency output of power supplies;
(c) resonant circuits that tune radios to particular frequencies.

Table 3: Physical properties of dielectrics and electrochemical performance of corresponding capacitors.[7]

	Specific dielectric dielectric constant	Dielectric strength (MV/m)	Max voltage (V)[8]
Waxed paper	1.2–2.6	10–40	400
Mica	5–7	118	100
Polymer			
Polypropylene	2.2–2.36	23.6	—
Polystyrene	2.5–2.7	19.7	—
Teflon	2.1	60	—
Ceramic			
Alumina	9.34	13.4	50
Porcelain	6–8	35–160	50

Dielectric capacitors can be easily stacked up for higher volumetric capacity. A special type of dielectric capacitor, *viz.* a ceramic capacitor, is constructed of alternating layers of metal and ceramic, with the ceramic material acting as the dielectric. Multilayer ceramic capacitors (MLCs) typically consist of ~100 alternate layers of metal electrode and dielectric ceramics sandwiched between two ceramic cover layers. They are fabricated by using screen-printing electrode layers on dielectric layers and co-sintering the laminate. Dielectric ceramic materials that have been identified and used include $BaTiO_3$, $CaZrO_3$, $MgTiO_3$, and $SrTiO_3$. Conductive Ag, Pd, Mn, Ca, etc. are used as internal electrodes. Ceramic capacitors are in widespread use in electronic equipment, providing high capacity and small size at low price compared to other types of capacitors.[9–11]

The capacitance of dielectric capacitors is often limited by the low surface to volume ratios of the electrodes. While the distance between electrodes is largely determined by the working voltage and the dielectric materials applied, to increase the capacitance, one has to increase the effective surface area or use high permittivity material.

3 Electrolytic Capacitors

Following dielectric capacitors, electrolytic capacitors were developed and designed as the next generation higher energy density devices, which use an electrolyte as an ion conductor between the dielectrics and a metal electrode. Aluminum (Al) and tantalum (Ta) capacitors are the two common types of electrolytic capacitors. The capacitances of these electrolytic capacitors are in the range of 0.1–10 μF with a voltage profile of 25–50 V.[8]

The typical construction of an aluminum electrolytic capacitor compromises two conducting aluminum foils, one of which is coated with an insulating oxide layer and the other one covered by a paper spacer soaked in electrolyte. The foil with the thin oxide layer acts as an anode while the liquid electrolyte and the second foil acts as a cathode (Fig. 3). The thin oxide layer on the anode acts as the dielectric. Since capacitance is inversely proportional to thickness, the small thickness of this dielectric layer allows a relatively high volumetric capacitance. An excellent dielectric material, aluminum oxide, has a relative permittivity of 10, which is several times higher than most common polymer insulators. It can withstand an electric field strength of the order of 25 megavolts per meter (MVm^{-1}), which is high but slightly lower than that of common polymers. This combination of high capacitance and reasonably high voltage results in high energy

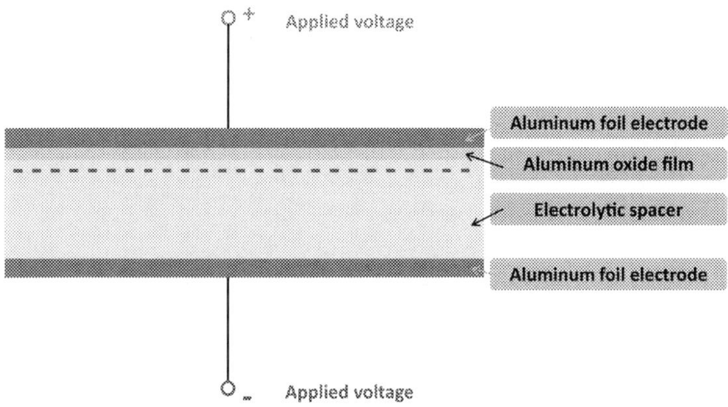

Fig. 3. Schematic of a typical aluminum electrolytic capacitor.

density. Electrolytic aluminum capacitors are mainly used in power supplies for automobiles, aircraft, space vehicles, computers, monitors, motherboards of personal computers, and other electronics. However, aluminum electrolytic capacitors have several drawbacks, which limit their use, such as relatively high leakage current and inductance, large capacitance variations from the defined specs (often ±100%), poor temperature range, and short lifetime.[1,12,13]

The construction of tantalum electrolytic capacitors is similar to that of aluminum electrolytic capacitors, with minor differences. The anode electrode is made of sintered tantalum grains, with an electrochemically deposited (by anodization) thin tantalum pentoxide layer acting as the dielectric. The small thickness of the oxide layer and the high surface area of the sintered grains contribute to a large volumetric capacitance. The cathode is formed either by a liquid electrolyte and a conductive metal current collector or by a conductive solid, such as a conductive polymer or a manganese dioxide coated with a conductive carbon layer, further connected with a conductive metal plate current collector. Compared to aluminum electrolytic capacitors, tantalum electrolytic capacitors have higher volumetric capacitance, higher charge–discharge rates and better cycle stability, less DC leakage and lower impedance at high frequencies, due to the higher relative permittivity of tantalum, the higher surface area of the porous sintered grains and the adoption of a solid electrolyte. However, the maximum voltage for tantalum electrolytic capacitors is lower than aluminum ones, which results in a lower energy density for tantalum capacitors.

In addition, the tantalum electrolytic capacitors are intolerant of positive or negative voltage spikes and can be destroyed (often exploding violently) if connected in the circuit backwards (reversed) or exposed to spikes above their voltage rating. Moreover, tantalum capacitors cost more than aluminum ones.

In order to enhance the capacitance of electrolytic capacitors, researchers have replaced cheaper oxide layers of aluminum and tantalum with more expensive oxides of niobium and titanium as dielectrics which have higher relative permittivity. Besides a considerable decrease in the size of the capacitors, the use of oxides of niobium and titanium also brings an increase in the range of working temperatures, an improvement in electrical characteristics, and an enhancement in the stability of the capacitors' performance.[2,14]

4 ECs

ECs, also called supercapacitors or ultracapacitors, are capacitors that achieve their capacitance by charge separation along a double layer of nanoscale thickness at the interface of the electrode and electrolyte, or reversible Faradaic charge transfer at the electrode. Their energy density (up to ~5 Wh kg^{-1} in commercial devices) is lower than that of batteries, but they achieve much higher power density (up to ~10 kW kg^{-1}) for short times (a few seconds in commercial devices and down to around a millisecond in ultra-thin electrode prototypes).[15]

ECs can be classified into several basic categories according to the charge storage mechanism (Table 1) as well as the active (ion storing) materials used (Table 4). EDLCs, one of the most common capacitors at present, achieve charge storage by charge separation along the Helmholtz double layer at the interface between electrode and electrolyte.[16–25] A second group of ECs, known as pseudocapacitors or redox supercapacitors, perform charge–discharge by fast and reversible surface or near-surface Faradaic reactions. Transition metal oxides as well as electrically conducting polymers are examples of pseudocapacitive active materials.[26,27]

Hybrid capacitors, combining a capacitive or pseudocapacitive electrode with battery electrode, are the latest kind of EC, which benefit from both the faster charge–discharge speed of the capacitor and the higher energy density of batteries.[28,29]

4.1 *Electrochemical Double-layer Capacitors*

Using carbon as the active material, EDLCs represent more than 80% of the commercially manufactured ECs.[30] EDLCs store the energy at the electrolyte–carbon interface via reversible ion adsorption on the carbon surface.[6] The double layer capacitance could be calculated similarly by Eq. (1), where C is the double layer capacitance. The absolute permittivity ε is the product of $\varepsilon_r\varepsilon_0$, where ε_r is the relative permittivity of the electrolyte, ε_0 is the permittivity of the vacuum, d is the charge separation distance (the thickness of electrical double layer in this case), and S is the electrode surface area. The thickness of the electrical double layer is typically on the scale of a nanometer, which results in a much larger

Table 4: Electrode materials explored in ECs.

Types of ECs	Electrode material
EDLCs	Activated carbons, templated carbons and carbide-derived carbons; Carbon fabrics, carbon fibers, carbon nanotubes; Carbon aerogels, carbon onions and carbon nanohorns; Graphene and graphene-based materials.
Pseudo-capacitors	Surface compounds such as hydrogen adsorption on platinum or lead adsorption on gold; High surface area porous carbons covered with redox-active functional groups Transition metal oxides such as RuO_2, Fe_3O_4, MnO_2, NiO_x, $Ni(OH)_2$, Co_3O_4, SnO_2, In_2O_3, Bi_2O_3, TiO_2, V_2O_5, $BiFeO_3$, $NiFe_2O_4$, and their mixtures, among others; Conductive polymers such as polyacetylene, polyparaphenylene, polyaniline, polypyrrole, polythiophene and their derivatives, often deposited onto or mixed with various high surface area porous carbons, carbon nanotubes and graphene; Nanostructured redox-active materials, such as nanocrystalline vanadium nitride (VN), manganese oxide, vanadium oxide and hydroxides, ruthenium oxides, nickel oxides and hydroxides, iron oxides and hydroxides on metal collectors, carbon nanotubes, nanofibers, porous carbons and graphene.
Hybrid capacitors	Pseudocapacitive metal oxides and hydroxides (based on nickel, manganese, vanadium, lead, ruthenium, and other transition metals) with a capacitive carbon electrode; Lithium-insertion electrodes (such as $Li_4Ti_5O_{12}$, $Li_4Mn_5O_{12}$, Li_2FeSiO_4, graphite and pre-lithiated high specific surface area (SSA) carbon anode) with a capacitive carbon electrode.

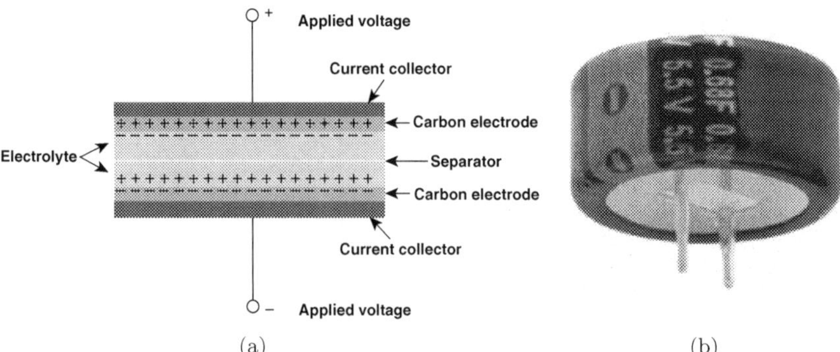

Fig. 4. (a) Schematic of an EDLC; (b) Panasonic double-layer capacitor (EEC-F5R5H105, 1F, 5.5V).

capacitance of EDLCs than conventional dielectric or electrolytic capacitors (Fig. 1).

Figure 4(a) presents the schematic of an EDLC. Charge separation occurs through polarization at the electrode–electrolyte interface. Figure 4(b) shows an example of a commercial EDLC (Panasonic).

With the capacitance of positive, C_+, and negative, C_-, electrodes and the maximum voltage, $V_{\max}^{\mathrm{EDLC}}$, at which the device can be operated, the energy of an EDLC can be calculated using Eq. (2):

$$E^{\mathrm{EDLC}} = \left(\frac{C_- \cdot C_+}{C_- + C_+}\right) \cdot (V_{\max}^{\mathrm{EDLC}})^2. \tag{2}$$

The energy density of EDLC is maximized when the capacitance (C) of both electrodes is identical:

$$E^{\mathrm{EDLC}} = \frac{1}{2} C \cdot (V_{\max}^{\mathrm{EDLC}})^2. \tag{3}$$

From Eq. (3), it is clear that to achieve high energy density of an EDLCs, one should maximize the capacitance of each electrode and increase the maximum allowed operational voltage.

The maximum voltage is generally determined by the stability window (from the lowest stable potential $V_{\mathrm{low}}^{\mathrm{stable}}$ to the highest stable potential $V_{\mathrm{high}}^{\mathrm{stable}}$) of the selected electrolyte (Fig. 5). The potential of each electrode in a fully discharged symmetric EDLC is the same and located between $V_{\mathrm{low}}^{\mathrm{stable}}$ and $V_{\mathrm{high}}^{\mathrm{stable}}$ (Fig. 5). The potential of each electrode changes linearly with time during galvanostatic charge–discharge. The negative

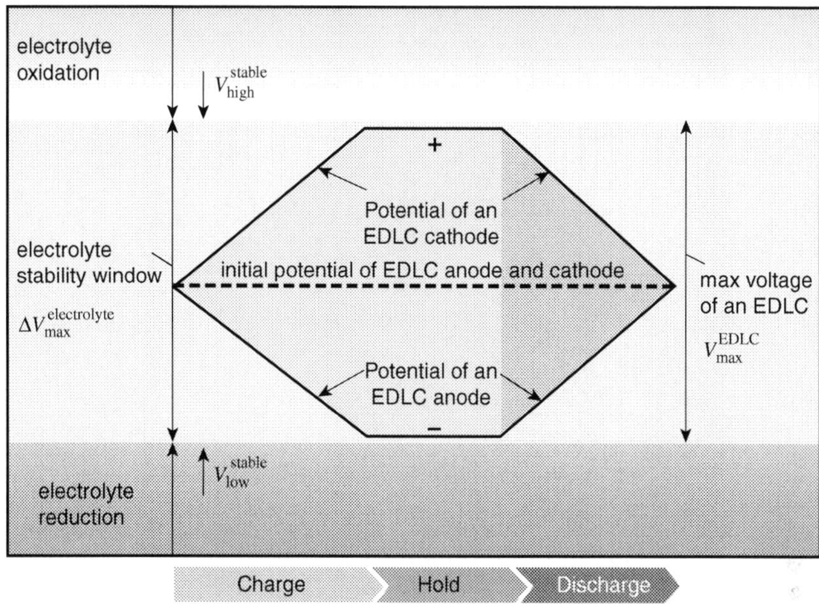

Fig. 5. Voltage profile of symmetric EDLC electrodes. The area of the dark zone (in step "discharge") is proportional to the energy stored.[31]

electrode decreases to V_{low}^{stable} at exactly the same time as the potential of the positive electrode increases to V_{high}^{stable}. In this case, the maximum EDLC voltage V_{max}^{EDLC} can approach the maximum electrolyte stability $\Delta V_{max}^{electrolyte} = V_{high}^{stable} - V_{low}^{stable}$. The area of the shaded zone in the diagram is proportional to the energy achieved.

According to Eq. (1), development of high SSA carbon is essential for EDLCs with large capacitance. High SSA and highly conductive carbon materials used in EDLCs include, but are not limited to, onion-like carbon (OLC), carbon nanotubes, graphene, activated carbons, zeolite- or silica-templated porous carbons and carbide-derived carbons. Activated carbons (Table 5) are the most widely used materials in commercial EDLCs today, because of their high SSA, moderate cost and well developed large-scale industrial production. Production of activated carbons usually involves carbonization (heat treatment) of carbon-rich organic precursors in an inert atmosphere with subsequent selective oxidation (activation) in CO_2, water vapor or strong acid/alkalis at elevated temperatures to promote porosity. A wide range of materials have been adopted as precursors for production

Table 5: Activated carbon materials for electrodes of EDLCs.

Carbon source	Activation method	S_{BET} (m^2 g^{-1})	Capacitance (Fg^{-1})	Electrolyte			Ref.
				Aqueous	Organic	IL	
Banana fibers	$ZnCl_2$	1097	74	1 M $NaSO_4$			37
Sucrose	CO_2	2102	163	1 M H_2SO_4			38
Sucrose	CO_2	1941	148			$EMImBF_4$	39
Starch	KOH	1510	194	30 wt% KOH			40
Wood	KOH	2967	236		1 M $TEABF_4$/AN		41
Corn grain	KOH	3199	257	6 M KOH			42
Seaweed		746	264	1 M H_2SO_4			43
Seaweed		270	198	1 M H_2SO_4			44
Pitch	KOH	2660	299	1 M H_2SO_4			45
Pitch	KOH	2171	140		1 M Et_4NBF_4/PC		46
Pitch	KOH	2860	130	1 M H_2SO_4			47
Pitch fiber	KOH	2436	224			DEME-BF_4	48
Pitch fiber	H_2O	880	28	1 M KCl			49
Rice husk	H_2SO_4		175	6 M KOH			50
Sugarcane bagasse	$ZnCl_2$	1788	300	1 M H_2SO_4			51
Coal	KOH	3150	317	1 M H_2SO_4			45
Coal	KOH	>2000	220		1 M $LiClO_4$/PC		52
Coconut shell	Melamine	804	230	1 M H_2SO_4			53
Apricot shell	NaOH	2335	339	6 M KOH			54
Coffee grounds	$ZnCl_2$	1019	368	1 M H_2SO_4			55
Wheat straw	KOH	2316	251		1.2 M $MeEt_3NBF_4$/AN		56
Polyfurfuryl alcohol	KOH	1140	160			PYR14TFSI	57
Polypyrrole	KOH	3432	300			$EMImBF_4$	58
Polyacrylonitrile		1340	66			1 M $LiPF_6$ in EC-DEC	59
Polyacrylonitrile		302	126	6 M KOH			60

Polyacrylonitrile		302	202	1 M H_2SO_4			60
Polyacrylonitrile	NaOH	3291	196			$LiN(SO_2CF_3)_2+C_3H_5NO_2$	61
Polyaniline	KOH	1976	455	6 M KOH			62
Polyaniline	K_2CO_3	917	210	6 M KOH			63
Polyaniline		400	125	6 M KOH			64
Poly (vinylidene chloride)	KOH	2050	38		1 M $TEABF_4$/PC		65
Polyvinyl alcohol	KOH	2218	218	30wt% KOH			66
Phenol resin	2M HNO_3		60	6 M KOH			67
Phenol resin		1232	40	1 M H_2SO_4			68
Phenol resin		1542	33	1 M H_2SO_4			68
Phenol resin/PVA		416	171	6 M KOH			69
Graphite	HNO_3/H_2SO_4	1071		0.1 M KOH			70
Rubber wood sawdust	CO_2	1232		1 M H_2SO_4			71
Poly (amide imide)	CO_2	1542		6 M KOH			72
Polystyrene	KOH	2350	258	6M KOH			73
Melamine mica	HNO_3	3487	148		1 M $TEABF_4$/PC		74

AN: acetonitrile.
PC: propylene carbonate.
DEME-BF_4: N,N-diethyl-N-methyl(2-methoxyethyl)ammonium tetrafluoroborate
$TEABF_4$/AN: tetraethylammonium tetrafluoroborate salt solution in acetonitrile.
Et_4NBF_4/PC: tetraethylammonium tetrafluoroborate in propylene carbonate.
PYR14TFSI: *N*-methyl-*N*-butyl-pyrrolidinium bis(trifluoromethanesulfonyl)imide.
EMImBF_4: 1-ethyl-3-methylimidazolium tetrafluoroborate.

of activated carbon, including both natural materials, such as coal, tar, pitch, nut shells, wood, corn grain, sucrose, banana fibers, and other agricultural and forest residues, as well as synthetic materials, such as polymers. The porosity of activated carbon is characterized by a broad distribution of pore size, ranging from micropores (<2 nm), mesopores (2–50 nm) to macropores (>50 nm). Longer activation time or higher temperature leads to larger mean pore size. The double layer capacitance of activated carbon reaches 70–250 Fg^{-1} in organic electrolytes, and 80–300 Fg^{-1} in ionic liquids (IL) (at elevated temperatures). This value can even exceed 150–300 Fg^{-1} in aqueous electrolytes (particularly in carbons containing redox active functional groups on the surface), but at a lower cell voltage because the stable voltage window of aqueous electrolytes is limited by water decomposition.

As previously mentioned (Table 4), many carbons have been tested for EDLC applications. The morphologies of some of these carbons are shown in Figs. 6(a)–6(d). The SSA and capacitances of various carbons are shown in Table 6. As a comparison, templated and carbide-derived carbons offer good pore size control and exhibit attractive properties, but their production is very limited. Activated carbon fabrics can reach the same capacitance as activated carbon powders, as they have similar SSA, but the higher price limits their use in price-sensitive applications.

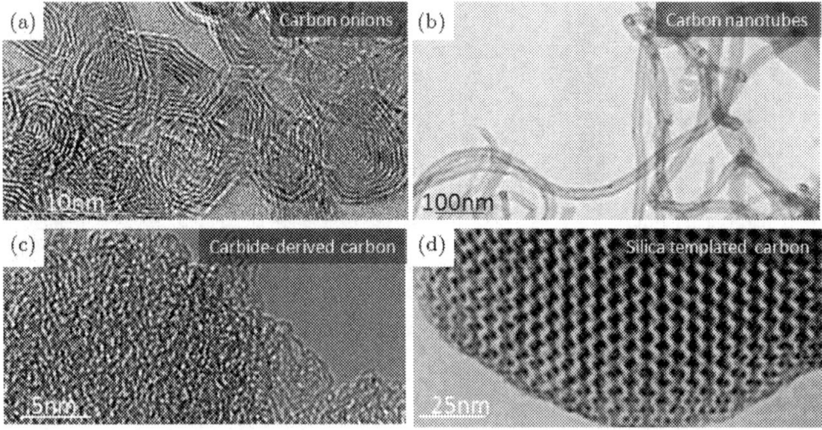

Fig. 6. Selected examples of structures of carbon used as EDLC electrode materials. (a) transmission electron microscopy (TEM) image of carbon onions[110]; (b) TEM image of carbon nanotubes[111]; (c) TEM image of carbide-derived carbon[85]; (d) TEM image of silica templated carbon.[112]

Table 6: Comparison of the EDLCs based on different carbon materials.

Carbon electrode material	S_{BET} (m^2g^{-1})	Capacitance (Fg^{-1})	Refs.
Activated carbon	1,000–3,500	70–450	14–30
Templated carbon	1,000–2,000	130–300	75–80
Carbide-derived carbon	600–3,000	100–200	17, 81–90
Carbon fabric	1,340	<100	59, 91, 92
Carbon fiber	1,000–3,000	100–200	61, 72, 93
Carbon nanotube	400–1,200	150–400	94–99
Carbon aerogel	600–1,600	90–190	100–102
Carbon onion	400–550	40–100	18, 103, 104
GBM	300–3,000	100–300	32–36, 105–109

Carbon nanotubes and macroporous carbon aerogels with large external surface area commonly offer high rate capability but low gravimetric and, more importantly, volumetric capacitance. Plasma treated, vertically aligned carbon nanotubes can deliver high specific capacitance and power density due to additional pseudocapacitance (which will be discussed in detail in the next chapter) contributed by the functional groups on their surface, but the preparation requires a complicated synthesis procedure and high stability of such materials has not been reached so far. Graphene, a two-dimensional nanosheet of graphite, has recently received much attention as a candidate electrode material for supercapacitors as it possesses superior electrical conductivity, a high theoretical surface area of over 2600 m^2g^{-1} and high chemical stability. However, due to the unavoidable aggregation of graphene nanosheets, the measured surface area of graphene is usually much lower than the theoretical value. More importantly, due to very high pore volume and thus low density, graphene electrodes offer moderate volumetric capacitance, commonly lower than 40 F cm^{-3}.

Other than single layer graphene, researchers have also studied the electrochemical performance of graphene-based materials (GBMs), such as graphene nanosheets, graphene oxide, etc. Although the SSAs of GBMs are usually much lower than that of a single-layer graphene because of the multilayer stacking of graphene sheets, many strategies have been applied to increase the porosities and SSAs of GBMs, and the capacitance of EDLCs with GBM electrodes can reach 300 Fg^{-1}.[32–36]

The double layer capacitance achieved by a certain electrode material also depends on the electrolyte used. Specific capacitance achieved with aqueous alkaline or acid solutions is generally higher than with organic electrolytes or ILs. In addition, the rate performance of aqueous solutions is better because the lower viscosity of aqueous solutions provides better charge shuttling kinetics. However, organic electrolytes are more widely used as they can sustain a higher operating voltage (up to 2.7 V).[15] Because the energy of an EDLC is proportional to the square of the maximum allowed operational voltage (Eq. (3)), replacing the aqueous electrolytes with organic electrolytes allows an increase in the allowed operational voltage ($V_{max}^{electrolyte}$) by up to a factor of 3–4, which results in an order of magnitude increase in the energy stored at the same capacitance. However, organic solvents do not meet the highly-desired requirements of environmental compatibility and safety, because organic solvents exhibit high vapor pressure, suffer from inherent flammability and potential explosion risks.

Recently, ILs have been explored as electrolytes in certain advanced EDLCs with improved energy and power densities, operational safety, and lifetime. This is because ILs have a large electrochemical window, wide liquid phase range, non-volatility, non-flammability, non-toxicity, and environmental compatibility with respect to conventional aqueous and organic electrolytes.[39,113]

4.2 *Pseudocapacitors*

Pseudocapacitors store electric charge via reversible Faradaic charge transfer during redox reactions in the electrode. The principal difference between a pseudocapacitor and a double-layer capacitor is that in the former there is a net ion exchange between the electrode and the electrolyte during the charge and discharge processes. It should be noted that the EDL capacitance in such systems always co-exists along with the pseudocapacitance. Nevertheless, in this case, the number of ions (such as protons) being exchanged between the two electrodes is dominant. In other words, the pseudocapacitance in the same system is much greater than the EDL capacitance. Several types of Faradaic processes occur in the electrodes of pseudocapacitors, for example, adsorption of hydrogen or lead on the surface of platinum or gold, redox reactions of transition metal oxides, and reversible processes of electrochemical doping/de-doping in electrodes based on conductive polymers. The specific pseudocapacitance far exceeds the specific double layer capacitance provided by the carbon materials, which is the reason for research interest in these systems. But because of the redox reactions involved in the charge storage, pseudocapacitors, like batteries, often suffer from lack of very long-term stability during cycling and rarely withstand >100,000 deep charge–discharge cycles.

4.2.1 *Pseudocapacitors with surface compounds*

One of the several mechanisms of reversible Faradaic surface redox reactions in pseudocapacitors is electrosorption or desorption on the electrode surface, with a charge transfer. Electrochemical processes of hydrogen adsorption on platinum and lead adsorption on gold are well studied.[114–116] In the case of the lead adsorption on gold, the following reaction takes place[2]:

$$\mathrm{Au} + x\mathrm{Pb}^{2+} + 2x\mathrm{e} \leftrightarrow \mathrm{Au}_x\mathrm{Pb}, \tag{4}$$

where, x represents the stoichiometry of chemisorbed lead atoms, which corresponds to a partial coverage of the surface by lead. The capacitance for reaction (4) is defined as follows:

$$C_\mathrm{F} = q_\mathrm{Pb} d\theta_\mathrm{Pb}/dE. \tag{5}$$

Here, $q_\mathrm{Pb} = 400\ \mu\mathrm{C\ cm}^{-2}$ is the minimum areal charge density required for full coverage of the gold surface by lead. The capacitance C_F, as calculated by (5), corresponds to a Faradaic reaction and is called a pseudocapacitance.

The existence of functional groups on the surface of high SSA carbon materials also often leads to pseudocapacitance.[53,74,117–120] These functional groups are involved in the redox processes with the electrolyte. This pseudocapacitance effect is most notably seen in aqueous electrolytes, but organic and IL electrolytes cannot be simply excluded as well. On a negative note, however, undesirable reactions between the functional groups and electrolyte have been found causing noticeable and highly undesired leakage currents,[19,23,121] which could be solvent-dependent.[110] Yushin *et al.* proposed that this is the reason for the strong correlation between device self-discharge and the presence of functional groups in aqueous solvent-based capacitors.[122] In an aqueous system, the local pH near the functional group is different from that of the electrolyte. Discharge-inducing undesirable redox reactions between functional groups and electrolyte take place when the local pH value shifts out of the stable window of the electrolyte. Other than the side reactions, one also needs to note the slower kinetics and therefore worse rate stability of pseudocapacitance contributed by functional groups.

4.2.2 *Pseudocapacitors with metal oxides*

A variety of metal oxides, such as RuO_2, MnO_2, NiO, In_2O_3, Co_3O_4, V_2O_5, Fe_3O_4, Bi_2O_3, IrO_2, $NiFe_2O_4$, $BiFeO_3$, etc., have been exploited as active materials for redox reactions in pseudocapacitors. Some of the metal oxide-based supercapacitors and their main electrochemical properties are summarized in Table 7.

As one of the popular pseudocapacitor electrode material, ruthenium oxide, RuO_2, has been widely studied for its high conductivity and three accessible distinct oxidation states within 1.2 V.[15] The pseudocapacitance of RuO_2 in acidic solutions is provided by a fast, reversible electron transfer together with an electro-adsorption of protons on the surface of RuO_2 particles, according to Eq. (6):

$$\mathrm{RuO_2} + x\mathrm{H}^+ + x\mathrm{e}^- \leftrightarrow \mathrm{RuO}_{2-x}(\mathrm{OH})_x, \qquad (6)$$

where $0 \le x \le 2$. The value of x gradually changes during proton insertion or de-insertion within a voltage window of about 1.2 V, which leads to a capacitive behavior with ion adsorption following a Frumkin-type isotherm. Specific capacitance of more than 600 Fg^{-1} has been achieved using RuO_2.[123,125] However, the high cost of RuO_2 limits its application as an electrode material for mass production. Among the less expensive alternative oxides, manganese oxide was found to be a promising candidate.[171] The charge storage mechanism based on surface adsorption of both electrolyte cations C^+ (K^+, $Na^+, \ldots$) and protons has been proposed:

$$\mathrm{MnO_2} + x\mathrm{C}^+ + y\mathrm{H}^+ + (x+y)e^- \leftrightarrow \mathrm{MnOOC}_x\mathrm{H}_y. \qquad (7)$$

MnO_2 films show a specific capacitance of about 700 Fg^{-1} in neutral aqueous electrolytes within a voltage window of <1V.[134]

Nanostructured transition metal oxides and nitrides have also been studied to enhance the capacitance.[170,172–175] Pseudocapacitance is primarily achieved close to the electrode surface, usually in the first few nanometers from the surface. Therefore, by decreasing the particle size, one can expect an increase in the utilization rate of the active material. Thin MnO_2 deposits of tens to hundreds of nanometers were produced on various substrates such as metal collectors, carbon nanotubes and activated carbons. Nanostructured MnO_2 was reported to achieve specific capacitances as high as 1,300 Fg^{-1}.[133] To improve the cycling stability and the specific capacitance of transition metal oxide nanoparticles, thin conducting polymer deposition on the surface of the electrode helps enhance proton exchange at the surface. Besides, in some studied nanostructured compounds, EDLC also makes a considerable contribution to the total capacitance. For example, the charging mechanism of nanocrystalline VN includes a combination of an electric double layer and a Faradaic reaction (II/IV) at the surface

Table 7: Metal oxide based supercapacitors and their main electrochemical properties.

Electrode material	Electrolyte	Voltage (V)	C_S (Fg^{-1})	Cycles	Refs.
RuO_2 film	0.5 M H_2SO_4	0–0.6	788	—	123
RuO_2 film	1 M H_2SO_4	0.1–0.9	650	4,000	124
RuO_2 thin film	0.5 M H_2SO_4	−0.1–0.6	1190	1,000	125
RuO_2 on CNTs	0.5 M H_2SO_4	0–1	1170	—	126
RuO_2 on CNTs	1 M H_2SO_4	0–1	667	—	127
Ru–Ir oxides	0.5 M H_2SO_4	0–1	367.6	—	128
Ru–Sn oxides	1 M H_2SO_4	0–1	930	1,000	129
$RuO_2{\cdot}xH_2O$	H_2SO_4	0–1	768	60,000	130
$RuO_2{\cdot}xH_2O$	1 M H_2SO_4	0–1	823	1,000	131
$RuO_2{\cdot}xH_2O$ thin film	Acrylic gel polymer	0–1	642	—	132
MnO_2 thin film	0.1 M Na_2SO_4	0–0.9	1370	—	133
MnO_2 film	0.1 M Na_2SO_4	0–0.9	698	1,500	134
MnO_2 nanostructure	1 M Na_2SO_4	0–0.8	120	—	135
MnO_2 nanocrystal on graphene	1 M Na_2SO_4	0–1	211	1,000	136
MnO_2 nanorods	2 M $(NH_4)_2SO_4$	0–0.9	400	1,000	137
Amorphous MnO_2	2 M KOH	−0.15–0.45	298.7	—	138
Tetrapropyl-ammonium/ Mn oxide	0.1 M $LiClO_4$	0–0.9	720	1,500	139
Mn–Ni–Cu oxides	6 M KOH	−0.4–0.4	490	500	140
Mesoporous Co_3O_4	2 M KOH	−0.2–0.5	202.5	1,000	141
Layered Co_3O_4	1 M KOH	−0.2–0.5	548	2,000	142
Co_3O_4 nanowire	1 M KOH	0.1–0.6	599	7,500	143
Co_3O_4 on r-GO	6 M KOH	0–0.5	159.8	1,000	144
Mesoporous α-MoO_3 film	1 M $LiClO_4$/PC	1.5–3.5	605	—	145
NiO film	1 M KOH	−0.1–0.5	200–278	—	146–149
NiO nanosphere	2 M KOH	0–0.8	612.5	1,000	150
NiO nanosheet	2 M KOH	0–0.6	989	1,000	151
Ni–Co oxides on CNTs	1 M KOH	−0.1–0.5	840	1,000	152
Ni–Co oxides	1 M NaOH+ 0.5 M Na_2SO_4	0–1.5	1840	500	153
SnO_2 film	0.1 M Na_2SO_4	0–1	285	1,000	154
SnO_2 film	0.5 M Na_2SO_4	—	66	—	155

(*Continued*)

Table 7: (*Continued*)

Electrode material	Electrolyte	Voltage (V)	C_S (Fg^{-1})	Cycles	Refs.
SnO_x on graphite	0.5 M KCl	0–1	298	1,000	156
SnO_2 on graphene	0.5M Na_2SO_4	0–2.5	294	2,000	157
In_2O_3 nanorods	1 M Na_2SO_4	−0.6–0.2	190	1,000	158
In_2O_3 nanorods	1 M Na_2SO_4	−0.9–0	105	—	159
Bi_2O_3 thin film	1 M NaOH	−1–0	98	1,000	160
Bi_2O_3 nanobelt	1 M Na_2SO_4	−1–−0.2	250	950	161
a-$V_2O_5{\cdot}nH_2O$	2 M KCl	0–0.8	350	100	162
$V_2O_5{\cdot}xH_2O$/CNT	1 M $LiClO_4$/PC	0.6–1.2	910	—	163
Fe_3O_4 thin film	1 M Na_2SO_3	−0.8–−0.1	170	—	164
$BiFeO_3$	1 M NaOH	−1–0.2	81	1,000	165
$NiFe_2O_4$	Na_2SO_3	—	354	1,000	27
Ti–V–W–O/Ti	0.5 M H_2SO_4	0–1.4	125	—	166
Ti/(RhO_x+ Co_3O_4)	0.5 M H_2SO_4	0.4–1.4	500–800	—	167
TiO_2+ NiO	2 M KOH	0–0.6	611	5,000	168
Co_xNi_{1-x} layered double hydroxides	1 M KOH	−1–0.6	2104	—	169
MnO_2/Co_3O_4 hybrid nanowire	1 M LiOH	−0.2–0.6	480	5,000	170

of the nanoparticles, leading to specific capacitance up to 1,200 Fg^{-1} at a scan rate of 2 mV s^{-1}.[176]

While academic research shows great interest in the high capacitance provided by metal oxides/nitrides, industrial R&D is more concerned about the long-term cycling stability of these electrodes, which is a technological issue that must be addressed for successful commercial development. There are still other issues that need to be examined carefully, such as self-discharge, corrosion of the current collector, low temperature performance, etc. Considering that metal oxide based supercapacitor technology is still in its infancy, future research, and development should ultimately yield high-performance, low cost, and safe energy storage devices.[27]

4.2.3 *Pseudocapacitors with conducting polymers*

One of the major recent achievements in the field of electrochemistry is the discovery of conducting polymers. The conductivity of conducting polymers relies on the conjugated bond system along the polymer backbone.

Conducting polymers are synthesized either through chemical oxidation of the monomer or electrochemical oxidation of the monomer. Two oxidation reactions occur simultaneously — the oxidation of the monomer and the oxidation of the polymer with the coincident insertion of a dopant/counter ion.[26] The pseudocapacitance provided by conducting polymers is based on the reversible redox reactions of their electrochemical doping/de-doping.[2,177] The process of reversible electrochemical doping/de-doping may be represented by the following reactions:

$$\mathrm{P}_m - xe^- + x\mathrm{A}^- \leftrightarrow \mathrm{P}_m^{x+}\mathrm{A}_x^-, \tag{8}$$

$$\mathrm{P}_m + ye^- + x\mathrm{M}^+ \leftrightarrow \mathrm{P}_m^{y-}\mathrm{M}_y^+, \tag{9}$$

where, P_m is a polymer with a network of conjugated double bonds, m is the polymerization degree, A^- denotes anions, and M^+ represents cations. In both Eqs. (8) and (9), the doping reactions proceed from the left to the right, and the de-doping reactions the other way around. Reaction (8) is a reaction of oxidative p-doping, and reaction (9) is a reaction of reductive n-doping. While most conducting polymers can only be p-doped, some of them (polyacetylene, polythiophene, and their derivatives) may be reversibly p-doped and n-doped. In the doped state, the conductive polymers have good intrinsic conductivities, from a few S cm^{-1} to 500 S cm^{-1} (Table 8).[26] They have relatively fast charge–discharge kinetics. Some previous studies on the application of conducting polymers as pseudocapacitive materials are summarized in Table 9.

Conducting polymers offer many advantages as supercapacitor electrodes. They are flexible, highly conductive, easy to process, and can be made into thin and thick porous films. As active materials, many conducting polymers exhibit high specific capacitances, while being able to charge and discharge at a relatively rapid rate. The major disadvantage of the conducting polymers used as supercapacitor electrodes is poor cycle life. In general, symmetric supercapacitors based on conducting polymers will have a lower cycle life than those based on carbon. This is unavoidable, because as anions or cations are doped or de-doped into the conducting

Table 8: Physical properties of common conducting polymers.[178]

Conducting polymer	Mw (g mol^{-1})	Dopant level	Potential range (V)	Conductivity (S cm^{-1})	Refs.
Polyaniline	93	0.5	0.7	0.1–5	179
Polypyrrole	67	0.33	0.8	10–50	180
Polythiophene	84	0.33	0.8	300–500	178
PEDOT	142	0.33	1.2	300–400	178

Table 9: Conducting polymer based supercapacitors and their main electrochemical properties.

Electrode material	Electrolyte	Voltage (V)	C_S (Fg^{-1})	Cycles	Refs.
Polyaniline	1 M Et_4 NBF_4/AN	−2–3	107	9,000	181
Polyaniline	1 M Et_4 NBF_4/AN	−0.5–0.5	150	1,000	182
Polyaniline/ graphene	4 M H_2SO_4	−1.4–1.4	243	600	183
Polyaniline/ graphene	2 M H_2SO_4	−0.2–0.8 vs. Ref	480	—	184
Polyaniline/CNT	1 M NaCl	−0.6–0.6	550	1,000	185
Polyaniline/ nanodiamond or onion like carbon	1 M H_2SO_4	−0.6–0.6	640	1,000	186
Polypyrrole	0.5 M Na_2SO_4	−0.5–0.4	254	—	187
Polypyrrole	3 M KCl	−0.4–0.4	400	100,000	188
Polypyrrole	1 M KCl	−0.8–0.8	480	1,000	189
Polypyrrole	PVdF-HFP based gel	−0.85–0.85	137	5,000	190
Polypyrrole/ Nafion	1 M Na_2SO_4	−0.8–0.5	380	5,000	191
Polypyrrole/CNT	1 M KCl	−0.4–0.6 vs. Ref	144	—	192
Polypyrrole/CNT	1 M H_2SO_4	−0.8–0.2 vs. Ref	190	100	193
Polypyrrole/ Fe_2O_3	1 M $LiClO_4$	−1–0.1	420	—	194
Polythiophene derivatives	1 M Et_4 NBF_4/AN	−1.3–0.6	160	2,000	195
Polythiophene-tartaric acid	PVdF-HFP in 1 M $LiPF_6$	−1–1	156	1,000	196
PEDOT	0.1 M $LiClO_4$	−0.2–0.8	124	—	197
PEDOT/ Polypyrrole	1 M KCl	−0.4–0.6 vs Ref	290	1,000	198
PEDOT/ Polypyrrole	1 M $LiClO_4$	−0.4–0.6 vs Ref	230	1,000	198
Irradiated HCl doped polyaniline	Gel polymer electrolyte	−1.2–1.2	243	10,000	199
Non-irradiated HCl doped polyaniline	Gel polymer electrolyte	−1.2–1.2	259	10,000	199

PVdF-HFP in 1 M $LiPF_6$: PVdF-co-HFP in 1 M $LiPF_6$ containing EC&PC (1:1 v/v)-based microporous polymer electrolyte.
PEDOT: poly (3,4-ethylenedioxythiophene).

polymer, there is a corresponding volume change of the electrode when compared with carbon supercapacitors, which involve only simple ion sorption and desorption.[26,200,201] In order to improve cycle-life, current research efforts on conducting polymers for supercapacitor applications are focused on polymer-ceramic[202–205] and polymer-carbon[184,206–216] composite materials as well as on hybrid systems. High specific capacitance of over 600 Fg^{-1} have been achieved by selected composite materials in symmetric two electrode configuration with no reduction in the capacitance after 10,000 charge–discharge cycles.[207]

4.3 *Hybrid Capacitors*

As the organic combination of conventional ECs and rechargeable batteries, hybrid capacitor systems are a promising approach to increase the energy density of capacitor devices without serious loss in their rate stability and cycling capability. This approach combines a battery-like electrode (Faradaic) with a capacitor-like electrode (commonly non-Faradaic) in the same cell, producing high working voltage and capacitance. Obviously, one needs to carefully balance the increase in the energy density contributed by the Faradaic electrode and the reduction in cycle stability for the same reason. In other words, it is important to avoid transforming a good supercapacitor into a mediocre battery.

So far, there are two major approaches to constructing hybrid capacitor systems: (i) pseudocapacitive metal oxides with a capacitive carbon electrode, such as MnO_2/AC, V_2O_5/AC, Fe_3O_4/AC, PbO_2/AC, RuO_2/AC, $Ni(OH)_2$/AC, TiO_2/AC, etc., and (ii) lithium-insertion electrodes with a capacitive carbon electrode, such as $Li_4Ti_5O_{12}$/AC, $LiNi_{0.5}Mn_{1.5}O_4$/AC, $LiCoPO_4$ nanoparticles/carbon nanofoam, $LiMn_2O_4$/AC, lithiated graphite/AC, etc. (Table 3).[15] Some of the previous work on these hybrid capacitor systems is summarized in Table 10. In the future, the advantages of nanomaterials as well as rapid progress in the field of rechargeable batteries will support advanced design of high-performance hybrid capacitor systems which will fill the gap between Li-ion batteries and EDLCs. These systems could be of particular interest in applications where high power, moderate energy, and medium cycle life are needed.

5 Promising Applications of ECs

Table 11 gives a partial summary of the state-of-the-art ECs that are commercially available. The recent development of low cost and highly

Table 10: Hybrid capacitor systems and their main electrochemical properties (AC: activated carbon; EC: ethylene carbonate; DMC: dimethyl carbonate; DEC: diethyl carbonate; PC: propylene carbonate.)

Cathode	Anode	Electrolyte	Voltage (V)	C_S^a (F g^{-1})	E_d^b (Wh kg^{-1})	P_d^c (kW kg^{-1})	Cycles	Refs.
MnO_2	AC	0.65 M K_2SO_4	0–2.2	105	10	3.6	10,000	217
Amorphous MnO_2	AC	1 M KCl	0–2	52	20.8–28.8	0.5–8	100	218
Amorphous MnO_2	AC	0.1 M K_2SO_4	0–2	21	10	16	195,000	219
MnO_2 nanoparticles	AC	0.1 M K_2SO_4	0–2.2	31	17.3	19	5000	220
δ-MnO_2 nanorods	AC	0.5 M Li_2SO_4	0–1.8	31	17	2	23,000	221
Amorphous V_2O_5	AC	2 M $NaNO_3$	−0.3–0.7	32.5	—	—	600	222
V_2O_5 nanoribbons	AC	0.5 M K_2SO_4	0–1.8	64.4	20.3–29	0.07–2	100	223
Fe_3O_4 nanoparticles	AC	6 M KOH	0–1.2	37.9	7.6	0.07	500	224
PbO_2 film	AC	1.28 g cm^{-3} H_2SO_4	0.8–1.8	34.7	11.7–7.8	0.02–0.26	200	225
Ru oxide	Modified carbon fabric	1 M H_2SO_4	0–1.3	159	26.7	17.3	—	226
$Co(OH)_2$	RuO_2	6 M KOH	0–1.4		23.7	8.1	—	227
CoAl double hydroxide	AC	6 M KOH	0.9–1.5	77	15.5	—	1,000	228
NiO nanoflakes	Porous carbon	6 M KOH	0–1.3	38	10	0.01–10	1,000	229
NiO/AC	AC	6 M KOH	−0.5–0.2 vs Ref	194	—	—	—	230
NiO	AC	5 M KOH/PVA aq.	0–1.5	73.4	26.1	—	800	231
$Ni(OH)_2$	AC	Polymer hydrogel	0.4–1.2	—	—	—	20,000	232
$Ni(OH)_2$/AC	AC	6 M KOH	0–0.5	530	—	—	—	233
$Ni(OH)_2$/MWNTs	AC	6 M KOH	0–1.5	96	32	1.5	2,000	234
PTMA[d]-AC	AC	1 M $LiClO_4$/PC	0–2.5	—	—	—	1,000	235
Polyaniline	AC	6 M KOH	1–1.6	380	18	1.25	4,000	236

MnO_2/porous carbon	V_2O_5 nanowires/MWNTs	1 M Na_2SO_4	0–1.6	45	5.5–16	0.075–3.75	100	237
MnO_2	FeOOH	1 M Li_2SO_4	0–1.85	51	12	3.7	2,000	238
MnO_2/RGO	MoO_3/RGO	1 M Na_2SO_4	0–2	307	42.6	0.276	1,000	239
MnO_2/MWNTs	SnO_2/MWNTs	2 M KCl	0–1.7	38	20.3	143.7	1,000	240
MnO_2/MWNTs	MWNTs	1 M $LiClO_4$	0–2.7	58	32.91	—	300	241
MnO_2 nanowire/SWNTs	In_2O_3 nanowire/SWNTs	1 M Na_2SO_4	0–2	184	25.5	50.3	—	242
AC	Zn	7.3 M KOH/0.7 M ZnO	0.4–1.4	120	20	0.1	—	243
AC	graphite	1 M $LiPF_6$/EC/DMC	1.5–5	—	103.8	0.11	3,000	244
AC	graphite	2 M LiTFSI/EC/DMC	1.5–4.2	150	80	—	1,000	245
AC	TiO_2	1 M $LiPF_6$/EC/DMC	1.2–3.5	44	30–80	0.35	600	246
MWNTs	TiO_2 nanowires	1 M $LiPF_6$/EC/DEC/DMC	0–2.8	11.5	8–12.5	0.3–1.2	600	247
MWNTs	Fe_2O_3/MWNTs	1 M $LiClO_4$/EC/DMC	0–2.8	80	50	1	500	248
Graphene	MnO_2-coated graphene	1 M KCl	0–0.9	328	11.4	25.8	1,300	249
Graphene/MnO_2	AC nanofiber	1 M Na_2SO_4	0–1.8	113.5	51.1	198	1,000	250
AC	$Li_4Ti_5O_{12}$	1 M $LiBO_4$/EC/DMC	1.5–3	—	85	—	5,000	251
AC	$Li_4Ti_5O_{12}$/[e]CNF	1 M $LiBF_4$/EC/DMC	1.3–2	—	40	7.5	9,000	252
$LiNi_{0.5}Mn_{1.5}O_4$	AC	1 M $LiPF_6$/EC/DMC	0–2.8	32	55	—	1,000	253
$LiMn_2O_4$	MnO_2/MWNTs	1 M $LiClO_4$/PC	0–2.5	60	26–56	0.3–2.4	—	254
$LiCoPO_4$ nanoparticles	Carbon nanofoam	1 M $LiClO_4$/EC/PC	0–2	21.9	11	0.2	1,000	255
$LiMn_2O_4$/AC	$Li_4Ti_5O_{12}$	1 M $LiPF_6$/EC/DEC/DMC	1.2–2.8	59	16	2.5	1,500	256

[a]C_S: Specific capacitance. [b]E_d: Energy density. [c]P_d: Power density. [d]PTMA: poly(2,2,6,6-tetramethylpiperidinyloxy methacrylate) nitroxide. [e]CNF: carbon nanofiber.

Table 11: Technical performance of some currently available ECs.

Company	Product	Voltage (V)	C (F)	ESR (mΩ)	Specific power ($W\ kg^{-1}$)	Specific energy ($Wh\ kg^{-1}$)	Cycle life	Mass (kg)
Maxwell	PC10 series	2.2–2.5	10	180	510–660	1.1–1.4	500,000	0.0063
Maxwell	HC series	2.7	1–150	14–700	1,100–3,400	0.9–4.7	500,000	0.0011–0.032
Maxwell	BC series	2.85	310–350	2.2–3.2	4,600–6,600	5.2–5.9	500,000	0.06
Maxwell	D cell series	2.7	310–350	2.2–3.2	4,600–6,600	5.2–5.9	500,000	0.06
Maxwell	K2 series	2.7	650–3,000	0.29–0.8	5,900–6,900	4.1–6.0	1,000,000	0.16–0.51
Maxwell	16 volt small modules	16	58	22	2,200	3.3	500,000	0.63
Maxwell	16 volt large modules	16	110–500	2.1–5.6	1,800–2,700	1.5–3.2	1,000,000	2.66–5.51
Maxwell	48 volt modules	48	83–165	6.3–10	2,700–3,300	2.6–3.9	1,000,000	10.3–13.5
Maxwell	56 volt UPS modules	56	130	8.1	2,600	3.1	1,000,000	18
Maxwell	75 volt power modules	75	94	13	2,100	2.9	1,000,000	25
Maxwell	125 volt transportation modules	125	63	18	1,700	2.3	1,000,000	60.5
Nesscap	Small EDLC cells	2.3–2.7	5–60	14–123	—	1.67–4.51	500,000	0.0022–0.0135
Nesscap	Medium EDLC cells	2.3–2.7	90–360	3.2–16	—	3.83–5.45	500,000	0.021–0.067
Nesscap	Large prismatic EDLC cells	2.7	600–5,000	0.25–0.64	—	2.9–5.44	500,000	0.21–0.93
Nesscap	Large cylindrical EDLC cells	2.7	650–3,000	0.22–0.5	—	3.13–5.73	1,000,000	0.205–0.535
Nesscap	Small pseudocapacitor cells	2.3	50–120	18–24	—	4.87–5.87	100,000	0.0076–0.015

Nesscap	Medium pseudocapacitor cells	2.3	220–300	12–14	—	7.03–8.73	100,000	0.023–0.0252
Nesscap	Multicell modules	5–125	1.5–500	1.6–110	—	1.05–3.83	1,000,000	0.0034–57
ApowerCap	Ultracapacitor	2.7	55	4	5,695	5.5	—	0.009
ApowerCap	Ultracapacitor	2.7	450	1.4	2,574	5.89	—	0.057
ApowerCap	Ultracapacitor	2.7	550	—	5,000–20,000	1.5–4.5	—	—
Panasonic	Multilayer coin type	3.6–5.5	0.10–1.5	—	—	—	—	—
Panasonic	Radial lead type	2.1–2.5	3.3–70	—	—	—	—	0.0015–0.0031
BatScap	Individual 2600F cell	2.7	2,680	0.20	2,050	4.2	—	0.50
Asahi Glass	Supercapacitor	2.7	1,375	2.5	390	4.9	—	0.21
LS Cable	LSUC 2.8V	2.8	100–3,000	0.25–9	—	4.8–5.2	500,000	0.023–0.025
LS Cable	LSUC 2.7V	2.7	650–3,000	0.26–0.6	—	5.68–5.79	1,000,000	0.525–0.535
LS Cable	LSUC 2.5V	2.5	110–2,800	0.65–14	—	3.57	500,000	0.028
LS Cable	LSHC 2.3–2.5V	2.3–2.5	220–5400	0.5–20	—	6.14–7.03	500,000	0.027–0.76
Power Sys.	Supercapacitor	2.7	1,350	1.5	650	4.9	—	0.21
Power Sys.	Supercapacitor	3.3	1,800	3.0	486	8.0	—	0.21
Fuji Heavy Industry	Supercapacitor	3.8	1,800	1.5	1,025	9.2	—	0.232
JSR Micro	Supercapacitor	3.8	2,000	1.9	1,038	12.1	—	0.206

capacitive ECs has enabled a wide range of new applications, such as the following:[8, 257]

(1) Industrial applications
   - Uninterruptible power supply (UPS) systems
   - Energy-efficient elevators, lift forks, and cranes
(2) Electric utility applications
   - Smart grids
   - Wind farms
(3) Automotive applications
   - Starting engines
   - Electric and hybrid vehicles
   - Heavy duty and large transport systems
(4) Electronic devices needing pulse-power (light flash, radio signal, and rapid heating)
(5) Specialized and military applications

Among these applications, the use of ECs as energy storage devices for various kinds of automobiles, including small range EV, hybrid electric vehicles (HEVs), and fuel cell vehicles, have received much attention. In this case, the EC serves as a short-time energy storage device with high power capability which facilitates the storage of energy obtained from regenerative braking. This energy could be reused in the next acceleration phase and thus boost the acceleration. It allows one to reduce the size of the primary power source (batteries (EV), internal combustion engine (HEV), fuel cell) and keep them running at an optimized operation point. Buses, delivery vans and cars in city areas with many go-stop intervals will benefit the most from this technology. Many more potential applications can be imagined. However, we emphasize that an EC is not a panacea for all applications concerning energy storage. Its working range is clearly labeled on the Ragone plot of Fig. 1, corresponding to a ratio of energy to power of about 1. For much larger and much smaller ratios, conventional capacitors and batteries, respectively, are more suitable, unless other criteria like cycling performance or environmental requirements become important.[8,258–261]

6 Self-discharge of Electrochemical Capacitive Storage

The shelf-life of an electrochemical capacitor (EC) is primarily determined by its self-discharge rate. Self-discharge of a supercapacitor refers to the gradual decrease in the voltage across the capacitor that occurs at open circuit.[262] There are several mechanisms for the self-discharge. Firstly, impurities in the supercapacitor materials[263] can cause redox-reactions which leads to charge transfer through the double layer. Secondly, inappropriate overcharge results in electrolyte decomposition, which also leads to self-discharge. Another reason for self-discharge is Ohmic leakage arising from inadvertent inter-electrode contacts or leaky bipolar electrodes.[264] Finally, in some cases, the observed voltage decay is believed to be caused by the redistribution of the charge carriers.[265] In an EDLC system, this process means the spatial redistribution of adsorbed charge carriers on the surface of the electrodes. In this process, the rate of voltage decay could be affected by the charge duration, charge history, the temperature, and the initial voltage, etc. In an aqueous pseudocapacitor system based on a transition metal oxide, as an example, this process means the proton redistribution from the electrode surface (highly charged) to the bulk (less charged), which involves the change of the local oxidation state of the transition metal. Besides external factors discussed in the case of EDLC, this process is also affected by the thickness of the oxide layer, the diffusion constant of protons in the solid electrode, etc.

7 Conclusions and Outlook

In this chapter, we have briefly reviewed the state-of-the-art capacitor technology including its development, classification, construction, materials, and applications. New knowledge in material science and electrochemistry is fundamental for the development of ECs with promising technical and economical properties, which are appropriate for a wide and growing range of applications. ECs are now available in multiple designs that cover almost all the performance range existing between conventional capacitors and batteries.

In addition to the large number of applications in consumer electronics, the new application areas described require larger capacitance, larger power or larger energy EC systems for various energy-intensive sectors: industrial, transport and electricity grids. The greater adoption of EC devices in these areas is also confirmed by market studies, forecasting an average yearly growth rate of about 27% from 2004 to 2014, with an increasing number (up to about 40%) of EC devices being used in large system applications.[258]

The overall prospects and the associated advantages of using EC devices in new applications will surely power near term as well as long term efforts in research, development, and even production processes. This in turn will further improve technical performance in a larger spectrum of applications and will also significantly reduce costs.

Acknowledgments

This work was partially supported by the US Air Force (AFOSR), Semiconductor Research Corporation (SRC) and the Petroleum Research Fund (PRF).

Glossary

AC/DC	Alternating current/direct current
ECs	Electrochemical capacitors
EDLC	Electrochemical double-layer capacitors
EV/HEV	Electric vehicle/Hybrid electric vehicle
GBMs	Graphene-based materials
IL	Ionic liquid
MLCs	Multilayer ceramic capacitors
OLC	Onion-like carbon
Redox reaction	Reduction-oxidation reaction
SSA	Specific surface area
TEM	Transmission electron microscopy

References

1. M. Jayalakshmi and K. Balasubramanian, Simple capacitors to supercapacitors — an overview, *Int. J. Electrochem. Sci.* **3** (2008), pp. 1196–1217.
2. Y. M. Vol'fkovich and T. M. Serdyuk, Electrochemical capacitors, *Russ. J. Electrochem.* **38** (2002), pp. 935–958.
3. T. R. J. Janet Ho, Overview of laminar dielectric capacitors, *IEEE Electr. Insul. M* **26** (2010), pp. 7–13.
4. K. Naoi, 'Nanohybrid capacitor': the next generation electrochemical capacitors, *Fuel Cells* **10** (2010), pp. 825–833.
5. B. E. Conway, Electrochemical supercapacitors, (Kluwer Academic/Plenum Publishers, New York, USA, 1999).
6. P. Sharma and T. S. Bhatti, A review on electrochemical double-layer capacitors, *Energy Convers. Manage.* **51** (2010), pp. 2901–2912.

7. R. C. Weast, M. J. Astle and W. H. Beyer, *CRC Handbook of Chemistry and Physics* (CRC Press, Boca Raton, FL, 1988).
8. A. Nishino, Capacitors: operating principles, current market and technical trends, *J. Power Sources* **60** (1996), pp. 137–147.
9. Y. Sakabe, Multilayer ceramic capacitors, *Solid State Mater. Sci.* **2** (1997), pp. 584–587.
10. J. C. Niepce, Multilayer ceramic capacitors, *Actual. Chim.* (2002), pp. 74–78.
11. H. Kishi, Y. Mizuno and H. Chazono, Base-metal electrode-multilayer ceramic capacitors: past, present and future perspectives, *JJAP* **42** (2003), pp. 1–15.
12. F. Trombetta, M. O. de Souza, R. F. de Souza and E. M. A. Martini, Electrochemical behavior of aluminum in 1-n-butyl-3-methylimidazolium tetrafluoroborate ionic liquid electrolytes for capacitor applications, *J. Appl. Electrochem.* **39** (2009), pp. 2315–2321.
13. S. Niwa and Y. Taketani, Development of new series of aluminium solid capacitors with organic semiconductive electrolyte (OS-CON), *J. Power Sources* **60** (1996), pp. 165–171.
14. A. Dehbi, W. Wondrak, Y. Ousten and Y. Danto, High temperature reliability testing of aluminum and tantalum electrolytic capacitors, *Microelectron. Reliab.* **42** (2002), pp. 835–840.
15. P. Simon and Y. Gogotsi, Materials for electrochemical capacitors, *Nat. Mater.* **7** (2008), pp. 845–854.
16. B. Xu, H. Zhang, G. P. Cao, W. F. Zhang and Y. S. Yang, Carbon materials for supercapacitors progress in chemistry, *Prog. Chem.* **23** (2011), pp. 605–611.
17. B. Daffos, P. L. Taberna, Y. Gogotsi and P. Simon, Recent advances in understanding the capacitive storage in microporous carbons, *Fuel Cells* **10** (2010), pp. 819–824.
18. C. Portet, G. Yushin and Y. Gogotsi, Electrochemical performance of carbon onions, nanodiamonds, carbon black and multiwalled nanotubes in electrical double layer capacitors, *Carbon* **45** (2007), pp. 2511–2518.
19. L. L. Zhang and X. S. Zhao, Carbon-based materials as supercapacitor electrodes, *Chem. Soc. Rev.* **38** (2009), pp. 2520–2531.
20. M. Inagaki, H. Konno and O. Tanaike, Carbon materials for electrochemical capacitors, *J. Power Sources* **195** (2010), pp. 7880–7903.
21. A. G. Pandolfo and A. F. Hollenkamp, Carbon properties and their role in supercapacitors, *J. Power Sources* **157** (2006), pp. 11–27.
22. P. Simon and Y. Gogotsi, Charge storage mechanism in nanoporous carbons and its consequence for electrical double layer capacitors, *Phil. Trans. R. Soc. A* **368** (2010), pp. 3457–3467.
23. E. Frackowiak and F. Beguin, Carbon materials for the electrochemical storage of energy in capacitors, *Carbon* **39** (2001), pp. 937–950.
24. F. B. Elzbieta Frackowiak, Electrochemical storage of energy in carbon nanotubes and nanostructured carbons, *Carbon* **40** (2002).
25. E. Frackowiak, Carbon materials for supercapacitor application, *PCCP* **9** (2007), pp. 1774–1785.

26. G. A. Snook, P. Kao and A. S. Best, Conducting-polymer-based supercapacitor devices and electrodes, *J. Power Sources* **196** (2011), pp. 1–12.
27. C. D. Lokhande, D. P. Dubal and O. S. Joo, Metal oxide thin film based supercapacitors, *Curr. Appl. Phys.* **11** (2011), pp. 255–270.
28. J. H. Chae, K. C. Ng and G. Z. Chen, Nanostructured materials for the construction of asymmetrical supercapacitors, *P. I. Mech. Eng. A-J. Pow. Energy* **224** (2010), pp. 479–503.
29. Y. Zhang, H. Feng, X. Wu, L. Wang, A. Zhang, T. Xia, H. Dong, X. Li and L. Zhang, Progress of electrochemical capacitor electrode materials: A review, *Int. J. Hydrogen Energy* **34** (2009), pp. 4889–4899.
30. Y. G. P. Simon, Capacitive energy storage in nanostructured carbon-electrolyte systems, *Acc. Chem. Res.* **46** (2013).
31. W. Gu and G. Yushin, Review of nanostructured carbon materials for electrochemical capacitor applications: advantages and limitations of activated carbon, carbide-derived carbon, zeolite-templated carbon, carbon aerogels, carbon nanotubes, onion-like carbon and graphene, *WIREs*, **3** (2013), pp. 435–457.
32. B. Xu, S. F. Yue, Z. Y. Sui, X. T. Zhang, S. S. Hou, G. P. Cao and Y. S. Yang, What is the choice for supercapacitors: graphene or graphene oxide? *Energ. Environ. Sci.* **4** (2011), pp. 2826–2830.
33. Y. Sun, Q. Wu and G. Shi, Graphene based new energy materials, *Energ. Environ. Sci.* **4** (2011), pp. 1113–1132.
34. M. Pumera, Graphene-based nanomaterials for energy storage, *Energ. Environ. Sci.* **4** (2011), pp. 668–674.
35. L. L. Zhang, R. Zhou and X. S. Zhao, Graphene-based materials as supercapacitor electrodes, *J. Mater. Chem.* **20** (2010), pp. 5983–5992.
36. M. Liang, B. Luo and L. Zhi, Application of graphene and graphene-based materials in clean energy-related devices, *Int. J. Energ. Res.* **33** (2009), pp. 1161–1170.
37. V. Subramanian, C. Luo, A. M. Stephan, K. S. Nahm, S. Thomas and B. Q. Wei, Supercapacitors from activated carbon derived from banana fibers, *J. Phys. Chem. C*, **111** (2007), pp. 7527–7531.
38. L. Wei and G. Yushin, Electrical double layer capacitors with activated sucrose-derived carbon electrodes, *Carbon*, **49** (2011), pp. 4830–4838.
39. L. Wei and G. Yushin, Electrical double layer capacitors with sucrose derived carbon electrodes in ionic liquid electrolytes, *J. Power Sources*, **196** (2011), pp. 4072–4079.
40. Q. Y. Li, H. Q. Wang, Q. F. Dai, J. H. Yang and Y. L. Zhong, Novel activated carbons as electrode materials for electrochemical capacitors from a series of starch, *Solid State Ion.* **179** (2008), pp. 269–273.
41. L. Wei, M. Sevilla, A. B. Fuertes, R. Mokaya and G. Yushin, Hydrothermal carbonization of abundant renewable natural organic chemicals for high-performance supercapacitor electrodes, *Adv. Energy Mater.* **1** (2011), pp. 356–361.
42. M. S. Balathanigaimani, W. G. Shim, M. J. Lee, C. Kim, J. W. Lee and H. Moon, Highly porous electrodes from novel corn grains-based activated

carbons for electrical double layer capacitors, *Electrochem. Commun.* **10** (2008), pp. 868–871.

43. E. Raymundo-Pinero, M. Cadek and F. Beguin, Tuning carbon materials for supercapacitors by direct pyrolysis of seaweeds, *Adv. Funct. Mater.* **19** (2009), pp. 1032–1039.
44. E. Raymundo-Piñero, F. Leroux and F. Béguin, A high performance carbon for supercapacitors obtained by carbonization of a seaweed biopolymer, *Adv. Mater.* **18** (2006), pp. 1877–1882.
45. K. Kierzek, E. Frackowiak, G. Lota, G. Gryglewicz and J. Machnikowski, Electrochemical capacitors based on highly porous carbons prepared by KOH activation, *Electrochim. Acta* **49** (2004), pp. 1169–1170.
46. D. Zhai, B. Li, H. Du, G. Wang and F. Kang, The effect of pre-carbonization of mesophase pitch-based activated carbons on their electrochemical performance for electric double-layer capacitors, *J. Solid State Electrochem.* **15** (2011), pp. 787–794.
47. T.-C. Weng and H. Teng, Characterization of high porosity carbon electrodes derived from mesophase pitch for electric double-layer capacitors, *J. Electrochem. Soc.* **148** (2001), p. A368.
48. Y. J. Kim, Y. Matsuzawa, S. Ozaki, K. C. Park, C. Kim, M. Endo, H. Yoshida, G. Masuda, T. Sato and M. S. Dresselhausc, High energy density capacitor based on the novel ammonium salt type ionic liquids and their mixing effect by propylene carbonate (PC), *J. Electrochem. Soc.* **152** (2005), pp. A710–A715.
49. H. Nakagawa, A. Shudo and K. Miura, High-capacity electric double-layer capacitor with high-density-activated carbon fiber electrodes, *J. Electrochem. Soc.* **147** (2000), pp. 38-42.
50. L. Ding, Z. Wang, Y. Li, Y. Du, H. Liu and Y. Guo, A novel hydrochar and nickel composite for the electrochemical supercapacitor electrode material, *Mater. Lett.* **74** (2012), pp. 111–114.
51. T. E. Rufford, D. Hulicova-Jurcakova, K. Khosla, Z. H. Zhu and G. Q. Lu, Microstructure and electrochemical double-layer capacitance of carbon electrodes prepared by zinc chloride activation of sugar cane bagasse, *J. Power Sources* **195** (2010), pp. 912–918.
52. D. Lozano-Castello, D. Cazorla-Amoros, A. Linares-Solano, S. Shiraishi, H. Kurihara and A. Oya, Influence of pore structure and surface chemistry on electric double layer capacitance in non-aqueous electrolyte, *Carbon* **41** (2003), pp. 1765–1775.
53. D. Hulicova-Jurcakova, M. Seredych, G. Q. Lu and T. J. Bandosz, Combined effect of nitrogen- and oxygen-containing functional groups of microporous activated carbon on its electrochemical performance in supercapacitors, *Adv. Funct. Mater.* **19** (2009), pp. 438–447.
54. B. Xu, Y. F. Chen, G. Wei, G. P. Cao, H. Zhang and Y. S. Yang, Activated carbon with high capacitance prepared by NaOH activation for supercapacitors, *Mater. Chem. Phys.* **124** (2010), pp. 504–509.
55. T. E. Rufford, D. Hulicova-Jurcakova, Z. Zhu and G. Q. Lu, Nanoporous carbon electrode from waste coffee beans for high performance supercapacitors, *Electrochem. Commun.* **10** (2008), pp. 1594–1597.

56. X. Li, C. Han, X. Chen and C. Shi, Preparation and performance of straw based activated carbon for supercapacitor in non-aqueous electrolytes, *Microporous Mesoporous Mater.* **131** (2010), pp. 303–309.
57. M. Lazzari, M. Mastragostino, A. G. Pandolfo, V. Ruiz and F. Soavi, Role of carbon porosity and ion size in the development of ionic liquid based supercapacitors, *J. Electrochem. Soc.* **158** (2011), pp. A22–A25.
58. L. Wei, M. Sevilla, A. B. Fuertes, R. Mokaya and G. Yushin, Polypyrrole-derived activated carbons for high-performance electrical double-layer capacitors with ionic liquid electrolyte, *Adv. Funct. Mater.* **22** (2011), pp. 827–834.
59. Q. Zhang, J. P. Rong, D. S. Ma and B. Q. Wei, The governing self-discharge processes in activated carbon fabric-based supercapacitors with different organic electrolytes, *Energ. Environ. Sci.* **4** (2011), pp. 2152–2159.
60. B. H. Kim, K. S. Yang, H. G. Woo and K. Oshida, Supercapacitor performance of porous carbon nanofiber composites prepared by electrospinning polymethylhydrosiloxane (PMHS)/polyacrylonitrile (PAN) blend solutions, *Synth. Met.* **161** (2011), pp. 1211–1216.
61. B. Xu, F. Wu, R. Chen, G. Cao, S. Chen and Y. Yang, Mesoporous activated carbon fiber as electrode material for high-performance electrochemical double layer capacitors with ionic liquid electrolyte, *J. Power Sources* **195** (2010), pp. 2118–2124.
62. J. Yan, T. Wei, W. Qiao, Z. Fan, L. Zhang, T. Li and Q. Zhao, A high-performance carbon derived from polyaniline for supercapacitors, *Electrochem. Commun.* 12 (2010), pp. 1279–1282.
63. X. Xiang, E. Liu, L. Li, Y. Yang, H. Shen, Z. Huang and Y. Tian, Activated carbon prepared from polyaniline base by K_2CO_3 activation for application in supercapacitor electrodes, *J. Solid State Electrochem.* **15** (2011), pp. 579–585.
64. K.-S. Kim and S.-J. Park, Easy synthesis of polyaniline-based mesoporous carbons and their high electrochemical performance, *Microporous Mesoporous Mater.* **163** (2012), pp. 140–146.
65. Y. J. Kim, Y. Horie, S. Ozaki, Y. Matsuzawa, H. Suezaki, C. Kim, N. Miyashita and M. Endo, Correlation between the pore and solvated ion size on capacitance uptake of PVDC-based carbons, *Carbon* **42** (2004), pp. 1491–1500.
66. F. W. Shuangling Guo, Hao Chen, He Ren, Rongshun Wang and Xiumei Pan, Preparation and performance of polyvinyl alcohol-based activated carbon as electrode material in both aqueous and organic electrolytes, *J. Solid State Electrochem.* **16** (2012).
67. X. Zhang, X. Wang, L. Jiang, H. Wu, C. Wu and J. Su, Effect of aqueous electrolytes on the electrochemical behaviors of supercapacitors based on hierarchically porous carbons, *J. Power Sources* **216** (2012), pp. 290–296.
68. M. Endo, T. Maeda, T. Takeda, Y. J. Kim, K. Koshiba, H. Hara and M. S. Dresselhaus, Capacitance and pore-size distribution in aqueous and nonaqueous electrolytes using various activated carbon electrodes, *J. Electrochem. Soc.* **148** (2001), p. A910.

69. C. Ma, Y. Song, J. Shi, D. Zhang, M. Zhong, Q. Guo and L. Liu, Phenolic-based carbon nanofiber webs prepared by electrospinning for supercapacitors, *Mater. Lett.* **76** (2012), pp. 211–214.
70. R. Nandhini, P. A. Mini, B. Avinash, S. V. Nair and K. R. V. Subramanian, Supercapacitor electrodes using nanoscale activated carbon from graphite by ball milling, *Mater. Lett.* **87** (2012), pp. 165–168.
71. M. D. E. Taer, I. A. Talib, A. Awitdrus, S. A. Hashmi and A. A. Umar, Preparation of a highly porous binderless activated carbon monolith from rubber wood sawdust by a multi-step activation process for application in supercapacitors, *Int. J. Electrochem. Sci.* **6** (2011), pp. 3301–315.
72. M. K. Seo and S. J. Park, Electrochemical characteristics of activated carbon nanofiber electrodes for supercapacitors, *Mater. Sci. Eng. B-Adv.* **164** (2009), pp. 106–111.
73. S.-J. Han, Y.-H. Kim, K.-S. Kim and S.-J. Park, A study on high electrochemical capacitance of ion exchange resin-based activated carbons for supercapacitor, *Curr. Appl. Phys.* **12** (2012), pp. 1039–1044.
74. D. Hulicova-Jurcakova, M. Kodama, S. Shiraishi, H. Hatori, Z. H. Zhu and G. Q. Lu, Nitrogen-enriched nonporous carbon electrodes with extraordinary supercapacitance, *Adv. Funct. Mater.* **19** (2009), pp. 1800–1809.
75. A. Kajdos, A. Kvit, F. Jones, J. Jagiello and G. Yushin, Tailoring the pore alignment for rapid ion transport in microporous carbons, *J. Am. Chem. Soc.* **132** (2010), pp. 3252–3253.
76. C. Portet, Z. Yang, Y. Korenblit, Y. Gogotsi, R. Mokaya and G. Yushin, Electrical double-layer capacitance of zeolite-templated carbon in organic electrolyte, *J. Electrochem. Soc.* **156** (2009), pp. A1–A6.
77. F. Lufrano and P. Staiti, Mesoporous carbon materials as electrodes for electrochemical supercapacitors, *Int. J. Electrochem. Sci.* **5** (2010), pp. 903–916.
78. M. Lazzari, F. Soavi and M. Mastragostino, Mesoporous carbon design for ionic liquid-based, double-layer supercapacitors, *Fuel Cells* **10** (2010), pp. 840–847.
79. T. Morishita, T. Tsumura, M. Toyoda, J. Przepiorski, A. W. Morawski, H. Konno and M. Inagaki, A review of the control of pore structure in MgO-templated nanoporous carbons, *Carbon* **48** (2010), pp. 2690–2707.
80. D. W. Wang, F. Li, M. Liu, G. Q. Lu and H. M. Cheng, 3D aperiodic hierarchical porous graphitic carbon material for high-rate electrochemical capacitive energy storage, *Angew. Chem. Int. Ed.* **47** (2008), pp. 373–376.
81. J. Eskusson, A. Janes, A. Kikas, L. Matisen and E. Lust, Physical and electrochemical characteristics of supercapacitors based on carbide derived carbon electrodes in aqueous electrolytes, *J. Power Sources* **196** (2011), pp. 4109–4116.
82. M. Rose, Y. Korenblit, E. Kockrick, L. Borchardt, M. Oschatz, S. Kaskel and G. Yushin, Hierarchical micro- and mesoporous carbide-derived carbon as a high-performance electrode material in supercapacitors, *Small* **7** (2011), pp. 1108–1117.

83. E. I. Shkol'nikov and D. E. Vitkina, Nanoporous structure characteristics of carbon materials for supercapacitors, *High Temperature* **48** (2010), pp. 815–822.
84. M. Oschatz, E. Kockrick, M. Rose, L. Borchardt, N. Klein, I. Senkovska, T. Freudenberg, Y. Korenblit, G. Yushin and S. Kaskel, A cubic ordered, mesoporous carbide-derived carbon for gas and energy storage applications, *Carbon* **48** (2010), pp. 3987–3992.
85. Y. Korenblit, M. Rose, E. Kockrick, L. Borchardt, A. Kvit, S. Kaskel and G. Yushin, High-rate electrochemical capacitors based on ordered mesoporous silicon carbide-derived carbon, *Acs Nano* **4** (2010), pp. 1337–1344.
86. E. N. Hoffman, G. Yushin, T. El-Raghy, Y. Gogotsi and M. W. Barsoum, Micro and mesoporosity of carbon derived from ternary and binary metal carbides, *Microporous Mesoporous Mater.* **112** (2008), pp. 526–532.
87. R. Dash, J. Chmiola, G. Yushin, Y. Gogotsi, G. Laudisio, J. Singer, J. Fischer and S. Kucheyev, Titanium carbide derived nanoporous carbon for energy-related applications, *Carbon* **44** (2006), pp. 2489–2497.
88. J. Chmiola, G. Yushin, R. Dash and Y. Gogotsi, Effect of pore size and surface area of carbide derived carbons on specific capacitance, *J. Power Sources* **158** (2006), pp. 765–772.
89. H. L. Wang and Q. M. Gao, Synthesis, characterization and energy-related applications of carbide-derived carbons obtained by the chlorination of boron carbide, *Carbon* **47** (2009), pp. 820–828.
90. G. N. Yushin, E. N. Hoffman, A. Nikitin, H. H. Ye, M. W. Barsoum and Y. Gogotsi, Synthesis of nanoporous carbide-derived carbon by chlorination of titanium silicon carbide, *Carbon* **43** (2005), pp. 2075–2082.
91. A. Lewandowski, A. Olejniczak, M. Galinski and I. Stepniak, Performance of carbon–carbon supercapacitors based on organic, aqueous and ionic liquid electrolytes, *J. Power Sources* **195** (2010), pp. 5814–5819.
92. K. S. Hung, C. Masarapu, T. H. Ko and B. Q. Wei, Wide-temperature range operation supercapacitors from nanostructured activated carbon fabric, *J. Power Sources* **193** (2009), pp. 944–949.
93. J. R. McDonough, J. W. Choi, Y. Yang, F. La Mantia, Y. G. Zhang and Y. Cui, Carbon nanofiber supercapacitors with large areal capacitances, *Appl. Phys. Lett.* **95** (2009), pp. 243109–243111.
94. G. Lota, K. Fic and E. Frackowiak, Carbon nanotubes and their composites in electrochemical applications, *Energ. Environ. Sci.* **4** (2011), pp. 1592–1605.
95. B. Q. Wei, Supercapacitors from carbon nanotubes, *Abstr. Pap. Am. Chem. S.* **241** (2011).
96. A. Izadi-Najafabadi, T. Yamada, D. N. Futaba, H. Hatori, S. Iijima and K. Hata, Impact of cell-voltage on energy and power performance of supercapacitors with single-walled carbon nanotube electrodes, *Electrochem. Commun.* **12** (2010), pp. 1678–1681.
97. H. Zhang, G. Cao and Y. Yang, Carbon nanotube arrays and their composites for electrochemical capacitors and lithium-ion batteries, *Energ. Environ. Sci.* **2** (2009), pp. 932–943.

98. W. Lu, L. Qu, K. Henry and L. Dai, High performance electrochemical capacitors from aligned carbon nanotube electrodes and ionic liquid electrolytes, *J. Power Sources* **189** (2009), pp. 1270–1277.
99. V. V. N. Obreja, On the performance of supercapacitors with electrodes based on carbon nanotubes and carbon activated material—a review, *Physica E Low Dimens. Syst. Nanostruct.* **40** (2008), pp. 2596–2605.
100. X. Wang, L. Liu, X. Wang, L. Bai, H. Wu, X. Zhang, L. Yi and Q. Chen, Preparation and performances of carbon aerogel microspheres for the application of supercapacitor, *J. Solid State Electrochem.* **15** (2011), pp. 643–648.
101. B. B. Garcia, S. L. Candelaria, D. W. Liu, S. Sepheri, J. A. Cruz and G. Z. Cao, High performance high-purity sol-gel derived carbon supercapacitors from renewable sources, *Renew. Energy* **36** (2011), pp. 1788–1794.
102. A. Halama, B. Szubzda and G. Pasciak, Carbon aerogels as electrode material for electrical double layer supercapacitors–synthesis and properties, *Electrochim. Acta* **55** (2010), pp. 7501–7505.
103. D. Pech, M. Brunet, H. Durou, P. Huang, V. Mochalin, Y. Gogotsi, P.-L. Taberna and P. Simon, Ultrahigh-power micrometre-sized supercapacitors based on onion-like carbon, *Nat. Nanotechnol.* **5** (2010), pp. 651–654.
104. E. G. Bushueva, P. S. Galkin, A. V. Okotrub, L. G. Bulusheva, N. N. Gavrilov, V. L. Kuznetsov and S. I. Moiseekov, Double layer supercapacitor properties of onion-like carbon materials, *Phys. Status Solidi B-Basic Solid State Phys.* **245** (2008), pp. 2296–2299.
105. J. J. Yoo, K. Balakrishnan, J. S. Huang, V. Meunier, B. G. Sumpter, A. Srivastava, M. Conway, A. L. M. Reddy, J. Yu, R. Vajtai and P. M. Ajayan, Ultrathin planar graphene supercapacitors, *Nano Letters* **11** (2011), pp. 1423–1427.
106. X. J. Lu, H. Dou, B. Gao, C. Z. Yuan, S. D. Yang, L. Hao, L. F. Shen and X. G. Zhang, A flexible graphene/multiwalled carbon nanotube film as a high performance electrode material for supercapacitors, *Electrochim. Acta* **56** (2011), pp. 5115–5121.
107. J. R. Miller, R. A. Outlaw and B. C. Holloway, Graphene double-layer capacitor with ac line-filtering performance, *Science* **329** (2010), pp. 1637–1639.
108. Y. W. Zhu, S. Murali, M. D. Stoller, K. J. Ganesh, W. W. Cai, P. J. Ferreira, A. Pirkle, R. M. Wallace, K. A. Cychosz, M. Thommes, D. Su, E. A. Stach and R. S. Ruoff, Carbon-based supercapacitors produced by activation of graphene, *Science* **332** (2011), pp. 1537–1541.
109. T. Y. K. T. Y. Kim, H. W. Lee, M. Stoller, D. R. Dreyer, C. W. Bielawski, R. S. Ruoff and K. S. Suh, High-performance supercapacitors based on poly(ionic liquid)-modified graphene electrodes, *ACS Nano* **5** (2011), pp. 436–442.
110. W. Gu, N. Peters and G. Yushin, Functionalized carbon onions, detonation nanodiamond and mesoporous carbon as cathodes in Li-ion electrochemical energy storage devices, *Carbon* **53** (2013), pp. 292–301.

111. B. Xu, F. Wu, Y. Su, G. Cao, S. Chen, Z. Zhou and Y. Yang, Competitive effect of KOH activation on the electrochemical performances of carbon nanotubes for EDLC: balance between porosity and conductivity, *Electrochim. Acta* **53** (2008), pp. 7730–7735.
112. S. H. J. Ryong Ryoo, Michal Kruk and Mietek Jaroniec, Ordered mesoporous carbons, *Adv. Mater.* **13** (2001), pp. 677–681.
113. M. Galinski, A. Lewandowski and I. Stepniak, Ionic liquids as electrolytes, *Electrochim. Acta* **51** (2006), 5567–5580.
114. G. S. Attard, P. N. Bartlett, N. R. B. Coleman, J. M. Elliott, J. R. Owen and J. H. Wang, Mesoporous platinum films from lyotropic liquid crystalline phases, *Science* **278** (1997), pp. 838–840.
115. F. G. Will and C. A. Knorr, Investigation of formation and removal of hydrogen and oxygen coverage on platinum by a new, nonstationary method, *Zeit. Elektrochem.* **64** (1960), p. 258.
116. B. E. Conway, V. Birss and J. Wojtowicz, The role and utilization of pseudocapacitance for energy storage by supercapacitors, *J. Power Sources* **66** (1997), pp. 1–14.
117. M. Seredych, D. Hulicova-Jurcakova, G. Q. Lu and T. J. Bandosz, Surface functional groups of carbons and the effects of their chemical character, density and accessibility to ions on electrochemical performance, *Carbon* **46** (2008), 1475–1488.
118. M. K. Denisa Hulicova, Hiroaki Hatori, Electrochemical performance of nitrogen-enriched carbons in aqueous and non-aqueous supercapacitors, *Chem. Mater.* **18** (2006), pp. 2318–2326.
119. C. O. Ania, V. Khomenko, E. Raymundo-Piñero, J. B. Parra and F. Béguin, The large electrochemical capacitance of microporous doped carbon obtained by using a zeolite template, *Adv. Funct. Mater.* **17** (2007), pp. 1828–1836.
120. T. J. B. H. Benaddi, J. Jagiello, J. A. Schwarz, J. N. Rouzaud, D. Legras, F. Beguin, Surface functionality and porosity of activated carbons obtained from chemical activation of wood, *Carbon* **38** (2000), pp. 669–674.
121. M. Nakamura, M. Nakanishi and K. Yamamoto, Influence of physical properties of activated carbons on characteristics of electric double-layer capacitors, *J. Power Sources* **60** (1996), pp. 225–231.
122. W. Gu, M. Sevilla, A. Magasinski, A. B. Fuertes and G. Yushin, Sulfur-containing activated carbons with greatly reduced content of bottle neck pores for double-layer capacitors: a case study for pseudocapacitance detection, *Energy Environ. Sci.* **6** (2013), pp. 2465–2476.
123. B. O. Park, C. D. Lokhande, H. S. Park, K. D. Jung and O. S. Joo, Performance of supercapacitor with electrodeposited ruthenium oxide film electrodes—effect of film thickness, *J. Power Sources* **134** (2004), pp. 148–152.
124. N. Soin, S. S. Roy, S. K. Mitra, T. Thundat and J. A. McLaughlin, Nanocrystalline ruthenium oxide dispersed few layered graphene (FLG) nanoflakes as supercapacitor electrodes, *J. Mater. Chem.* **22** (2012), pp. 14944–14950.

125. V. D. Patake, S. M. Pawar, V. R. Shinde, T. P. Gujar and C. D. Lokhande, The growth mechanism and supercapacitor study of anodically deposited amorphous ruthenium oxide films, *Curr. Appl. Phys.* **10** (2010), pp. 99–103.
126. I. H. Kim, J. H. Kim, Y. H. Lee and K. B. Kim, Synthesis and characterization of electrochemically prepared ruthenium oxide on carbon nanotube film substrate for supercapacitor applications, *J. Electrochem. Soc.* **152** (2005), pp. A2170–A2178.
127. H.-T. Fang, M. Liu, D.-W. Wang, X.-H. Ren and X. Sun, Outstanding performance of activated graphene based supercapacitors in ionic liquid electrolyte from -50 to 80°C, *Nano Energy* **2** (2013), pp. 403–411.
128. C. C. Hu, Y. H. Huang and K. H. Chang, Annealing effects on the physicochemical characteristics of hydrous ruthenium and ruthenium-iridium oxides for electrochemical supercapacitors, *J. Power Sources* **108** (2002), pp. 117–127.
129. S. L. Kuo and N. L. Wu, Composite supercapacitor containing tin oxide and electroplated ruthenium oxide, *Electrochem. Solid-State Lett.* **6** (2003), pp. A85–A87.
130. J. P. Zheng and T. R. Jow, High energy and high power density electrochemical capacitors, *J. Power Sources* **62** (1996), pp. 155–159.
131. X. Wu, Y. Zeng, H. Gao, J. Su, J. Liu and Z. Zhu, Template synthesis of hollow fusiform $RuO_2{\cdot}xH_2O$ nanostructure and its supercapacitor performance, *J. Mater. Chem. A* **1** (2013) p. 469.
132. K. M. Kim, J. H. Nam, Y.-G. Lee, W. I. Cho and J. M. Ko, Supercapacitive properties of electrodeposited RuO_2 electrode in acrylic gel polymer electrolytes, *Curr. Appl. Phys.* **13** (2013), pp. 1702–1706.
133. M. Toupin, T. Brousse and D. Belanger, Charge storage mechanism of MnO_2 electrode used in aqueous electrochemical capacitor, *Chem. Mater.* **16** (2004), pp. 3184–3190.
134. S. C. Pang, M. A. Anderson and T. W. Chapman, Novel electrode materials for thin-film ultracapacitors: comparison of electrochemical properties of sol-gel-derived and electrodeposited manganese dioxide, *J. Electrochem. Soc.* **147** (2000), pp. 444–450.
135. P. Yu, X. Zhang, D. Wang, L. Wang and Y. Ma, Shape-controlled synthesis of 3D hierarchical MnO_2 nanostructures for electrochemical supercapacitors, *Crystal Growth and Design* **9** (2008), pp. 528–533.
136. S. Chen, J. Zhu, X. Wu, Q. Han and X. Wang, Graphene oxide-MnO_2 nanocomposites for supercapacitors, *ACS Nano* **4** (2010), pp. 2822–2830.
137. C. Ye, Z. M. Lin and S. Z. Hui, Electrochemical and capacitance properties of rod-shaped MnO_2 for supercapacitor, *J. Electrochem. Soc.* **152** (2005), pp. A1272–A1278.
138. M.-W. Xu, D.-D. Zhao, S.-J. Bao and H.-L. Li, Mesoporous amorphous MnO_2 as electrode material for supercapacitor, *J. Solid State Electrochem.* **11** (2007), pp. 1101–1107.
139. S. F. Chin, S. C. Pang and M. A. Anderson, Material and electrochemical characterization of tetrapropylammonium manganese oxide thin films

as novel electrode materials for electrochemical capacitors, *J. Electrochem. Soc.* **149** (2002), pp. A379–A384.

140. D. L. Fang, Z. D. Chen, B. C. Wu, Y. Yan and C. H. Zheng, Preparation and electrochemical properties of ultra-fine Mn-Ni-Cu oxides for supercapacitors, *Mater. Chem. Phys.* **128** (2011), pp. 311–316.
141. D. W. Wang, Q. H. Wang and T. M. Wang, Morphology-controllable synthesis of cobalt oxalates and their conversion to mesoporous Co_3O_4 nanostructures for application in supercapacitors, *Inorg. Chem.* **50** (2011), pp. 6482–6492.
142. S. K. Meher and G. R. Rao, Ultralayered Co_3O_4 for high-performance supercapacitor applications, *J. Phys. Chem. C* **115** (2011), pp. 15646–15654.
143. X.-h. Xia, J.-p. Tu, Y.-j. Mai, X.-l. Wang, C.-d. Gu and X.-b. Zhao, Three-dimentional porous nano-Ni/$Co(OH)_2$ nanoflake composite film: a pseudocapacitive material with superior performance, *J. Mater. Chem.* **21** (2011), pp. 9319–9325.
144. K. SengáTan and C. MingáLi, Fabrication of Co_3O_4-reduced graphene oxide scrolls for high-performance supercapacitor electrodes, *Phys. Chem. Chem. Phys.* **13** (2011), pp. 14462–14465.
145. T. Brezesinski, JohnWang, S. H. Tolbert and B. Dunn, Ordered mesoporous α-MoO_3 with iso-oriented nanocrystalline walls for thin-film pseudocapacitors, *Nat. Mater.* **9** (2010), pp. 146–151.
146. K. W. Nam, W. S. Yoon and K. B. Kim, X-ray absorption spectroscopy studies of nickel oxide thin film electrodes for supercapacitors, *Electrochim. Acta* **47** (2002), pp. 3201–3209.
147. E. E. Kalu, T. T. Nwoga, V. Srinivasan and J. W. Weidner, Cyclic voltammetric studies of the effects of time and temperature on the capacitance of electrochemically deposited nickel hydroxide, *J. Power Sources* **92** (2001), pp. 163–167.
148. V. Srinivasan and J. W. Weidner, Studies on the capacitance of nickel oxide films: effect of heating temperature and electrolyte concentration, *J. Electrochem. Soc.* **147** (2000), pp. 880–885.
149. K. C. Liu and M. A. Anderson, Porous nickel oxide/nickel films for electrochemical capacitors, *J. Electrochem. Soc.* **143** (1996), pp. 124–130.
150. Z. Yang, F. Xu, W. Zhang, Z. Mei, B. Pei and X. Zhu, Controllable preparation of multishelled NiO hollow nanospheres via layer-by-layer self-assembly for supercapacitor application, *J. Power Sources* **246** (2014), pp. 24–31.
151. K. K. Purushothaman, I. Manohara Babu, B. Sethuraman and G. Muralidharan, Nanosheet-assembled NiO Microstructures for High-Performance Supercapacitors, *ACS Appl. Mater. Interfaces* **5** (2013), pp. 10767–10773.
152. H. Kuan-Xin, W. Quan-Fu, Z. Xiao-Gang and W. Xin-Lei, Electrodeposition of nickel and cobalt mixed oxide/carbon nanotube thin films and their charge storage properties, *J. Electrochem. Soc.* **153** (2006), pp. A1568–A1574.
153. G. Wang, L. Zhang, J. Kim and J. Zhang, Nickel and cobalt oxide composite as a possible electrode material for electrochemical supercapacitors, *J. Power Sources* **217** (2012), pp. 554–561.

154. K. R. Prasad and N. Miura, Electrochemical synthesis and characterization of nanostructured tin oxide for electrochemical redox supercapacitors, *Electrochem. Commun.* **6** (2004), pp. 849–852.
155. S. Pusawale, P. Deshmukh and C. Lokhande, Chemical synthesis of nanocrystalline SnO_2 thin films for supercapacitor application, *Appl. Surf. Sci.* **257** (2011), pp. 9498–9502.
156. M. Q. Wu, L. P. Zhang, D. M. Wang, C. Xiao and S. R. Zhang, Cathodic deposition and characterization of tin oxide coatings on graphite for electrochemical supercapacitors, *J. Power Sources* **175** (2008), pp. 669–674.
157. Z. Li, T. Chang, G. Yun, J. Guo and B. Yang, 2D tin dioxide nanoplatelets decorated graphene with enhanced performance supercapacitor, *J Alloy. Compd.* **586** (2014), pp. 353–359.
158. K. R. Prasad, K. Koga and N. Miura, Electrochemical deposition of nanostructured indium oxide:? high-performance electrode material for redox supercapacitors, *Chem. Mater.* **16** (2004), pp. 1845–1847.
159. J. Chang, W. Lee, R. S. Mane, B. W. Cho and S.-H. Han, morphology-dependent electrochemical supercapacitor properties of indium oxide, *Electrochem. Solid-State Lett.* **11** (2008), pp. A9–A11.
160. T. P. Gujar, V. R. Shinde, C. D. Lokhande and S. H. Han, Electrosynthesis of Bi_2O_3 thin films and their use in electrochemical supercapacitors, *J. Power Sources* **161** (2006), pp. 1479–1485.
161. F.-L. Zheng, G.-R. Li, Y.-N. Ou, Z.-L. Wang, C.-Y. Su and Y.-X. Tong, Synthesis of hierarchical rippled Bi_2O_3 nanobelts for supercapacitor applications, *Chem. Commun.* **46** (2010), pp. 5021–5023.
162. H. Y. Lee and J. B. Goodenough, Ideal supercapacitor behavior of amorphous $V_2O_5{\cdot}nH_2O$ in potassium chloride (KCl) aqueous solution, *J. Solid State Chem.* **148** (1999), pp. 81–84.
163. I. H. Kim, J. H. Kim, B. W. Cho, Y. H. Lee and K. B. Kim, Synthesis and electrochemical characterization of vanadium oxide on carbon nanotube film substrate for pseudocapacitor applications, *J. Electrochem. Soc.* **153** (2006), pp. A989–A996.
164. W. Xunhong and W. Kuaishe, Microstructure and properties of friction stir butt-welded AZ31 magnesium alloy, *Mater. Sci. Eng. A* **431** (2006), pp. 114–117.
165. C. D. Lokhande, T. P. Gujar, V. R. Shinde, R. S. Mane and S. H. Han, Electrochemical supercapacitor application of pervoskite thin films, *Electrochem. Commun.* **9** (2007), pp. 1805–1809.
166. Y. Takasu, S. Mizutani, M. Kumagai, S. Sawaguchi and Y. Murakami, Ti-V-W-O/Ti oxide electrodes as candidates for electrochemical capacitors, *Electrochem. Solid-State Lett.* **2** (1999), pp. 1–2.
167. A. R. de Souza, E. Arashiro, H. Golveia and T. A. F. Lassali, Pseudocapacitive behavior of $Ti/RhO_x + Co_3O_4$ electrodes in acidic medium: application to supercapacitor development, *Electrochim. Acta* **49** (2004), pp. 2015–2023.
168. J. B. Wu, R. Q. Guo, X. H. Huang and Y. Lin, Construction of self-supported porous TiO_2/NiO core/shell nanorod arrays for electrochemical capacitor application, *J. Power Sources* **243** (2013), pp. 317–322.

169. V. Gupta, S. Gupta and N. Miura, Potentiostatically deposited nanostructured Co_xNi_{1-x} layered double hydroxides as electrode materials for redox-supercapacitors, *J. Power Sources* **175** (2008), pp. 680–685.
170. J. Liu, J. Jiang, C. Cheng, H. Li, J. Zhang, H. Gong and H. J. Fan, Co_3O_4 nanowire@ MnO_2 ultrathin nanosheet core/shell arrays: a new class of high-performance pseudocapacitive materials, *Adv. Mater.* **23** (2011), pp. 2076–2081.
171. J. W. Long, A. L. Young and D. R. Rolison, Spectroelectrochemical characterization of nanostructured, mesoporous manganese oxide in aqueous electrolytes, *J. Electrochem. Soc.* **150** (2003), p. A1161.
172. J. Yan, Z. Fan, T. Wei, J. Cheng, B. Shao, K. Wang, L. Song and M. Zhang, Carbon nanotube/MnO_2 composites synthesized by microwave-assisted method for supercapacitors with high power and energy densities, *J. Power Sources* **194** (2009), pp. 1202–1207.
173. J. Hu, A.-B. Yuan, Y.-Q. Wang and X.-L. Wang, Improved cyclability of nano-MnO_2/CNT composite supercapacitor electrode derived from room-temperature solid reaction, *Acta Phys-Chim. Sin.* **25** (2009), pp. 987–993.
174. W. Sugimoto, H. Iwata, Y. Yasunaga, Y. Murakami and Y. Takasu, Preparation of ruthenic acid nanosheets and utilization of its interlayer surface for electrochemical energy storage, *Angew. Chem. Int. Ed.* **42** (2003), pp. 4092–4096.
175. T. Brezesinski, J. Wang, S. H. Tolbert and B. Dunn, Next generation pseudocapacitor materials from sol-gel derived transition metal oxides, *J. Sol-Gel Sci. Technol.* **57** (2010), pp. 330–335.
176. D. Choi, G. E. Blomgren and P. N. Kumta, Fast and reversible surface redox reaction in nanocrystalline vanadium nitride supercapacitors, *Adv. Mater.* **18** (2006), pp. 1178–1182.
177. B. E. Conway, Transition from "supercapacitor" to "battery" behavior in electrochemical energy storage, *J. Electrochem. Soc.* **138** (1991), pp. 1539–1548.
178. K. Lota, V. Khomenko and E. Frackowiak, Capacitance properties of poly (3, 4-ethylenedioxythiophene)/carbon nanotubes composites, *J. Phys. Chem. Solids* **65** (2004), pp. 295–301.
179. K. S. Ryu, K. M. Kim, Y. J. Park, N.-G. Park, M. G. Kang and S. H. Chang, Redox supercapacitor using polyaniline doped with Li salt as electrode, *Solid State Ion.* **152** (2002), pp. 861–866.
180. F. Faverolle, A. Attias, B. Bloch, P. Audebert and C. Andrieux, Highly conducting and strongly adhering polypyrrole coating layers deposited on glass substrates by a chemical process, *Chem. Mater.* **10** (1998), pp. 740–752.
181. K. S. Ryu, K. M. Kim, N. G. Park, Y. J. Park and S. H. Chang, Symmetric redox supercapacitor with conducting polyaniline electrodes, *J. Power Sources* **103** (2002), pp. 305–309.
182. F. Fusalba, P. Gouerec, D. Villers and D. Belanger, Electrochemical characterization of polyaniline in nonaqueous electrolyte and its evaluation as electrode material for electrochemical supercapacitors, *J. Electrochem. Soc.*, **148** (2001), pp. A1–A6.

183. X. Jiang, S. Setodoi, S. Fukumoto, I. Imae, K. Komaguchi, J. Yano, H. Mizota and Y. Harima, An easy one-step electrosynthesis of graphene/polyaniline composites and electrochemical capacitor, *Carbon* **67** (2014), pp. 662–672.
184. K. Zhang, L. L. Zhang, X. S. Zhao and J. Wu, Graphene/polyaniline nanofiber composites as supercapacitor electrodes, *Chem. Mater.* **22** (2010), pp. 1392–1401.
185. J. Benson, I. Kovalenko, S. Boukhalfa, D. Lashmore, M. Sanghadasa and G. Yushin, Multifunctional CNT-polymer composites for ultra-tough structural supercapacitors and desalination devices, *Adv. Mater.* **25** (2013), pp. 6625–6632.
186. I. Kovalenko, D. G. Bucknall and G. Yushin, Detonation nanodiamond and onion-like-carbon-embedded polyaniline for supercapacitors, *Adv. Funct. Mater.* **20** (2010), pp. 3979–3986.
187. C. Shi and I. Zhitomirsky, Electrodeposition and capacitive behavior of films for electrodes of electrochemical supercapacitors, *Nanoscale Res. Lett.* **5** (2010), pp. 518–523.
188. J. Wang, Y. L. Xu, F. Yan, J. B. Zhu and J. P. Wang, Template-free prepared micro/nanostructured polypyrrole with ultrafast charging/discharging rate and long cycle life, *J. Power Sources* **196** (2011), pp. 2373–2379.
189. L. Z. Fan and J. Maier, High-performance polypyrrole electrode materials for redox supercapacitors, *Electrochem. Commun.* **8** (2006), pp. 937–940.
190. S. Tripathi, A. Kumar and S. Hashmi, Electrochemical redox supercapacitors using PVdF-HFP based gel electrolytes and polypyrrole as conducting polymer electrode, *Solid State Ion.* **177** (2006), pp. 2979–2985.
191. B. C. Kim, C. O. Too, J. S. Kwon, J. M. Bo and G. G. Wallace, A flexible capacitor based on conducting polymer electrodes, *Synth. Met.* **161** (2011), pp. 1130–1132.
192. J. Wang, Y. Xu, X. Chen and X. Sun, Electromagnetic interference shielding effect of nanocomposites with carbon nanotube and shape memory polymer, *Compos. Sci. Technol.* **67** (2007), pp. 2981–2985.
193. V. Khomenko, E. Frackowiak and F. Beguin, Determination of the specific capacitance of conducting polymer/nanotubes composite electrodes using different cell configurations, *Electrochim. Acta* **50** (2005), pp. 2499–2506.
194. M. Mallouki, F. Tran-Van, C. Sarrazin, P. Simon, B. Daffos, A. De, C. Chevrot and J.-F. Fauvarque, Polypyrrole-Fe_2O_3 nanohybrid materials for electrochemical storage, *J. Solid State Electrochem.* **11** (2007), pp. 398–406.
195. F. Fusalba, H. A. Ho, L. Breau and D. Belanger, Poly(cyano-substituted diheteroareneethylene) as active electrode material for electrochemical supercapacitors, *Chem. Mater.* **12** (2000), pp. 2581–2589.
196. S. R. P. Gnanakan, N. Murugananthem and A. Subramania, Organic acid doped polythiophene nanoparticles as electrode material for redox supercapacitors, *Polymer. Adv. Tech.* **22** (2011), pp. 788–793.
197. J. H. Huang and C. W. Chu, Achieving efficient poly(3,4-ethylene dioxythiophene)-based supercapacitors by controlling the polymerization kinetics, *Electrochim. Acta* **56** (2011), pp. 7228–7234.

198. J. Wang, Y. L. Xu, X. Chen and X. F. Du, Electrochemical supercapacitor electrode material based on poly(3,4-ethylenedioxythiophene)/polypyrrole composite, *J. Power Sources* **63** (2007), pp. 1120–1125.
199. A. M. P. Hussain, A. Kumar, F. Singh and D. K. Avasthi, Effects of 160MeV Ni^{12+} ion irradiation on HCl doped polyaniline electrode, *J. Phys. D: Appl. Phys.* **39** (2006), pp. 750–755.
200. L. L. Tu and C. Y. Jia, Conducting polymers as electrode materials for supercapacitors, *Prog. Chem.* **22** (2010), pp. 1610–1618.
201. D. Aradilla, F. Estrany and C. Aleman, Symmetric supercapacitors based on multilayers of conducting polymers, *J. Phys. Chem. C* **115** (2011), pp. 8430–8438.
202. Jaidev, R. I. Jafri, A. K. Mishra and S. Ramaprabhu, Polyaniline-MnO_2 nanotube hybrid nanocomposite as supercapacitor electrode material in acidic electrolyte, *J. Mater. Chem.* **21** (2011), p. 17601.
203. L. Chen, L.-J. Sun, F. Luan, Y. Liang, Y. Li and X.-X. Liu, Synthesis and pseudocapacitive studies of composite films of polyaniline and manganese oxide nanoparticles, *J. Power Sources* **195** (2010), pp. 3742–3747.
204. F.-J. Liu, Electrodeposition of manganese dioxide in three-dimensional poly(3,4-ethylenedioxythiophene)-poly(styrene sulfonic acid)-polyaniline for supercapacitor, *J. Power Sources* **182** (2008), pp. 383–388.
205. X. Zhang, L. Ji, S. Zhang and W. Yang, Synthesis of a novel polyaniline-intercalated layered manganese oxide nanocomposite as electrode material for electrochemical capacitor, *J. Power Sources* **173** (2007), pp. 1017–1023.
206. V. Gupta and N. Miura, Polyaniline/single-wall carbon nanotube (PANI/SWCNT) composites for high performance supercapacitors, *Electrochim. Acta* **52** (2006), pp. 1721–1726.
207. I. Kovalenko, D. Bucknall and G. Yushin, Detonation nanodiamond and onion-like-carbon-embedded polyaniline for supercapacitors, *Adv. Funct. Mater.* **20** (2010), p. 3979.
208. G.-m. Zhou, D.-W. Wang, F. Li, L.-l. Zhang, Z. Weng and H.-M. Cheng, The effect of carbon particle morphology on the electrochemical properties of nanocarbon/polyaniline composites in supercapacitors, *New Carbon Mater.* **26** (2011), pp. 180–186.
209. L. Zheng, X. Wang, H. An, X. Wang, L. Yi and L. Bai, The preparation and performance of flocculent polyaniline/carbon nanotubes composite electrode material for supercapacitors, *J. Solid State Electrochem.* **15** (2010), pp. 675–681.
210. W.-x. Liu, N. Liu, H.-h. Song and X.-h. Chen, Properties of polyaniline/ordered mesoporous carbon composites as electrodes for supercapacitors, *New Carbon Mater.* **26** (2011), pp. 217–223.
211. S. Konwer, Studies on conducting polypyrrole/graphene oxide composites as supercapacitor electrode, *J. Electronic Mater.* **40** (2011), p. 2248.
212. J. Ge, G. Cheng and L. Chen, Transparent and flexible electrodes and supercapacitors using polyaniline/single-walled carbon nanotube composite thin films, *Nanoscale* **3** (2011), pp. 3084–3088.

213. C. Yuan, L. Shen, F. Zhang, X. Lu and X. Zhang, Reactive template fabrication of uniform core-shell polyaniline/multiwalled carbon nanotube nanocomposite and its electrochemical capacitance, *Chem. Lett.* **39** (2010), pp. 850–851.
214. M. Yang, B. Cheng, H. Song and X. Chen, Preparation and electrochemical performance of polyaniline-based carbon nanotubes as electrode material for supercapacitor, *Electrochim. Acta* **55** (2010), pp. 7021–7027.
215. H. Gómez, M. K. Ram, F. Alvi, P. Villalba, E. Stefanakos and A. Kumar, Graphene-conducting polymer nanocomposite as novel electrode for supercapacitors, *J. Power Sources* **196** (2011), pp. 4102–4108.
216. J. Yan, T. Wei, B. Shao, Z. Fan, W. Qian, M. Zhang and F. Wei, Electrochemical properties of graphene nanosheet/carbon black composites as electrodes for supercapacitors, *Carbon* **48** (2010), pp. 487–493.
217. T. Brousse, M. Toupin and D. Belanger, A hybrid activated carbon-manganese dioxide capacitor using a mild aqueous electrolyte, *J. Electrochem. Soc.* **151** (2004), pp. A614–A622.
218. M. S. Hong, S. H. Lee and S. W. Kim, Use of KCl aqueous electrolyte for 2 V manganese oxide/activated carbon hybrid capacitor, *Electrochem. Solid-State Lett.* **5** (2002), pp. A227–A230.
219. T. Brousse, P. L. Taberna, O. Crosnier, R. Dugas, P. Guillemet, Y. Scudeller, Y. Zhou, F. Favier, D. Belanger and P. Simon, Long-term cycling behavior of asymmetric activated carbon/MnO 2 aqueous electrochemical supercapacitor, *J. Power Sources* **173** (2007), pp. 633–641.
220. T. Cottineau, M. Toupin, T. Delahaye, T. Brousse and D. Belanger, Nanostructured transition metal oxides for aqueous hybrid electrochemical supercapacitors, *Appl. Phys. A* **82** (2006), pp. 599–606.
221. Q. Qu, P. Zhang, B. Wang, Y. Chen, S. Tian, Y. Wu and R. Holze, Electrochemical performance of MnO_2 nanorods in neutral aqueous electrolytes as a cathode for asymmetric supercapacitors, *J. Phys. Chem. C* **113** (2009), pp. 14020–14027.
222. L. M. Chen, Q. Y. Lai, Y. J. Hao, Y. Zhao and X. Y. Ji, Investigations on capacitive properties of the AC/V_2O_5 hybrid supercapacitor in various aqueous electrolytes, *J Alloy. Compd.* **467** (2009), pp. 465–471.
223. Q. T. Qu, Y. Sh, L. L. Li, W. L. Guo, Y. P. Wu, H. P. Zhang, S. Y. Guan and R. Holze, $V_2O_5 \cdot q0.6H_2O$ nanoribbons as cathode material for asymmetric supercapacitor in K_2SO_4 solution, *Electrochem. Commun.* **11** (2009), pp. 1325–1328.
224. X. Du, C. Wang, M. Chen, Y. Jiao and J. Wang, Electrochemical performances of nanoparticle Fe_3O_4/activated carbon supercapacitor using KOH electrolyte solution, *J. Phys. Chem. C* **113** (2009), pp. 2643–2646.
225. N. Yu, L. Gao, S. Zhao and Z. Wang, Electrodeposited PbO_2 thin film as positive electrode in PbO_2/AC hybrid capacitor, *Electrochim. Acta* **54** (2009), pp. 3835–3841.
226. Z. Algharaibeh, X. Liu and P. G. Pickup, An asymmetric anthraquinone-modified carbon/ruthenium oxide supercapacitor, *J. Power Sources* **187** (2009), pp. 640–643.

227. X. F. Wang, Z. You and D. B. Ruan, A hybrid metal oxide supercapacitor in aqueous KOH electrolyte, *Chin. J. Chem.* **24** (2006), pp. 1126–1132.
228. Y.-G. Wang, L. Cheng and Y.-Y. Xia, Electrochemical profile of nano-particle CoAl double hydroxide/active carbon supercapacitor using KOH electrolyte solution, *J. Power Sources* **153** (2006), pp. 191–196.
229. D. W. Wang, F. Li and H. M. Cheng, Hierarchical porous nickel oxide and carbon as electrode materials for asymmetric supercapacitor, *J. Power Sources* **85** (2008), pp. 1563–1568.
230. G.-h. Yuan, Z.-h. Jiang, A. Aramata and Y.-Z. Gao, Electrochemical behavior of activated-carbon capacitor material loaded with nickel oxide, *Carbon* **43** (2005), pp. 2913–2917.
231. C. Yuan, X. Zhang, Q. Wu and B. Gao, Effect of temperature on the hybrid supercapacitor based on NiO and activated carbon with alkaline polymer gel electrolyte, *Solid State Ion.* **177** (2006), pp. 1237–1242.
232. S. Nohara, T. Asahina, H. Wada, N. Furukawa, H. Inoue, N. Sugoh, H. Iwasaki and C. Iwakura, Hybrid capacitor with activated carbon electrode, $Ni(OH)_2$ electrode and polymer hydrogel electrolyte, *J. Power Sources* **157** (2006), pp. 605–609.
233. J. H. Park, O. O. Park, K. H. Shin, C. S. Jin and J. H. Kim, An electrochemical capacitor based on a $Ni(OH)_2$/activated carbon composite electrode, *Electrochem. Solid State Lett.* **5** (2002), pp. H7–H10.
234. Y. G. Wang, L. Yu and Y. Y. Xia, Hybrid aqueous energy storage cells using activated carbon and lithium-intercalated compounds, *J. Electrochem. Soc.* **153** (2006), pp. A743–A748.
235. H.-q. Li, Y. Zou and Y.-y. Xia, A study of nitroxide polyradical/activated carbon composite as the positive electrode material for electrochemical hybrid capacitor, *Electrochim. Acta* **52** (2007), pp. 2153–2157.
236. J. H. Park and O. O. Park, Hybrid electrochemical capacitors based on polyaniline and activated carbon electrodes, *J. Power Sources* **111** (2002), pp. 185–190.
237. Z. Chen, Y. C. Qin, D. Weng, Q. F. Xiao, Y. T. Peng, X. L. Wang, H. X. Li, F. Wei and Y. F. Lu, Design and Synthesis of Hierarchical Nanowire Composites for Electrochemical Energy Storage, *Adv. Funct. Mater.* **19** (2009), pp. 3420–3426.
238. W.-H. Jin, G.-T. Cao and J.-Y. Sun, Hybrid supercapacitor based on MnO_2 and columned FeOOH using Li_2SO_4 electrolyte solution, *J. Power Sources* **175** (2008), pp. 686–691.
239. J. Chang, M. Jin, F. Yao, T. H. Kim, V. T. Le, H. Yue, F. Gunes, B. Li, A. Ghosh and S. Xie, Asymmetric supercapacitors based on graphene/MnO_2 nanospheres and graphene/MoO_3 nanosheets with high energy density, *Adv. Funct. Mater.* **23** (2013), pp. 5074–5083.
240. K. C. Ng, S. Zhang, C. Peng and G. Z. Chen, Individual and bipolarly stacked asymmetrical aqueous supercapacitors of CNTs/SnO_2 and CNTs/MnO_2 nanocomposites, *J. Electrochem. Soc.* **156** (2009), pp. A846–A853.

241. G.-X. Wang, B.-L. Zhang, Z.-L. Yu and M.-Z. Qu, Manganese oxide/MWNTs composite electrodes for supercapacitors, *Solid State Ion.* **176** (2005), pp. 1169–1174.
242. P. C. Chen, G. Z. Shen, Y. Shi, H. T. Chen and C. W. Zhou, Preparation and characterization of flexible asymmetric supercapacitors based on transition-metal-oxide nanowire/single-walled carbon nanotube hybrid thin-film electrodes, *ACS Nano* **4** (2010), pp. 4403–4411.
243. H. Inoue, T. Morimoto and S. Nohara, Electrochemical characterization of a hybrid capacitor with Zn and activated carbon electrodes, *Electrochem. Solid-State Lett.* **10** (2007), pp. A261–A263.
244. V. Khomenko, E. Raymundo-Piñero and F. Béguin, High-energy density graphite/AC capacitor in organic electrolyte, *J. Power Sources* **177** (2008), pp. 643–651.
245. C. Decaux, G. Lota, E. Raymundo-Pinero, E. Frackowiak and F. Béguin, Electrochemical performance of a hybrid lithium-ion capacitor with a graphite anode preloaded from lithium bis(trifluoromethane)sulfonimide-based electrolyte, *Electrochim. Acta* **86** (2012), pp. 282–286.
246. T. Brousse, R. Marchand, P. L. Taberna and P. Simon, TiO_2(B)/activated carbon non-aqueous hybrid system for energy storage, *J. Power Sources* **158** (2006), pp. 571–577.
247. Q. Wang, Z. H. Wen and J. H. Li, A hybrid supercapacitor fabricated with a carbon nanotube cathode and a TiO_2-B nanowire anode, *Adv. Funct. Mater.* **16** (2006), pp. 2141–2146.
248. X. Zhao, C. Johnston and P. S. Grant, A novel hybrid supercapacitor with a carbon nanotube cathode and an iron oxide/carbon nanotube composite anode, *J. Mater. Chem.* **19** (2009), pp. 8755–8760.
249. Q. Cheng, J. Tang, J. Ma, H. Zhang, N. Shinya and L. C. Qin, Graphene and nanostructured MnO_2 composite electrodes for supercapacitors, *Carbon* **49** (2011), pp. 2917–2925.
250. Z. J. Fan, J. Yan, T. Wei, L. J. Zhi, G. Q. Ning, T. Y. Li and F. Wei, Asymmetric supercapacitors based on graphene/MnO_2 and activated carbon nanofiber electrodes with high power and energy density, *Adv. Funct. Mater.* **21** (2011), pp. 2366–2375.
251. G. G. Amatucci, F. Badway, A. Du Pasquier and T. Zheng, An asymmetric hybrid nonaqueous energy storage cell, *J. Electrochem. Soc.* **148** (2001), pp. A930–A939.
252. K. Naoi, S. Ishimoto, Y. Isobe and S. Aoyagi, High-rate nano-crystalline $Li_4Ti_5O_{12}$ attached on carbon nano-fibers for hybrid supercapacitors, *J. Power Sources* **195** (2010), pp. 6250–6254.
253. H. Q. Li, L. Cheng and Y. Y. Xia, A hybrid electrochemical supercapacitor based on a 5 V Li-ion battery cathode and active carbon, *Electrochem. Solid-State Lett.* **8** (2005), pp. A433–A436.
254. S.-B. Ma, K.-W. Nam, W.-S. Yoon, X.-Q. Yang, K.-Y. Ahn, K.-H. Oh and K.-B. Kim, A novel concept of hybrid capacitor based on manganese oxide materials, *Electrochem. Commun.* **9** (2007), pp. 2807–2811.

255. R. Vasanthi, D. Kalpana and N. G. Renganathan, Olivine-type nanoparticle for hybrid supercapacitors, *J. Solid State Electrochem.* **12** (2008), pp. 961–969.
256. X. Hu, Z. Deng, J. Suo and Z. Pan, A high rate, high capacity and long life ($LiMn_2O_4$ + AC)/$Li_4Ti_5O_{12}$ hybrid battery–supercapacitor, *J. Power Sources* **187** (2009), pp. 635–639.
257. M. Al Sakka, H. Gualous, J. Van Mierlo and H. Culcu, Thermal modeling and heat management of supercapacitor modules for vehicle applications, *J. Power Sources* **194** (2009), pp. 581–587.
258. M. Conte, Supercapacitors technical requirements for new applications, *Fuel Cells* **10** (2010), pp. 806–818.
259. R. Kotz and M. Carlen, Principles and applications of electrochemical capacitors, *Electrochim. Acta* **45** (2000), pp. 2483–2498.
260. A. Burke, Ultracapacitor technologies and application in hybrid and electric vehicles, *Int. J. Energ. Res.* **34** (2010), pp. 133–151.
261. A. Hammar, P. Venet, R. Lallemand, G. Coquery and G. Rojat, Study of accelerated aging of supercapacitors for transport applications, *IEEE T. Ind. Electron.* **57** (2010), pp. 3972–3979.
262. B. Ricketts and C. Ton-That, Self-discharge of carbon-based supercapacitors with organic electrolytes, *J. Power Sources* **89** (2000), pp. 64–69.
263. F. Rafik, H. Gualous, R. Gallay, A. Crausaz and A. Berthon, Frequency, thermal and voltage supercapacitor characterization and modeling, *J. Power Sources* **165** (2007), pp. 928–934.
264. T. Liu, W. Pell and B. Conway, Self-discharge and potential recovery phenomena at thermally and electrochemically prepared RuO_2 supercapacitor electrodes, *Electrochim. Acta* **42** (1997), pp. 3541–3552.
265. M. Kaus, J. Kowal and D. U. Sauer, Modelling the effects of charge redistribution during self-discharge of supercapacitors, *Electrochim. Acta* **55** (2010), pp. 7516–7523.

Chapter 6

Fuel Cells and the Hydrogen Economy

John T.S. Irvine, Gael P.G. Corre and Xiaoxiang Xu

School of Chemistry, University of St. Andrews, Fife, UK KY16 9ST

The present chapter summarizes the current state and perspectives of fuel cells and hydrogen energy. First of all the relevant fuels and fuel chemistry are discussed focusing on hydrogen, hydrocarbons and oxygenates, then the basics of fuel cell technology and applications are discussed. The different types of fuel cells are presented with particular attention on solid oxide fuel cells, and thermodynamics and the factors influencing efficiency reviewed.

1 Introduction

The fuel cell concept, ascribed to Sir Humphrey Davy, dates from the beginning of the 19th century. The first hydrogen–oxygen cell was successfully operated by Sir William Grove in 1839[1] and is generally referred to as the first fuel cell. While investigating the electrolysis of water, Grove observed that when the current was switched off, a small current flowed through the circuit in the opposite direction, as a result of a reaction between the electrolysis products, hydrogen and oxygen, catalysed by the platinum electrodes. Grove recognized the possibility of combining several of these in series to form a gaseous voltaic battery,[2] and also made the crucially important observation that there must be a "notable surface of action" between the gas, the electrolyte, and the electrode phases in a cell. Maximizing the area of contact between these three phases remains at the forefront of fuel cell research and development. Some 50 years after Grove's "gas battery", Mond and Langer introduced the term fuel cell[3] to describe their device which had a porous platinum black electrode structure, and used a diaphragm made of a porous non-conducting substance to hold the electrolyte.

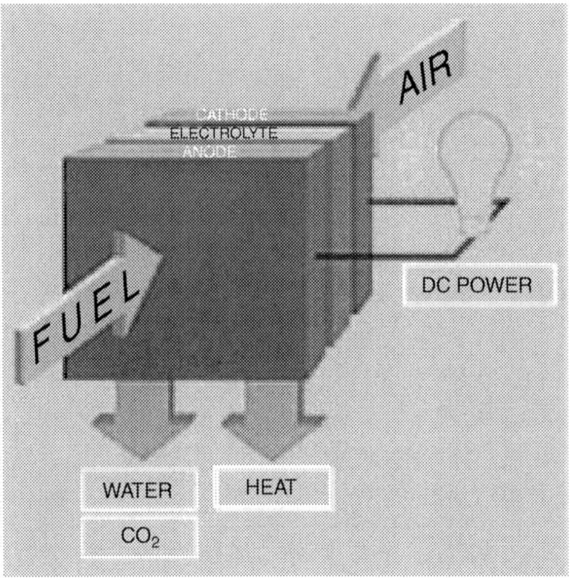

Fig. 1. Principle of an individual fuel cell.[4]

Despite the fact that the fuel cell was discovered over 160 years ago, with the high efficiencies and environmental advantages that it offered, it is only now that fuel cells are approaching commercial reality.

A fuel cell is essentially an energy conversion device that produces electricity, and may be viewed as a battery with external fuel supply. A fuel cell consists of four essential components: two electrodes, the *anode* and the *cathode*, separated by an *electrolyte*, and connected by an external circuit or *interconnect*, as shown in Fig. 1. These may be further subdivided in relation to function. Fuel is fed to the anode where it is oxidized, releasing electrons to the external circuit. Oxidant is fed to the cathode where it is reduced using the electrons delivered by the external circuit. The electrons flow through the interconnect from the anode to the cathode, producing direct current (DC) electricity. In theory, any gas capable of electrochemical oxidation and reduction can be used as fuel or oxidant in a fuel cell. Air is the most common oxidant for fuel cells since it is readily and economically available from the atmosphere. Hydrogen, which offers high electrochemical reactivity, is the most obvious fuel. However, fuel cells can be developed to work with alternative fuels to hydrogen.

The essential feature of a fuel cell is its high energy conversion efficiency. Because fuel cells convert the chemical energy of the fuel directly to electrical energy without the intermediate of thermal energy (unlike indirect conversion in conventional systems), their conversion efficiency is not subject to the Carnot limitation. Efficiency can be further improved to levels as high as 80% when the produced heat is used in combined heat and power, or gas turbine applications. Besides their high energy conversion efficiency, fuel cells offer several other advantages over conventional methods of power generation. They offer a much lower production of pollutants, certainly at point of use. A fuel cell fueled with H_2 and air only produces water. Other significant advantages offered are modular construction and size flexibility, which makes them well suited for decentralized applications, high efficiency at part load, fuel flexibility and remote/unattended operation. Moreover, their vibration free operation reduces noise usually associated with conventional power generation systems, although there is still some noise due to compressors and pumps. More details concerning general features of fuel cells can be found in Refs. 4 and 5

2 Fuel Cell Types and Applications

There is a wide range of fuel cells in different stages of development. Although all types of fuel cells have the same basic operating principle, they have different characteristics that stem from the nature of the electrolyte involved. The six main types are summarized in Table 1. The nature

Table 1: Different types of fuel cells and characteristics.[4]

Fuel cells	Electrolyte	Charge carrier	Working temperatures (°C)	Anticipated scales	Anticipated electrical efficiency (hhv) (%)
Proton exchange membrane fuel cells (PEMFCs)	Ion exchange membranes	H^+	80–120	1–100 kWe	40–50
SAFCs	Solid acids	H^+	20–250		
AFCs	KOH solution	OH^-	80	10–100 kWe	60
PAFCs	H_3PO_4	H^+	180–200	100–500 kWe	40
MCFCs	Immobilized liquid molton carbonates	CO_3^{2-}	–650	300 kWe–3 MWe	45–50
SOFCs	Ceramics	O^{2-} or H^+	600–1000	1 kWe–2 MWe	60

of the electrolyte determines the mobile ions transferred and the direction of this transport, which in turn determines on which side of the electrolyte water is produced. Moreover, each electrolyte must be operated in a specific temperature range, which is a major difference in characteristics between different types of fuel cells. Molten carbonate fuel cells (MCFCs) and solid oxide fuel cells (SOFCs) have elevated operating temperature, compared to much lower operating temperature for alkaline fuel cell (AFC), polymer electrolyte membrane fuel cell (PEMFC), solid acid fuel cell (SAFC) and phosphoric acid fuel cells (PAFCs). The operating temperature dictates in turn the physicochemical and thermomechanical properties of materials to be used as cell component as well as the type of fuel the cell can be operated on. Moreover, this difference in operating temperatures has a number of implications for the applications for which particular fuel cell types are most suited.

Large differences exist in application, design, size, cost, and operating range for the different type of fuel cells. Of the available fuel cell technologies, PEMFCs, and SOFCs are thought to have the most potential to achieve cost and efficiency targets for widespread use in power generation, and have been the most investigated types.

In general, high temperature fuel cells exhibit higher efficiencies and are less sensitive to fuel composition or impurities. PEMFC systems require a pure H_2 fuel stream because the precious metal anode catalysts are poisoned by even low levels of CO or other compounds such as those containing sulfur. Current PEMFC systems operate below 100°C, but there is a great deal of research to find polymer electrolytes that can operate at higher temperatures since increasing the operating temperature relaxes the fuel-purity requirements relative to catalyst poisoning. In contrast, due to their high operating temperature, CO is rather a fuel than a poison for SOFC. Hence, high temperature fuel cells can be operated on fuels other than H_2.

2.1 *Fuel Cell Applications*

The potential applications of fuel cells in society are ever increasing, driven by the various benefits that the implementation of fuel cells would bring over current technologies, such as environmental and efficiency improvements. Applications being considered range all the way from very small scale, requiring only a few watts to larger-scale distributed power generation of hundreds of megawatts.

The small-scale power supply market is well suited for fuel cells. Indeed, fuel cells offer significantly higher power densities than batteries, as well as being smaller and lighter and having much longer lifetimes. Hence, an increasing number of applications are emerging where only a few watts are required, such as palmtop and laptop computers, mobile phones, and other portable electronic devices.

Their potential high reliability and low maintenance coupled to their quiet operation and modular nature makes fuel cells well suited to localized "off grid" power generation, either for high quality uninterrupted power supplies, or remote applications. High temperature fuel cells (MCFC and SOFC) are suitable for continuous power production, where the cell temperature can be maintained. If the released heat is used to drive a gas turbine to produce extra energy, the system efficiency can be increased to levels as high as 80%, significantly higher than any conventional electricity generation process. Moreover, the produced heat makes SOFCs particularly suited to combined heat and power (CHP) applications ranging from less than 1 kW to several MW, which covers individual households, larger residential units, and business and industrial premises, providing all the power and hot water from a single system.

The combination of their high efficiency (approaching 50% for Hydrogen PEMFC) and significantly reduced emissions of pollutants mean that fuel cell powered vehicles are a very attractive proposition, especially in heavily populated urban areas. The efficiency is to be compared with about 20% for a combustion engine. Low temperature fuel cells, in particular PEMFC, are the most suited to transport applications, because of the need for short warm-up. The concept of a fuel cell powered vehicle running on hydrogen, the so-called "zero emission vehicle", is a very attractive one and is currently an area of intense activity for almost all the major motor manufacturers. A major advance has been the recent introduction of vehicles such the Toyota Mirai into the commercial sphere. As a further example, fuel cell powered buses, running on compressed hydrogen are successfully operated in several cities around the world.[6]

3 Fuels

3.1 *Hydrogen*

Although fuel cells are intimately linked with hydrogen, a major advantage of fuel cells is their potential to use a wide range of fuels. This feature is particularly important for high temperature fuel cells such as MCFCs and

SOFCs although other types of fuel cells such as PEMFCs are well suited to operate on fuels such as alcohol.

Hydrogen remains however the preferred fuel for many reasons. In terms of power output, hydrogen is the most attractive fuel since it shows the highest energy density per unit of mass and its oxidation requires only the breakage of the H–H bond, allowing for fast oxidation kinetics. Fuel cells typically produce higher power output when fueled with H_2 than with any other fuels. In terms of the environmental impact, the oxidation of hydrogen produces only water, allowing for pollution-free electricity generation. Finally, hydrogen does not contain any carbon. Therefore, any concern related to solid carbon deposition on the anode does not apply to hydrogen.

Because they can efficiently convert hydrogen to electricity, fuel cells are considered a keystone of a potential future economy based on hydrogen. In the hydrogen economy, hydrogen would be used as a widespread energy carrier, providing the energy source for motive power, stationary power and portable electronic devices. Due to their high efficiency and the range of application covered, fuel cells would then play a major role in converting the hydrogen to CO_2, free electricity and possibly valuable heat in the case of high temperature fuel cells. Nonetheless, many technological hurdles need to be overcome before the transition to any hydrogen-based economy can be considered. Important advances are required mainly in the transport, storage and production of hydrogen.

Since molecular hydrogen is not available on earth in convenient natural reservoirs (most of the hydrogen on earth is bonded to oxygen in water), it needs to be produced and hydrogen is hence referred to as an energy vector or as energy carrier. The vast majority of available hydrogen, 96%,[6] is currently produced via the reforming of hydrocarbons, which has two important consequences. The feedstock used is non-renewable and the production process emits CO_2.

Hydrogen can be produced through the electrolysis of water. In this process, electricity is used to split water molecules into hydrogen and oxygen. Pollution-free electricity generated using renewable sources such as solar and wind energy can be used, meaning such a process is well suited for decentralized hydrogen production. In this respect, an interesting example is the Shetland Islands in Scotland, where wind power is used to produce hydrogen with the aim of fulfilling the community's energy needs. This project was the first off-grid renewable hydrogen system in Europe and the first community owned hydrogen production plant in the world.[7] On a larger scale, the use of nuclear energy could be considered to provide the electricity required for water electrolysis but uranium is non-renewable and the development of this kind of energy is opposed in many countries. High temperature electrolysis (HTE) is another process being considered. In this process, both electricity and heat are used to convert water to hydrogen.

Another interesting alternative would be the reforming of biofuels. Whilst the current reforming of fossil fuels uses non-renewable feedstock, replacing fossil fuels in this process with renewable biofuels could provide a sustainable mean of hydrogen production. The first hydrogen production plants reforming bio-ethanol have recently been built in the USA. However, if the same biofuels could be efficiently oxidized in fuel cells, both efficiency and system simplicity could be gained. Other means of hydrogen production from biomass include fermentative production and biological production.

Besides its production, many issues associated with hydrogen storage and transport need to be addressed. A transition to a hydrogen economy requires the ability to safely and efficiently transport and store molecular hydrogen. The difficulties associated with storage and transport stem from hydrogen's low energy density per unit of volume at ambient conditions. Despite showing a high energy per unit of mass, hydrogen has a low molecular weight. To be stored for practical applications such as on board a vehicle for motive power, hydrogen must be pressurized or liquefied to provide sufficient driving range. Achieving the high pressures required to obtain reasonable storage volumes necessitates high use of energy to power the compression. Alternatively, the use of liquid hydrogen is being considered. Liquid hydrogen possesses a higher volumetric energy density but is cryogenic and boils at −250°C. While being attractive in terms of weight, cryogenic storage is energy consuming. Moreover, liquefied hydrogen has lower energy density by volume than gasoline by approximately a factor 4, due to the low density of liquid hydrogen (there is actually more hydrogen in a liter of gasoline (116 g) than there is in a liter of pure liquid hydrogen (71 g)).

The inherent difficulties related to both compressed and liquefied hydrogen storage have led to the investigation of alternative storage methods. Hydrogen can be stored as a chemical hydride. In this storage method, hydrogen gas is reacted with a solid material to produce the hydride, which is much easier to transport. The hydride is then made to decompose at the point of use, yielding hydrogen gas. Current barriers to practical storage systems stem from the high pressure and temperature conditions needed for hydride formation and hydrogen release. For many potential systems, hydriding and dehydriding kinetics and heat management are also problems that need to be overcome. Another method is to absorb molecular hydrogen into a solid storage material such as carbon nanotubes. Unlike hydrides, the hydrogen does not dissociate/recombine upon charging/discharging the storage system and hence does not suffer from kinetics limitations.

The problems related to hydrogen transport basically stem from the same causes as for storage. Although the infrastructure used for the transport of natural gas could be used, hydrogen embrittlement of steel requires the pipes to be coated on the inside or new pipelines to be installed. Although expensive to install, once in place, pipelines are the cheapest way to transport hydrogen. Transport can be achieved using compressed hydrogen tanks or liquid hydrogen tanks, but involve the energy costs previously mentioned.

To summarize, the hydrogen economy could offer an interesting alternative to the current fossil fuel-based economy. The implementation of a hydrogen economy will undoubtedly accelerate the research and development of fuel cells. But before the technological hurdles mentioned above are overcome, the use of hydrogen as an energy carrier will be limited to small-scale and community projects such as the Shetlands community, the hydrogen highway in California or public transport in big cities.[8–10]

3.2 *Fuel Processing*

The mismatch between the desired fuel, hydrogen, and available ones, has contributed to the limited commercial implementation of fuel cells so far. The important drawbacks that hinder the widespread use of hydrogen as an energy carrier have led to the development of fuel cell systems that rely on practical fuels. Figure 2 illustrates the general concepts and requirements of processing gaseous, liquid and solid fuels for fuel cell applications.[11] Different strategies must be applied according to the type of fuel cell involved. Higher temperatures release the requirements on fuel purities, and typically, less processing is required with higher temperatures.

Low temperature fuel cells, such as PEMFC and PAFC require many fuel processing steps. The fuel has to be converted to a fairly pure hydrogen rich gas. The conversion of hydrocarbon-related fuels to hydrogen can be done mainly by steam reforming, partial, or complete oxidation, and is performed in a set of reactors external to the fuel cell stack. Catalysts are poisoned by CO at low temperatures. The CO content of the fuel entering the fuel cell stack needs to be reduced to the purity levels required by the

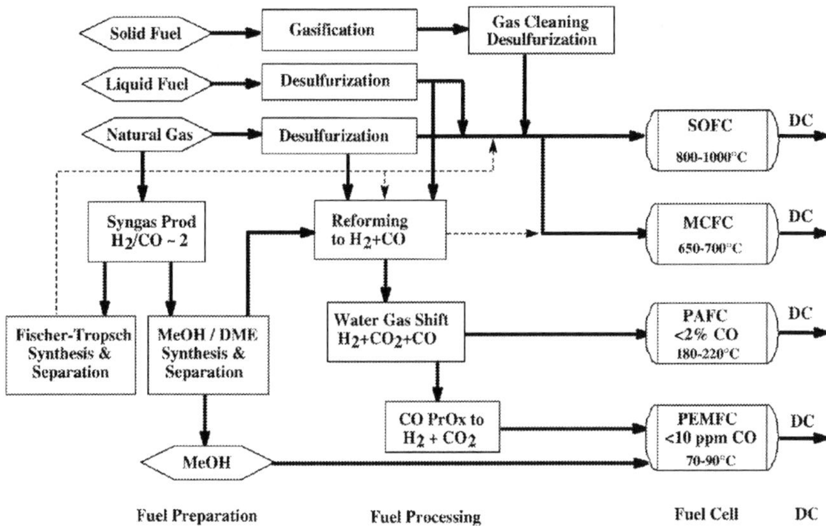

Fig. 2. The concepts and steps for fuel processing of gaseous, liquid and solid fuels for high temperature and low temperature fuel cells.[11]

type of cell involved. Requirements can be as strict as a few of ppm for PEMFC.

The onsite pre-processing of the fuel provides a suitable solution for supplying hydrogen to fuel cells. Transport and storage are no longer subjects of concern, since hydrogen is stored chemically in liquid fuels or gases that can be more easily transported. However, the addition of external reactors leads to significant complications in the system, by increasing system complexity and costs and decreasing efficiency.

As opposed to low temperature cells, CO is not a poison for SOFCs but can be oxidized and therefore acts as a fuel. SOFCs are the most flexible fuel cells with respect to their multifuel capability so that not only hydrogen and carbon monoxide but also various kind of hydrocarbon-related species could be used in the feed. Purity requirements are less stringent for SOFCs due to their higher temperature.[45] The desulfurization step is still required but as can be seen in Fig. 2, the route from chemicals to electricity could be greatly simplified if SOFCs were to be fed directly on hydrocarbons.

Due to the high temperature, feeding SOFCs with hydrocarbons leads to multiple and complex physicochemical processes that will considerably modify the feed composition before any electrochemical reaction take place. Taking advantage of the high temperature, the conversion of the fuel to more reactive species such as hydrogen and carbon monoxide can be performed *in situ* on the anode, which serves both as a catalyst and electrocatalyst, eliminating the need for external converters. This approach requires the fuel to be diluted with steam, carbon dioxide, or oxygen to provide reactants for steam reforming and/or partial oxidation reactions and so lessens the overall efficiency by diluting the fuel.

Processes occurring within the high temperature environment are both homogeneous and heterogeneous. Homogeneous processes occur regardless of the nature of the materials involved, while heterogeneous processes involve material surfaces, hence making the choice of anode materials and indeed surfaces of prime importance.

The main problem when feeding an SOFC with hydrocarbon is carbon deposition, which is a serious problem with any processes that involve hydrocarbons at high temperature. Two different types of carbon deposition can occur. The first mechanism results from reactions over a catalyst and thus belongs to the heterogeneous chemistry. This is a well-know process, which has been intensively studied over Ni, Fe, and Co catalysts,[12–17] leading to the precipitation of carbon as a graphite fiber at some surface of the metal particle.[18] This deposition results in irreversible damage to the anode catalyst, but can be prevented with addition of reforming agents in sufficient amounts. Thermodynamic studies reporting the amount of oxidant to be added to prevent carbon deposition have been performed for various fuels.

The second mechanism for carbon deposition occurs in the absence of a catalyst via free radical gas phase reactions.[19] Those reactions, usually initiated by the C–C bond scission, form, polyaromatic compounds, perhaps best referred to as tars and quite different in form from the graphitic fibers. These polyaromatic compounds have very low vapor pressure and hence easily deposit on surfaces.[20] While the catalytic deposition damages irreversibly the active surface, the carbon deposits from gas-phase reactions do not chemically interact with the anode surface. Carbon deposition by free-radical, gas-phase reaction is probably unavoidable for most hydrocarbon fuels other than methane above 700°C as shown by Gorte and co-workers in a series of studies addressing the use of hydrocarbons on copper-ceria anodes.[21–24] However, the deposit layer can be controlled by appropriate choice of temperature and materials, such as ceria-based oxides, which can oxidize the deposits. Hence, a steady-state layer of deposited carbon can be achieved, which in turn can improve the performance through improved connectivity in the anode conductive phase.

Besides carbon formation, both homogeneous and heterogeneous chemistry will contribute to modify the fuel composition. The parent hydrocarbon may undergo pyrolysis. Under the effect of high temperature, hydrocarbons decompose, without any interaction either with catalytic surfaces or with additional species in the inlet stream. The importance of homogeneous pyrolysis in the operation of hydrocarbon fueled SOFCs has been demonstrated by Dean *et al.*,[19,25] in studies of a variety of fuels under typical SOFCs conditions. As an example, at 800°C, butane undergoes full conversion for a residence time of 5 seconds. Oxygenates, such as ethanol, show a higher conversion than hydrocarbons, due to the presence of the OH group that leads to a weakening of the C–H bonds. Besides, the decomposition is more efficient, as large amounts of hydrogen and CO are produced.

When the gaseous mixture reaches the surface of porous anodes, catalytic reactions will occur and play a definite role in further converting the fuel mixture before it participates in electrochemical reactions. The heterogeneous processes likely to occur are steam reforming, CO_2 reforming, partial oxidation and auto-thermal reforming. These are basically the processes commonly used in the upstream processing of the fuel presented in Fig. 2. The nature of the ones occurring along the anode chamber will be dictated by the nature of the catalysts present in the anode chamber, and the reactants. Since hydrocarbons are often diluted with steam rather than with O_2 or CO_2, steam reforming will be the dominant conversion catalytic process within the anode. Even if the feed is a dry hydrocarbon, gas phase reactions have been shown to produce steam, while electrochemical reactions at the TPB will also produce steam and CO_2 during operation. Hence, only a small fraction of the hydrocarbon fuel, if any at all, will be available for electrochemical reactions. H_2 and CO are likely to be the oxidized species.

Specific energies of different fuels are shown in Table 2. Based on the availability of natural gas (methane with small amounts of other hydrocarbons) most stationary fuel cells have been designed for this fuel.

4 Proton Conducting Electrolyte Fuel Cells

In Grove's prototype fuel cells, sulfuric acid was used as the electrolyte.[1] Although high efficiency and environmental benefits were promised by fuel cell concepts, developing the early scientific experiments into commercially industrial products proved difficult. The main problems are associated with developing appropriate materials and manufacturing techniques that enables fuel cells to become competitive with existing power generation

Table 2: Theoretical energy densities of different fuels (HHV).

Fuel	Gravimetric energy density (kWh/kg)	Volumetric energy density (kWh/L)
Hydrogen	39.41	0.003
Compressed hydrogen (700 bars)	39.41	1.25
Methane	15.45	0.011
Natural gas	14.89	0.010
CNG (250 bars)	14.89	2.50
Gasoline	12.89	9.50
Diesel	12.83	10.36
Propane	13.78	7.03
Methanol	6.36	5.04
Ethanol	8.33	6.67
Coal	9.02	20.11

methods in terms of cost per kWh. Three variants of proton conducting electrolyte are considered here, polymer membrane, solid acid, and phosphoric acid. All of these are currently being developed commercially.

4.1 *Proton Exchange Membrane Fuel Cells (PEMFCs)*

PEMFCs are composed of an ion exchange membrane sandwiched between two electrode sheets. Various proton conducting polymers could be fabricated as electrolytes in PEMFCs such as Nafion and SPEEK. The only liquid in this kind of fuel cell is water so that corrosion problems are minimized. Water management in these systems is crucial since ion exchange membranes rely on water to be conductive but electrodes prefer a relatively "dry" condition for fuel transportation. A typical working temperature for PEMFCs is around 80°C and could be increased to 120°C at high pressure conditions; this is mainly limited by the water balance in the membrane and electrodes. Noble metals such as platinum are used as catalysts due to their high catalytic activation at low temperatures and high purity hydrogen is chosen as fuel. Impurities such as H_2S and CO are detrimental to platinum catalysts and would lead to a marked degradation of fuel cell performance. This is due to the poisoning effect of CO to Pt as CO is more stably absorbed on Pt surface than H_2 at low temperatures (Fig. 3). The main problems for PEMFCs are high catalyst loading (Pt on anodes and cathodes), the requisite for cooling systems and for purified systems. Recent work mainly focused on developing high temperature ion exchange membranes so that fuel cells could be operated at higher temperatures. This gives the benefit of decreasing or even replacing the loading of noble metals

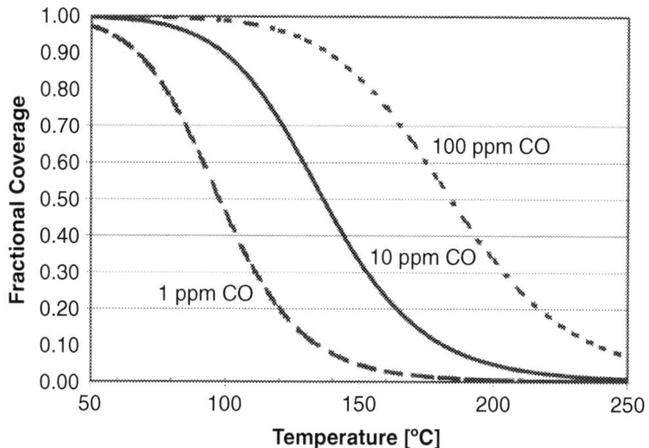

Fig. 3. CO coverage on a platinum surface as a function of temperature and CO concentration. H_2 partial pressure is 0.5 bar.[26]

and enables the fuel cells to be self-sustaining, avoiding the requirement of cooling and purifying systems.

4.2 *SAFCs*

SAFCs use solid acids as electrolytes. The charge carriers within the fuel cells are protons in most cases so that fuels and products can be well separated by the electrolyte, a great advantage over oxygen ion conductors. However, some problems still need to be solved. One is the solubility of most solid acids (such as $H_3PW_{12}O_{40} \cdot xH_2O$) in the water product that may lead to the collapse of the fuel cell infrastructures. Another is the requisite of humidified environments during fuel cell operations in many cases. Other technical problems such as thin membrane fabrication and fuel cell sealing were also encountered during fuel cell manufacturing. Recent studies have focused on synthesizing and developing stable solid acids that are independent of humidity at elevated temperatures. Figure 4 is a schematic display of a solid acid fuel cell based on CsH_2PO_4.[27] Thanks to advanced fabricating techniques, fuel cells could work with various fuels to provide promising applications in the future.

4.3 *PAFCs*

PAFCs are the first commercialized and widely used type of fuel cells. Hydrogen or a hydrogen rich gas mixture is used as fuel. The electrolyte,

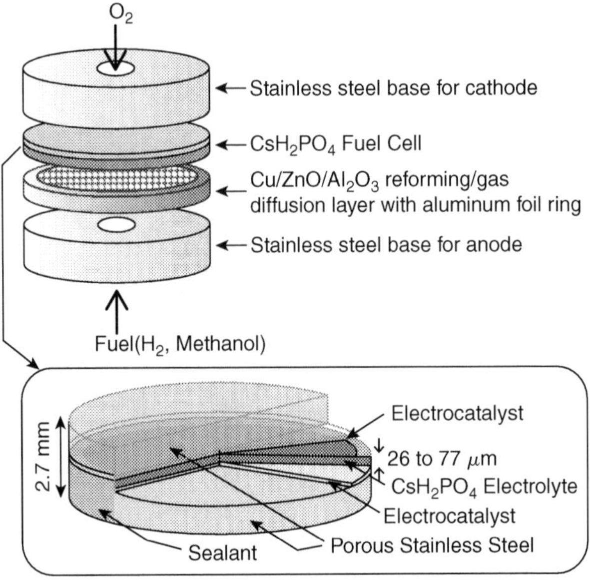

Fig. 4. Schematic display of the constitution of solid acid fuel cells.[27]

primarily composed of phosphoric acid (H_3PO_4) is a pure proton conductor with good thermal, chemical and electrochemical stability. The operating temperature of the PAFCs is typically between 150°C and 200°C and is a compromise between the electrolyte conductivity (increases with temperature) and the cell life (decreases with temperature).[28] The electrolyte, however, is highly corrosive so that only noble metals such as Pt could be used. Presumably, the large capital cost of PAFCs was expected and was only improved when Pt/carbon or graphite electrodes were deployed in the 1960s.[29] Evolution and cell component for PAFCs are listed in Table 3. However, one of the major problems in such fuel cells is the accelerated carbon corrosion and Pt dissolution when cell voltages are above 0.8 V. Therefore, long time, hot idles at open circuit have to be avoided. Another problem usually encountered is the electrode flooding and drying and this has been recognized as one of the major causes of declining fuel cell performance. Migrations of phosphoric acid between the matrix and the electrodes during cell load cycling are responsible for this. An alternative matrix that is capable of maintaining acid is still under development.[29] Recently, introducing phosphoric acid into polymers bearing basic groups such as ether, alcohol, imine, amide, or imide groups has attracted much interest. The

Table 3: Evolution and cell component technology for PAFCs.[30]

Component	ca.1965	ca. 1975	Current status
Anode	PTFE-bonded Pt black	PTFE-bonded Pt/C Vulcan XC-72[a]	PTFE-boned Pt/C Vulcan XC-72[a]
	9 mg/cm^2	0.25 mg Pt/cm^2	0.1 mg Pt/cm^2
Cathode	PTFE-bonded Pt black	PTFE-bonded Pt/C Vulcan XC-72	PTFE-bonded Pt/C Vulcan XC-72
	9 mg/cm^2	0.5 mg Pt/cm^2	0.5 mg Pt/cm^2
Electrode support	Ta mesh screen	Carbon paper	Carbon paper
Electrolyte support	Glass fiber paper	PTFC-bonded SiC	PTFC-bonded SiC
Electrolyte	85% H_3PO_4	95% H_3PO_4	100% H_3PO_4

[a]Conductive oil furnace black, product of Cabot Corp. Typical properties: 002 d-spacing 3.6 Å by X-ray diffusion, surface area of 220 m^2/g by nitrogen adsorption, and average particle size of 30 μm by electron microscopy.

resulting acid–base complex systems not only exhibit a high conductivity but also possess reasonable mechanical stability at elevated temperatures (>100°C) such as polybenzimidazole (PBI)/H_3PO_4 systems.[30]

5 SOFCs

5.1 *Basic Definitions*

A SOFC is defined by its solid ceramic electrolyte, which is a non-porous metal oxide. Such electrolytes are oxygen-ion (O^{2-}) conductors, impervious to gas flow and have negligible electronic conductivity. Solid oxide electrolytes require a high operating temperature to display suitable conductivities, typically in the range 700–1000°C.[31] SOFCs involve multiple complex physico-chemical processes. The principle of an SOFC, involving hydrogen as a fuel is illustrated in Fig. 5.

Oxygen is electrochemically reduced at the cathode–electrolyte–gas interface. Electrons are delivered to the cathode through the interconnect where they react with oxygen molecules in the gas phase to deliver oxygen ions to the electrolyte via a charge-transfer reaction:

$$\frac{1}{2}O_2(g) + 2e^-(c) \Leftrightarrow O^{2-}(e). \tag{1}$$

In this equation, the three phases are denoted as (g) for the gas, (c) for the cathode and (e) for the electrolyte. Oxygen ions migrate through the electrolyte via a vacancy hopping mechanism towards the

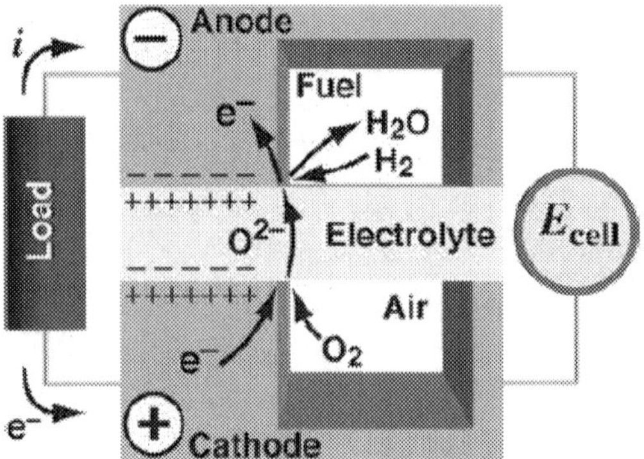

Fig. 5. Principle of a SOFC involving H_2 as a fuel.[10]

anode–electrolyte–fuel interface where they participate in the fuel oxidation, written as follows for hydrogen:

$$H_2(g) + O^{2-}(e) \Leftrightarrow H_2O(g) + 2e^-(a), \tag{2}$$

where (a) denotes the anode. The hydrogen in the gas phase reacts with the oxygen ions provided by the electrolyte to deliver electrons to the anode. As long as a load is connected between the anode and the cathode, the electrons from the anode will flow through the load back to the cathode, and electrical current will flow through the circuit.

The overall electrochemical cell reaction in a SOFC, based on oxygen and hydrogen, is written as follows:

$$\int O_2 + 2H_2 \Leftrightarrow 2H_2O. \tag{3}$$

Another accurate way to represent the overall reaction occurring in the cell, regardless of the fuel involved, is to describe the oxygen transfer:

$$O_2(c) \Leftrightarrow O_2(a). \tag{4}$$

An SOFC can therefore be considered as an oxygen pump. The amount of oxygen transported to the anode will depend on the type of fuel used and the reactions occurring at the anode.

Solid oxide electrolytes are the basis of two important energy technologies, SOFCs and solid oxide electrolyser cells (SOECs) that are jointly referred to as Solid Oxide Cells. SOFCs offer efficient alternative for electricity production and, similarly, SOECs offer enhanced efficiency in the conversion of steam. These technologies are very flexible with a wide-ranging scale of applications. When practical fuels are used, the environmental impact is better than for combustion technologies with less CO_2 and NO_x produced per unit of power generated. The high quality exhaust heat released during operation can be used as a valuable energy source, e.g. to drive a gas turbine when pressurized or the achievement of high efficiency electrical to chemical conversion.

5.2 *History of SOFC*

The solid oxide-fuel cell was first conceived following the discovery of solid oxide electrolytes in 1899 by Nernst.[32] Nernst discovered that the very high electrical resistance of pure solid oxides could be greatly reduced by the addition of certain other oxides. The most promising of these mixtures consisted mainly of zirconia (ZrO_2) with small amounts of added yttria (Y_2O_3). This is still the most widely used electrolyte material in the SOFC.

The first working SOFC was demonstrated by Baur and Preis in 1937, using stabilized zirconia as the electrolyte and coke and magnetite, respectively as a fuel and oxidant.[33] Unfortunately, the high-operating temperature and the reducing nature of the fuel led to serious materials problems and despite very significant efforts by Baur and other researchers, the search for suitable materials was unsuccessful.

This effectively hindered the development of SOFCs until the 1960s, when a first period of intensive activity in SOFC development began. Intensive research programs that were driven by new energy needs for military, space and transport applications, addressed mainly the electrolyte conductivity improvement and the first steps in SOFC technology. A second period of intense activity began in the mid-1980s and still continues today. These research efforts have brought SOFC commercialization close to reality. Different companies have developed different concepts, and several demonstration units have been operated for significant amounts of times.

5.3 *Characteristics*

The advantages of the high operating temperature include the possibility of running directly on practical hydrocarbon fuels without the need for a complex and expensive external fuel reformer and purification system. Internal reforming can be performed at high temperatures and SOFCs are not poisoned by CO, which can be oxidized at the anode and act as a fuel. When practical fuels are used, the environmental impact is better than for combustion technologies, in the sense that less CO_2 and NO_x are produced per unit of power generated. Looking at the overall system efficiency, the high quality exhaust heat released during operation can be used as a valuable energy source, either to drive a gas turbine when pressurized or for combined heat and power applications.[34]

Originally, SOFCs have been developed for operation primarily in the temperature range of 900–1000°C, which is beneficial for the fuel reforming, electrochemistry kinetics and the added value of the exhaust heat. However, some important drawbacks stem from such elevated temperatures. The materials that can be used are limited with respect to their chemical stability in oxidizing and/or reducing environments and their chemical and thermomechanical compatibility with adjacent components. Hence, there are considerable efforts to lower the operating temperature by 200°C or more which would allow the use of a broader set of materials, with less demands on seals and balance-of-plant components, simplifying thermal management, aiding in faster start-up and cool down, and resulting in less degradation of cell and stack components.[5] Because of these advantages, activity in the development of SOFCs capable of operating in the temperature range of 600–800°C has increased dramatically in the last few years. However, at lower temperatures, electrolyte conductivity and electrode kinetics decrease significantly. In terms of applications, the length of time that is generally required to heat up and cool down the system restricts the use of SOFCs in applications that require rapid temperature fluctuations. This is a consequence of the need to use a relatively weak, brittle component as the substrate material and because of problems associated with thermal expansion mismatches. This restriction applies particularly for transport applications, where a rapid transport start-up is essential.

Reducing the cost of SOFCs is a crucial issue for their commercialization. Currently, the high cost-to-performance ratio, limits SOFC introduction on the energy market. In this respect, lower operation temperature also makes possible the use of inexpensive metallic interconnections in place of lanthanum chromite-based ceramic interconnections.

5.4 *Design*

For a given electrolyte material and thickness, performance is largely determined by the electrode processes and these can be subdivided into Faradaic processes, largely charge transfer and non-Faradaic, largely mass transport. The non-Faradaic processes are strongly dependent upon the macroscopic and microscopic features of the electrode as a whole, whereas the Faradaic processes are very much dependent upon the nanoscale features of the interface between the electrode and electrolyte. This region which typically only extends a few microns from the electrolyte is often considered the "true electrode" with the remainder of the electrode having mechanical, electron, and gas transport functions.[35] The components, structure, and evolution of this interface are central to many of the recent exciting advances in the development and understanding of an SOFC device.

A wide range of material functions and architectures can be implemented to enhance the activity of this electrode–electrolyte interface and its electrochemistry. An ideal cell microstructure would offer an optimized contact between the electrolyte and the electrode, while being dimensionally, mechanically, chemically, and thermally stable during operation. The electrochemically active region of a SOFC electrode must support three essential functionalities which allow for electrochemical reactions to occur effectively. These are high electrocatalytic activity towards desirable reactions such as H_2O splitting or O_2 reduction, ionic conduction and electronic conduction. Often these are provided from a single material, as a mixed ionic and electronic conductor (MIEC). Since electrode reactions occur exclusively at discreet locations where these functionalities converge in the presence of reactants, these locations also constitute the active areas within the electrode structure. Electrodes must be designed with an extended active surface area through adequate choice of materials, microstructures, and porosity.

The solid state character of all SOFC components means that, in principle, there is no restriction on the cell configuration. Instead, it is possible to shape the cell according to criteria such as overarching design or applications issues. As for other fuel cell concepts, it is necessary to stack SOFCs to increase the voltage and the power produced. A stack can in principle comprise any number of cells depending on the desired power, and a fuel cell plant can be designed in modules of stacks in series- and parallel connections. To construct an electric generator, individual cells are connected in both electrical parallel and series to form a semi-rigid bundle that becomes the basic building block of a generator.

The most two common designs of SOFCs, the tubular and the planar, are represented in Fig. 6. In the tubular cells, the cell components are deposited in the form of thin layers on a cathode tube. In the planar design, the cell components are configured as thin, flat plates. The interconnection, which is ribbed on both sides, forms gas flow channels and serves as a current conductor.

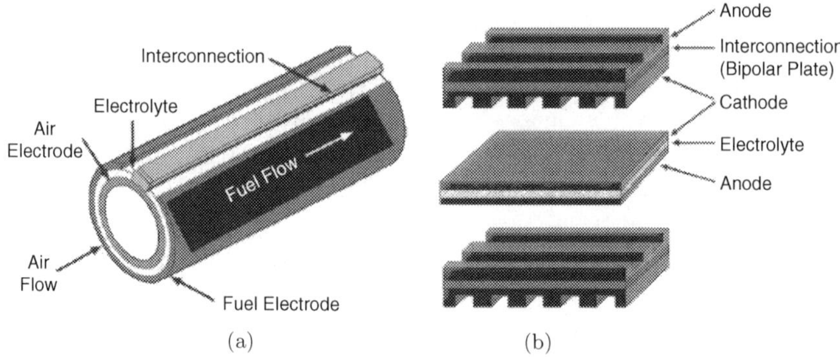

Fig. 6. The two most common SOFC designs: tubular (a) and planar (b).[4]

Alternative designs have been proposed, such as the planar segmented design developed at Rolls Royce,[36] or the SOFC roll developed at the University of St. Andrews.[37] The latter is an innovative design that takes advantage of both planar and tubular designs.

5.5 *Materials*

Electrodes, electrolyte, and interconnect materials for SOFCs are limited in choice by the several important requirements brought by the high temperature. All materials must be non-reactive with adjacent components at the high operating temperature, and must have compatible thermal expansion coefficient. Interconnects and electrolytes must be impermeable to gas, show high conductivities to minimize losses (electronic and ionic, respectively), and be stable in both reducing and oxidizing atmospheres. Electrodes must show high electrocatalytic activity and must be designed with an extended active surface area (triple phase boundary (TPB) points) since electrochemical reactions will occur only on sites possessing conductivities for those three phases detailed in Eqs. (1) and (2). Electrodes must fulfill some important requirements to ensure high and durable power output. To extend the TPB area, electrodes are fabricated as MIEC porous ceramics or ceramic–metallic composites. An ideal microstructure would offer the highest TPB length for electrochemical reactions, an optimized contact between the electrolyte and the electrode, and be dimensionally stable during operation (mechanically, chemically, and thermally).

The most commonly used electrolyte in SOFCs is yttria-stabilized zirconia (YSZ) as it possesses all the required characteristics. ZrO_2, in its pure form, does not serve as a good electrolyte because its ionic conductivity is too low. The addition of certain aliovalent oxides stabilizes the cubic fluorite structure of ZrO_2 from room temperature to its melting point and, at the same time, creates a large concentration of oxygen vacancies by charge compensation. The properties of stabilized zirconia have been extensively studied and several reviews dedicated to this material have been published.[38–40] Conventional zirconia-based SOFCs generally require an operating temperature above 850°C. This high operating temperature places severe demands on the material used as interconnects and for manifolding and sealing, and necessitate the use of expensive ceramic materials and specialist metal alloys. There is therefore considerable interest in lowering the operating temperature of SOFCs to below 750°C to enable the use of cheaper materials, such as stainless steel, and reduce fabrication costs, whilst maintaining high power outputs. Reducing the electrolyte thickness will obviously allow a reduction of the operating temperature but this approach is limited. The alternative route consists in developing new electrolyte materials showing higher conductivity than doped zirconia. SOFC electrolyte materials have been reviewed by Goodenough[41] and Skinner and Kilner,[42] and are detailed in reviews addressing SOFCs.[8,31] Among those various materials, two promising alternative electrolytes to YSZ are gadolinia-doped ceria[43,44] and lanthanum gallate-based perovskites.[45] Both these electrolytes offer the possibility of lower temperature operation for SOFCs between 500°C and 700°C. Scandia-doped zirconia has also received particular attention, since it has similar properties to YSZ but exhibits higher ionic conductivities, though it is also more expensive.

Lanthanum chromite, $LaCrO_3$, was at one stage considered as an important candidate for interconnects; however, metallic interconnects, such as chrome-based alloys, are now preferred except in some special ceramic geometries with integrated interconnect. The main drawback associated with lanthanum chromite interconnects is their manufacturing costs due their difficult sinter ability. For SOFCs operating in the intermediate temperature range, 500–750°C, it becomes feasible to use certain ferritic stainless-steel composites which fulfill the necessary criteria for the SOFC interconnect.

The standard anode material used in SOFC is the Ni/YSZ cermet, which was introduced by Spacil as a response to the failure of all-metal anodes.[46] These cermets have been extensively studied and their performance optimized. Nickel serves as an excellent reforming catalyst and electrocatalyst for electrochemical oxidation of hydrogen and the intrinsic charge-transfer resistance that is associated with the electrocatalytic activity at Ni/YSZ boundary is low. Despite being used in most SOFC applications Ni/YSZ anodes suffer from a few significant limitations. A severe limitation is their inability to operate on hydrocarbons, except when diluted with steam. The possibility to use practical fuels in SOFCs, without the need of a pre-reforming step and improved redox stability of the anode would be a definite advantage for the development of practical systems. Therefore, much research has focused on the development of alternative anode materials that are catalytically active for the oxidation of methane and higher hydrocarbons, and inactive for cracking reactions that lead to carbon deposition. Other desirable properties include the tolerance to sulfur, which would allow for the use of practical fuels such as natural gas and biogas.

Cu-based SOFC cermet anodes have successfully synthesized and demonstrated the direct electrochemical utilization of a large variety of hydrocarbon fuels with little carbon deposition.[47] GDC anodes ($Ce_{0.6}Gd_{0.4}O_{1.8}$) have been successfully operated with steam diluted CH_4 as a fuel.[48] Perovskites have also been widely investigated as potential SOFC anode materials. Among these materials, chromites and titanates are promising SOFC anode materials.[49,50] Another important double perovskite is $Sr_2MgMoO_{6-\delta}$, which has recently been shown to offer good performance in CH_4 and good sulfur tolerance.[51]

Materials suitable for a SOFC cathode (SOFC) have to fulfill the following key requirements: high electronic conductivity, stability in oxidizing atmospheres at high temperature, thermal expansion match with other cell components, compatibility with different cell components and sufficient porosity to allow transport of the fuel gas to the electrolyte–electrode interface.

Strontium-doped lanthanum manganite ($La_{0.85}Sr_{0.15}$) MnO_3, a *p*-type semiconductor, is most commonly used for the cathode material. Although adequate for most SOFCs, other materials may be used, particularly attractive being p-type conducting perovskite structures that exhibit mixed ionic and electronic conductivity. This is especially important for lower-temperature operation since the polarization of the cathode increases significantly as the SOFC temperature is lowered to, e.g. around 650°C. Important alternative cathodes are the perovskites, lanthanum strontium ferrite, lanthanum strontium cobalite, and lanthanum strontium cobalt ferrite, which are better electrocatalysts than the state-of-the-art lanthanum strontium manganite because they are mixed conductors.[52]

6 Molten Carbonate Fuel Cell (MCFC)

The second, and more commercially developed, main type of high temperature fuel cell system is the MCFC. The MCFC uses a eutectic mixture of alkali carbonates, Li_2CO_3 and K_2CO_3 immobilized in a $LiAlO_2$ ceramic matrix as the electrolyte. At high operating temperature, e.g. 650–700°C, the alkali carbonates form a highly conductive molten salt, with carbonate ions,CO_3^{2-}, providing ionic conduction. Carbon dioxide, CO_2 and oxygen, O_2, must be supplied to the cathode to be converted to carbonate ions, which provide the means of ion transfer between the cathode and the anode. The anode and the cathode reactions are as follows:

$$\text{Anode: } H_2 + CO_3^{2-} \rightarrow CO_2 + H_2O + 2e^-, \tag{5}$$

$$\text{Cathode: } \frac{1}{2}O_2 + CO_2 + 2e^- \rightarrow CO_3^{2-}. \tag{6}$$

In the MCFC, CO_2 is produced at the anode and consumed at the cathode. Therefore, MCFC systems generally feed the CO_2 from the anode to the cathode. The electrodes are nickel based, the anode usually consisting of a nickel–chromium alloy while the cathode is made of a lithiated nickel oxide. At both electrodes, the nickel phase provides catalytic activity and conductivity. At the anode, the chromium additions maintain high porosity and protect against corrosion.[52]

Because they operate at high temperatures, MCFCs have flexibility in the chosen fuel, there is no need to use precious metals as a catalyst, and they have high quality waste heat for cogeneration applications. There are also some disadvantages: a CO_2 recycling system must be implemented, the molten electrolyte is corrosive which gives rise to degradation issues, and the materials are relatively expensive. The main application area for this technology is distributed power generation, often utilizing biogas.

6.1 *Direct Carbon Fuel Cell (DCFC)*

Another promising application of MCFCs that is currently being developed is the direct oxidation of carbon (DCFCs). DCFCs are one of the most efficient electrochemical conversion system with practical overall achievable electrical efficiencies of 80%.[53] The overall reaction of this system is shown in Eq. (7).

$$C_{(s)} + O_{2(g)} \rightarrow CO_{2(g)}. \tag{7}$$

ΔG for this reaction is -395 kJ mol^{-1} and the ratio $\Delta G/\Delta H = 1.003$, meaning that this process theoretically offers 100% efficiency for converting chemical energy to electricity, more than twice that typically obtained from thermal conversion. Practical efficiency should be as high as 80%, which is a major improvement on the current inefficient traditional coal-fired power plants, using MCFC or SOFCs running on hydrogen or natural gas (where the efficiency is 40–60%). Moreover, DCFC systems are scalable, and therefore suitable for decentralized electricity production. Opportunities for DCFC applications are further strengthened by the abundance of available fuels, which include both fossil fuels, such as petroleum coke, or coal (the most abundant fossil fuel on earth), and renewable fuels, such as biomass (e.g. wood or nut shells, switchgrass, corn stover, palm, rice, and algae) or even other sources of fuel (food waste, wood waste). This technology enhances the biomass conversion, which is an important long-term consideration in producing electricity from renewable biomass sources. The dominant potential application is in the conversion of coal, especially in China, where there is significant new commercial opportunities.

7 Efficiency

7.1 *Thermodynamics of Fuel Cells*

A comprehensive study of the fuel cell thermodynamics has been performed by Kee *et al.*[10] The thermodynamic limit on fuel cell performance can be understood by considering the energy and entropy accounting associated with a generic steady-flow process.

The rate at which work is done by a system, $\dot{W}$, can be obtained by combining the first and second law of thermodynamics.

$$\dot{W} = -\dot{m}(\Delta h - T_0 \Delta S) - T_0 \dot{P}_s, \tag{8}$$

where $\Delta h = h_{\text{out}} - h_{\text{in}}$ and $\Delta S = S_{\text{out}} - S_{\text{in}}$ are respectively the net enthalpy and entropy differences associated with the flow streams entering and leaving the system. Entropy is produced within the system due to internal irreversible processes at a rate $\dot{P}_s$. Since $\dot{P}_s$ is required to be positive by the second law of thermodynamics, the greatest power is produced by a reversible process and equals:

$$\dot{W}_{\text{rev}} = -\dot{m}(\Delta h - T_0 \Delta S). \tag{9}$$

While this general expression applies to any steady-flow process, a relevant case for fuel cells is one in which the temperature remains fixed at T_0 and the pressure is constant, but the composition changes due to internal chemical reactions. In this case, the greatest work production rate achievable is:

$$\dot{W}_{\text{rev}} = -\sum_k \Delta(\dot{N}_k \mu_k), \tag{10}$$

where μ_k and N_k are respectively the species chemical potentials and molar flow rate, and the sum runs over all species. If depletion effects are small enough, so that $\mu_{k,\text{in}} = \mu_{k,\text{out}}$ and if $\dot{N}_k = v_k \dot{N}$, where ν_k is the stoichiometric coefficient of species k and N is a rate of progress variable for a global oxidation reaction, Eq. (10) reduces to:

$$\dot{W}_{\text{rev}} = -\dot{N} \sum_k v_k \mu_k. \tag{11}$$

If the fluid stream is an ideal gas mixture, the reversible work production rate can be written as:

$$\dot{W}_{\text{rev}} = -\dot{N}[-\Delta G^0 - RT \ln \Pi_k P_k^{vk}], \tag{12}$$

where ΔG^0 represents the free-energy change between reactants and products in the global reaction, and p_k is the partial pressure. This expression also holds in the case where one or more reactants are supplied in separate streams. All that is required is to evaluate the partial pressures of each species in the stream in which it is present.

Since the cell potential can be expressed as $E_{\text{cell}} = \dot{W}/I$, Eq. (12) can be used to determine the reversible cell potential. The electric current generated in a fuel cell as a direct consequence of the reactions that result in oxidation of the fuel, is given by:

$$I = nF\dot{N}, \tag{13}$$

where F is Faraday's constant and n the number of exchanged electrons. Hence, the potential developed by a reversible cell is:

$$E_{\text{rev}} = -\frac{\Delta G^0}{nF} - \frac{RT}{nF} \ln \prod_k P_k^{v_k}. \tag{14}$$

When using Eq. (4), accounting for the oxygen transfer, the reversible potential simplifies to:

$$E_{\text{rev}} = -\frac{RT}{4F} \ln \frac{P_{O_2(a)}}{P_{O_2(c)}}. \tag{15}$$

This reversible potential is known as the Nernst potential. This ideal potential depends on the electrochemical reactions that occur with different fuels and oxygen. The Nernst equation provides a relationship between the ideal standard potential E^0 for the cell reaction and the ideal equilibrium potential (E) at other temperatures and partial pressures of reactants and products.

7.2 Fuel Cell Efficiency

Different efficiencies must be combined to produce the overall efficiency of a fuel cell. The overall efficiency is defined by the product of the electrochemical efficiency ε_E, and the heating efficiency ε_H. The electrochemical efficiency is in turn, the product of the thermodynamic efficiency ε_T, the voltage efficiency ε_V and the current or Faradic efficiency ε_J[4,10,54]:

$$\varepsilon_{FC} = \varepsilon_E \varepsilon_H = \varepsilon_T \varepsilon_V \varepsilon_J \varepsilon_H. \tag{16}$$

7.2.1 *Heating efficiency*

The heating efficiency applies to cases where the fuel contains more species than the electrochemically active ones, such as gases, impurities and other combustibles. The heating efficiency, ε_H, is defined as:

$$\varepsilon_H = \frac{\Delta H^0}{\Delta H_{\text{com}}}, \tag{17}$$

where ΔH^0 represents the amount of enthalpy in the electrochemically active species and ΔH_{com} represents the amount of enthalpy included in all combustible species in the fuel gases fed to the fuel cell. A pure fuel will obviously give a heating efficiency of 100%.

7.2.2 *Thermodynamic efficiency*

The thermodynamic efficiency of a process measures how efficiently chemical energy extracted from the fuel stream is converted to useful power, rather than heat:

$$\varepsilon_t = \frac{\dot{W}}{\dot{W} + \dot{Q}} \quad \text{or,} \quad \varepsilon_t = \frac{\dot{W}}{\dot{m}|\Delta h|}, \tag{18}$$

where $\dot{Q}$ is the heat rate production of the cell. Using the maximum work production rate achievable, which is the one of a reversible process (Eq. (8)), one can write the maximum theoretical efficiency of a system as:

$$\varepsilon_t = \frac{\Delta(h - T_0 S)}{\Delta h}. \tag{19}$$

The thermodynamic efficiency is an extremely important feature when analyzing a fuel cell. This efficiency justifies the need for fuel cell development. Indeed, since the chemical energy is transferred directly to electricity, the free enthalpy change of the cell reaction may be totally converted to electrical energy. In a conventional heat engine, where only the temperature changes, ε_{rev} is limited to the familiar Carnot efficiency:

$$\varepsilon_{t,\text{carnot}} = 1 - \frac{T_0}{T}. \tag{20}$$

For a constant-temperature fuel cell, the intrinsic maximum thermodynamic efficiency is given by:

$$\varepsilon_{\text{rev}} = \frac{\Delta G}{\Delta H} = 1 - \frac{T\Delta S}{\Delta H} = \frac{\Delta G^0}{\Delta H^0} + \frac{RT}{\Delta H^0} \ln \prod_k p_k^{v_k}. \tag{21}$$

7.2.3 *Current efficiency*

The current efficiency can be commonly expressed as the fuel utilization efficiency. The efficiency of a SOFC drops if all the reactants are not converted to reaction products. For a 100% conversion of a fuel, the amount of current density, i_F, produced is given by Faraday's law:

$$i_F = zF\left(\frac{df}{dt}\right), \tag{22}$$

where (df/dt) is the molar flow rate of the fuel. For the amount of fuel actually consumed, the current density produced is given by:

$$i = zF\left(\frac{df}{dt}\right)_{\text{consumed}}. \tag{23}$$

The current efficiency, ε_J, is the ratio of the actual current produced to the current available for complete electrochemical conversion of the fuel:

$$\varepsilon_J = \frac{i}{i_F}. \tag{24}$$

This efficiency can be expressed as well in terms of fuel consumption:

$$\varepsilon_J = \frac{h_{\text{in}} - h_{\text{out}}}{h_{\text{in}} - h_{\text{ox}}}, \tag{25}$$

where h_{ox} corresponds to the enthalpy change when all the useful fuel has been consumed.

7.2.4 *Voltage efficiency*

When an electrical current is drawn from a fuel cell, part of the chemical potential available must be used to overcome the irreversible internal losses. Hence, the actual cell potential is decreased from its equilibrium potential meaning that in an operating SOFC, the cell voltage is always less than the reversible voltage. The voltage efficiency, ε_V, is defined as the ratio of the operating cell voltage under load, E, to the reversible cell voltage, ε_R, and is given as:

$$\varepsilon_V = \frac{E}{E_r}. \tag{26}$$

A voltmeter connecting the anode and cathode can measure the cell electrical potential E, which depends on the current flow. Indeed, many irreversibilities in a fuel cell scale with current density. If no current flows, the cell voltage is the open circuit potential or open circuit voltage (OCV). In most cases, the OCV will equal the potential developed by a reversible cell. The difference between the operating cell voltage and the expected reversible voltage is termed polarization or overpotential and is represented as η. This cell overpotential comprises the total Ohmic losses for the cell, and the polarization losses associated with the electrodes. The useful voltage under load conditions can therefore be expressed as:

$$V = E^0 - IR - \eta_{\text{an}} - \eta_{\text{cath}}. \tag{27}$$

In this equation, I is the current passing through the cell. The electrical resistance R encompasses the Ohmic resistance of all components. The polarization resistances, η_{an} and η_{cath} respectively for the anode and the cathode account for non Ohmic losses in each electrode. The polarization loss of each electrode is composed of: (i) activation overpotential due to energy barriers to charge-transfer reactions, (ii) concentration overpotential associated with gas-phase species diffusion resistance through the electrodes, (iii) contact resistance which is caused by poor adherence between electrode and the electrolyte.[51] Although polarizations cannot be eliminated, material choice and electrode designs can contribute to their minimization. Figure 7 shows a typical voltage–current polarization curve for a SOFC. The voltage loss increases with the current density. Activation overpotentials contribute the most at low current, while at high currents, concentration polarizations become important. Ohmic losses dominate the losses in the intermediate currents zone.

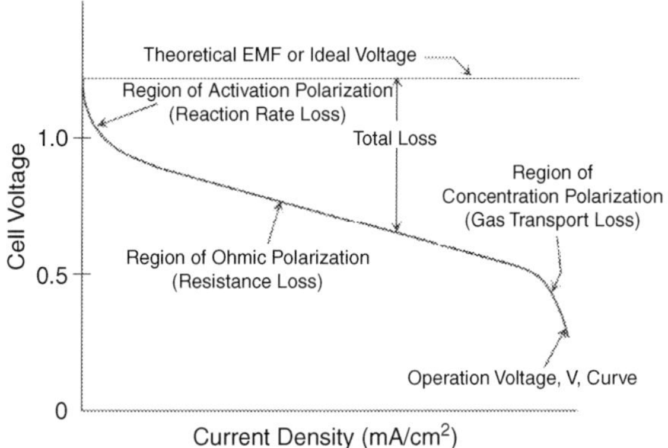

Fig. 7. Ideal and actual fuel cell voltage-current characteristics.[4]

The concave portion at low current density corresponds to activation-related losses. When the current density increases, the losses are dominated by the Ohmic polarization. When the current density approaches its highest values, losses are dominated by concentration polarizations that cause the steep drop in the cell voltage. Each of the different types of losses is described hereafter.

7.2.5 *Internal resistance*

The size of the voltage drop due to Ohmic losses is simply proportional to the current:

$$V = IR. \tag{28}$$

The internal resistance R encompasses the contribution from the electrodes, electrolyte, interconnect and bipolar plates:

$$R = R_{\text{Electronic}} + R_{\text{ionic}} + R_{\text{contact}}. \tag{29}$$

In most fuel cells, the electrolyte contribution to this resistance is the most important, due to the ionic nature of its conductivity. The interconnect and bipolar plates contribution can be important as well. To minimize the Ohmic losses, the preferred practice is to fabricate dense, gas-tight electrolyte membranes as thin as possible.

7.2.6 *Charge transfer or activation polarization*

The activation polarization is related to the charge-transfer processes occurring during the electrochemical reactions on electrode surfaces. The losses are caused by the slowness of the reactions taking place on the surface of the electrodes. Electrochemical reactions involve an energy barrier that must be overcome by the reacting species. A proportion of the voltage generated is lost therefore in driving the electron transfer. This energy barrier, called the activation energy, results in activation or charge-transfer polarization, η_A. Activation polarization is related to current density, i, by the Butler–Volmer equation:

$$i = i_0 \exp\left[\frac{(1-\beta)\eta_A F}{RT}\right] - i_0 \exp\left[-\frac{\beta\eta_A F}{RT}\right], \tag{30}$$

where β is the symmetry coefficient and i_0 the exchange current density. The symmetry coefficient is considered as a fraction of the change in polarization which leads to a change in the reaction rate constant. The exchange current density is related to the balanced forward and reverse electrode reaction rates at equilibrium. A high exchange current density means a high electrochemical reaction rate and, in that case, a good fuel- cell performance is expected.

The exchange current density can be determined experimentally by extrapolating plots of log i vs. η to $\eta = 0$. For large values of η (either negative or positive) one of the bracketed terms in Eq. (29) becomes negligible. After rearranging one obtains,

$$\eta_A = a \pm b \log i, \tag{31}$$

which is usually referred to as the Tafel equation. Parameters a and b are constants which are related to the applied electrochemical material, type of electrode reaction and temperature.

The constant a (in the form $v = a \ln(i/i_0)$ is higher for an electrochemical reaction which is slow. The constant i_0 is higher if the reaction is faster. The current density i_0 can be considered as the current density at which the overvoltage begins to move from zero. The smaller i_0, the greater is the voltage drop.

The exchange current density is a crucial factor in reducing the activation overvoltage. The cell performance can be improved through an increase of the exchange current density. This can be done in the following ways[4]:

— raising the cell temperature
— using more effective catalysts
— increasing the roughness of the electrodes
— increasing reactant concentration, e.g. pure O_2 instead of air
— increasing the pressure

In low- and medium-temperature fuel cells, activation overvoltage is the most important irreversibility and cause of voltage drop, and occurs mainly at the cathode. Activation overvoltage can be important at the anode when fuels other than hydrogen are involved.

7.2.7 *Diffusion or concentration polarization*

Concentration polarization, η_D, is related to the transport of gaseous species through the porous electrodes and thus its magnitude is dictated by the microstructure of the electrode, specifically, the volume percent porosity, the pore size, and the tortuosity factor. It becomes significant when the electrode reaction is hindered by mass transport effects, i.e. when the supply of reactant and/or the removal of reaction products by diffusion to or from the electrode is slower than that corresponding to the charging–discharging current i. When the electrode process is governed completely by diffusion, the limiting current, i_L, is reached. In such a case, the demand for reactants exceeds the capacity of the porous anode to supply them by gas diffusion mechanisms. High tortuosity (bulk diffusion resistance) is often assumed to explain this behavior.

The voltage drop due to the mass transport limitations can be expressed as:

$$\Delta V = \frac{RT}{2F} \ln\left(1 - \frac{i}{i_l}\right), \tag{32}$$

where i_L is postulated to be the limiting current density at which the fuel is used up at a rate equal to its maximum supply speed. The current density cannot rise above this value because the fuel gas cannot be supplied.

8 Summary

In general, high-temperature fuel cells exhibit higher efficiencies and are less sensitive to fuel composition. PEMFC systems require a pure H_2 fuel stream because the precious metal anode catalysts are poisoned by even low levels of CO or other compounds such as those containing sulfur. These PEMFCs are the most suited to transport applications, because of the need for short warm-up. The concept of a fuel cell powered vehicle running on hydrogen, the so-called "zero emission vehicle", is a very attractive one and is currently an area of intense activity for almost all the major motor manufacturers. The potential high reliability and low maintenance coupled to their quiet operation and modular nature makes fuel cells well suited to localized "off grid" power generation, either for high quality uninterrupted power supplies, or remote applications. High temperature fuel cells (MCFC and SOFC) are suitable for continuous power production.

References

1. W. R. Grove, *Philos. Mag.* **14** (1839), p. 127.
2. W. R. Grove, *Philos Mag.* **21** (1843), p. 417.
3. L. Mond and C. Langer, *Proc. Roy. Soc. London.* **46** (1889), p. 296.
4. *Fuel Cell Handbook* (7th edition) (US Department of Energy, Southwest Washington D.C., 2004).
5. S. C. Singhal, *Solid State Ion.* **152–153** (2002), p. 405.
6. IEA Technology essentials, Hydrogen production and distribution, OECD/IEA (2007).
7. Available at: www.pure.shetland.co.uk.
8. M. Ormerod, *Chem. Soc. Rev.* **32** (2003), pp. 17–28.
9. A. Atkinson, S. Barnett, R. J. Gorte, J. T. S. Irvine, A. J. McEvoy, M. Mogensen, S. C. Singhal and J. Vohs., *Nat. Mater.* **3** (2004), pp. 17–27.
10. R. J. Kee, H. Zhu and D. G. Goodwin, *P. Combus. In.* **30** (2005), pp. 2379–2404.
11. C. Song, *Catal. Today* **77** (2002), pp. 17–49.
12. R. T. K. Baker, M. A. Barber, P. S. Harris, F. D. Feates and R. J. Waite, *J. Catal.* **26** (1972), p. 51.
13. R. T. K. Baker, P. S. Harris, J. Henderson and R. B. Thomas, *Carbon* **13** (1975), p. 17.
14. R. T. K. Baker, P. S. Harris and S. Terry, *Nature* **253** (1975), p. 37.
15. C. W. Keep, R. T. K. Baker and J. A. France, *J. Catal.* **47** (1977), p. 232.
16. C. H. Bartholomew, *Catal. Rev. Sci. Eng.* **24** (1982), p. 67.
17. R. T. K. Baker, *Carbon* **27** (1989), p. 315.
18. M. L. Toebes, J. H. Bitter, A. J. van Dillen and K. P. de Jong, *Catal. Today* **76** (2002), pp. 33–42.
19. C. Y. Sheng and A.M. Dean, *J. Phys. Chem. A* **108** (2004), p. 3772.
20. C. H. Toh, P. R. Munroe, D. J. Young and K. Foger, *Mater. High Temp.* **20** (2003), p. 129.
21. S. McIntosh, H. He, S. -I. Lee, O. Costa-Nunes, V. V. Krishnan, J. M.Vohs and R. J. Gorte, *J. Electrochem. Soc.* **151** (2004), pp. A604–A608.
22. T. Kim, G. Liu, M. Boaro, S.-I. Lee, J. M. Vohs, R. J. Gorte, O. H. Al-Madhi, B. O. Dabboussi, *J. Power Sources* **155** (2006), pp. 231–238.
23. S. McIntosh, J. M. vohs and R. J. Gorte, *J. Electrochem. Soc.* **150**(4) (2003), pp. A470–A476.
24. H. He, J. M Vohs and R. J. Gorte, *J. Power Sources* **144** (2005), pp. 135–140.
25. K. M. Walters, A. M. Dean, H. Zhu and R. J. Kee, *J. Power Sources* **123** (2003), pp. 182–189.
26. C. Yang, P. Costamagna, S. Srinivasan, J. Benziger and A. B. Bocarsly, *J. Power Sources* **103** (2001), pp. 1–9.

27. T. Uda, D. A. Boysen, C. R. I. Chisholm and S. M. Haile, *Electrochem. Solid-State Lett.* **9**(6), pp. A261–A264.
28. N. Sammes, R. Bove and K. Stahl, *Curr. Opin. Solid-State Mater. Sci.* **8** (2004), pp. 372–378.
29. J. H. Hirschenhofer, D. B. Stauffer, R. R. Engleman and M. G. Klett, *Fuel Cell Handbook* (4th edition) (Parsons Corporation, Reading PA, 1998).
30. Q. F. Li, R. H. He, J. O. Jensen and N. J. Bjerrum, *Chem. Mater.* **15** (2003), pp. 4896–4915.
31. N. Q. Minh, *J. Am. Ceram. Soc.* **76** (1993), p. 563.
32. W. Nernst, *Z .Electrochem.* **6** (1899), p. 41.
33. E. Baur and H. Z. Preis, Über Brennstoff-ketten mit FestLeitern, *Z. Electrochem.* **43** (1937), p. 727.
34. J. T. S. Irvine, D. Neagu, M. C. Verbraeken, C. Chatzichristodoulou, C. Graves and M. B. Mogensen. *Nat. Energy.* **1** (2016), p. 15014.
35.
36. F. J. Gardner, M. J. Day, N. P. Brandon, M. N. Pashley and M. Cassidy, *J. Power Sources*, **86** (2000).
37. F. G. E. Jones, P. A. Connor, A. J. Feighery, J. Nairn, J. Rennie and J. T. S. Irvine, *J. Fuel Cell Sci. Tech.*, **4** (2007), p. 1.
38. R. Stevens, *An Introduction to Zirconia* (Magnesium Elektron, London, 1986).
39. H. Etsell and S. N. Flengas, *Chem. Rev.* **70** (1970), pp. 339–376.
40. E. C. Subbarao and H. S. Maiti, *Solid-State Ion.*, **11** (1984), p. 317.
41. J. B. Goodenough, *Ann. Rev. Mater. Res.*, **33** (2003), p. 91.
42. S. J. Skinner and J. A. Kilner, *Mater. Today* **6** (2003), p. 30.
43. M. Mogensen, N. M. Sammes and G. A. Tompsett, *Solid State Ion.* **129** (2000), p. 63.
44. B. C. H. Steele, *Solid State Ion.* **129** (2000), p. 95.
45. B. C. H. Steele and A. Heinzel, *Nature* **414** (2001), p. 345.
46. S. Spacil, Electrical device including nickel-containing stabilized zirconia electrode. US Patent 3,558,360 (1970).
47. H. Kim, S. Park, J. M. Vohs and R. J. Gorte, *J. Electrochem. Soc.* **148** (2001), p. A693.
48. O. A. Marina, C. Bagger, S. Primdahl and M. Mogensen, *Solid State Ion.* 123 (1999), p. 199.
49. S. Primdahl, J. R, Hansen, L. Grahl-Madsen and P. H. Larsen. *J. Electrochem. Soc.* **148** (2001), p. A74.
50. G. Pudmich, B. A. Boukamp, M. Gonzalez-Cuenca, W. Jungen and F. Zipprich Tietz, *Solid State Ion.* **135** (2000), p. 433.
51. Y. H. Huang, R. I. Dass, Z. L. Xing and J. B. Goodenough, *Science* **312** (2006), p. 254.
52. J. Larminie and A. Dicks, *Fuel Cells Systems Explained* (2nd edition) (John Wiley & sons Ltd., London, 2005).
53. D. Cao, Y. Sun and G. Wang, *J. Power Sources* **167** (2007), pp. 250–257.
54. B. de Boer, PhD thesis, (1998), University of Twente "SOFC Anode: Hydrogen oxidation at porous nickel and nickel/yttria stabilised zirconia cermet electrodes".

Chapter 7

Flywheels

Donald Bender

Sandia National Laboratories Livermore, CA 94550
bender@alum.mit.edu

Ubiquitous in rotating machinery, flywheels smooth the flow of energy in rotary systems ranging from small engines to large reciprocating machines. A new class of standalone flywheel energy storage system has been in use since the late 20th century. These systems are electrically connected to the applications that they serve. They may be found in applications as diverse as uninterruptible power supplies, racecars, and large physics research facilities. This chapter presents the theoretical and practical foundations of flywheel design, the history of flywheels from ancient times to the present, flywheel subsystem technology, and a survey of applications where flywheel energy storage systems are currently in service.

1 Introduction

Flywheels store kinetic energy (the energy of motion) in a rotating mass. In use since ancient times, the mass may be affixed directly to a rotating machine such as a mill or steam engine. In contrast, a class of modern advanced flywheel systems employs a rotor spinning at high speed in an evacuated enclosure that is charged and discharged electrically.

During charging, a torque is applied to the flywheel in the direction of rotation, accelerating the rotor to a higher speed. Discharge is accomplished by applying a braking torque that decelerates the flywheel while performing useful work. When used in engines or industrial equipment, torque magnitude may vary significantly and torque direction may reverse one or more times per revolution. In this application, spin speed of the flywheel varies only slightly between pulses. The flywheel serves to damp out speed variations caused by a pulsed motive source such as an engine or an intermittent load such as a mechanical punch press.[1] A standalone flywheel developed expressly for energy storage will experience much longer charge and discharge intervals and may be operated over a speed range of greater than 2:1 between charged and discharged states.

Flywheel rotors have been built in a wide range of shapes. The oldest configurations were simple stone disks. From the time of the industrial revolution, "wagon wheel" iron and steel rims from a few pounds to tens of tons have been used in a variety of stationary machines. High strength, mass-produced metal rotors are found in modern industrial equipment. These rotors are generally disk-like with a geometry tailored for the requirements of the application.

In the recent development of standalone flywheel energy storage systems, numerous unusual configurations have been explored. These include flexible subcircular rims, straight fibers oriented along the diameter ("brush" rotors), tapered rotors such as the constant stress Stodola hub, and many others.[2] After decades of development, the configurations of choice for flywheel energy storage systems in use today comprise disks, solid cylinders, and thick-walled cylinders constructed from high strength steel or carbon and glass composite materials.

2 Kinetic Energy, Stress, and Cost of Flywheel Rotors

The kinetic energy of a rotating object is expressed as:

$$E_k = \frac{1}{2} I \omega^2, \tag{1}$$

where E_k is kinetic energy, I is moment of inertia, and ω is angular velocity. While many rotor shapes have been used, the most common shapes can be classified as disk-like and drum-like. In this equation and those that follow the terms disk or disk-like cover a range of shapes from a thin circular plate to a solid cylinder of extended length. For a disk, the moment of inertia is given by:

$$I = \frac{1}{2} m r^2,$$

where m is the mass of the disk and r is its radius. For a thin-walled cylinder, the moment of inertia is given by:

$$I = m r^2.$$

More specifically:

$$I = 2\pi r^3\, l\, t\, \rho,$$

where l is the length of the cylinder, t is the thickness of the cylinder, and ρ is the density of the material in mass per unit volume. In terms of weight rather than mass:

$$\rho = \frac{\delta}{g},$$

where δ is the weight per unit volume and g is the acceleration due to gravity.

For a thick-walled cylinder, referred to throughout as a drum-like configuration, the moment of inertia is given by:

$$I = \frac{1}{2} m(r_i^2 + r_o^2),$$

where r_i and r_o are the inner and outer radii.

The maximum speed at which a flywheel may operate is limited by the strength of the rotor material. The stress experienced by the rotor must remain below the strength of the rotor material while maintaining a suitable safety margin.

A thin-walled cylinder is a cylinder for which the inner and outer radii are nearly equal and are expressed as r. Stress in a rotating thin-walled cylinder is given by:[3]

$$\sigma_\theta = \rho r^2 \omega^2, \tag{2}$$

where σ_θ is the stress in the circumferential or hoop direction.

For a uniform disk or solid cylinder the maximum stress occurs at the center and has a value given by[3]:

$$\sigma_{\max} = \frac{1}{8} \rho r_o^2 \omega^2 (3 + \nu), \tag{3}$$

where $\sigma_{\max}$ is the maximum stress and ν is the Poisson ratio[a] of the rotor material.

[a]Poisson's ratio, ν, is defined as the ratio of the transverse contraction strain to the longitudinal extension strain in the direction of the stretching force. Strain is the change in length divided by the original length.

For a thick-walled cylinder, maximum circumferential stress occurs at the bore of the cylinder and is given by[3]:

$$\sigma_{\theta-\max} = \frac{1}{4}\rho\omega^2((3+\nu)r_i^2 + (1-\nu)r_o^2). \tag{4}$$

Two parameters of interest in flywheel design are specific energy of the rotor and cost per unit of stored energy of the rotor. Specific energy is significant when the weight of the system is important, such as in mobile applications.

Since the surface speed of a flywheel is $V = r\omega$, the specific energy of a flywheel rotor can be expressed as:

$$\frac{E_k}{m} = K_{\text{sev}} V^2, \tag{5}$$

where K_{sev} is a shape factor with a value of 0.25 for a disk and 0.5 for a thin-walled cylinder. In order to make best use of the rotor material the flywheel will generally be designed to operate at the highest surface speed allowed by the rotor material. High performance composite flywheel rotors have a maximum operating surface speed in the range of 500–1000 m/s while high performance steel rotors have a maximum operating surface speed in the range of 200–400 m/s.

Specific energy may also be expressed in terms of rotor material properties:

$$\frac{E_k}{m} = K_{\text{se}\sigma} \frac{\sigma}{\rho},$$

where $K_{\text{se}\sigma}$ is a shape factor. For any particular rotor geometry, the maximum specific energy that can be attained for that geometry is a function of the maximum allowable stress in the material and the density of the material. A light strong material such as a carbon composite stores considerably more energy per unit mass than a heavy strong material such as high strength steel.

An important concept to consider in stating the capacity of a flywheel system is the distinction between total stored energy and extractable energy. Equation (1) gives the total stored energy of a rotating mass. In practice, it is difficult to access all of the stored energy in the rotor. For instance, since $P = T\omega$ where P is power and T is torque, discharging the flywheel at constant power from full speed to a complete stop would lead to infinite torque as the rotor approaches zero speed.

In practice, an energy storage flywheel is operated over a speed range where the maximum speed is set by the allowable stress of the rotor material and the minimum speed is determined by economic or technical limits for the deposition or extraction of energy. A typical standalone flywheel energy storage system may operate over a 2:1 speed range. Since energy scales with the square of speed, a flywheel discharged to half speed will retain 25% of the energy stored at full speed and will have an extractable energy equal to 75% of the total stored energy at full speed.

A further consideration in the design of a flywheel is its cost. In applications, where the flywheel is connected directly to rotating machinery, relatively little energy is stored and incremental cost per unit energy storage is not a critical factor in total system cost. However, for the particular case where flywheels store relatively large amounts of energy (>1 kWh) rotor cost is a significant consideration in the design of the system. Additionally, irrespective of rotor material selection, non-rotor energy-related costs, power-related costs, and balance of system costs are typically the dominant elements in the total cost of a flywheel system. Rotor material selection has an appreciable impact on the cost of the remainder of the system.

The cost of energy storage is commonly expressed in terms of \$/kWh. For a flywheel rotor:

$$C_r = \frac{K_{\text{sec}}}{K_{\text{ex}}} \frac{p\rho}{\sigma_{\text{allowable}}},$$

where C_r is the cost of the rotor in \$/kWh, K_{ex} is the fraction of total stored energy that is extractable (commonly 0.75), K_{sec} is a shape factor, p is the price of the fabricated rotor material in \$ per unit mass, and $\sigma_{\text{allowable}}$ is the allowable stress at the end of the design life of the rotor. For any particular rotor geometry, the most cost-effective rotor material will have the optimal combination of lowest cost per unit mass, lowest density, and the highest allowable operating stress that affords adequate safety margins at the end of life.

When an allowance is made for a loss in strength due to cyclic fatigue over 10^6 cycles, the following table gives an approximation of the incremental cost of rotor material.

In Table 1, the stress figures in the first column correspond to the yield strength of various grades of steel when new. The carbon composite values are based on a filament wound construction using 700,000 psi fiber with 65% fiber fraction and proven safety factors for high cycle life. It is important to recognize that this metric applies only to the incremental cost of increasing the mass of a rotor to store more energy. This metric does not reflect the

Table 1: Flywheel rotor material cost per unit energy.

Material	$/kWh	mass/kWh
Carbon composite	1,200	1
260,000 psi steel	1,800	$7x$
160,000 psi steel	2,000	$12x$
90,000 psi steel	4,000	$24x$

balance of system costs such as the motor, bearings, and the housing that are generally greater than the cost of the rotor itself.

The third column of Table 1 indicates the ratio of mass for a steel rotor normalized to the mass of a carbon composite rotor storing the same amount of energy. For example, a rotor constructed from heat-treated steel with a tensile strength of 260,000 psi will weigh $7x$ more than a carbon composite rotor storing the same amount of energy when both are designed for a life of 10^6 cycles. A heavier rotor requires higher capacity bearings and a heavier, more costly housing. Therefore, not only does a carbon composite rotor have lower incremental cost per unit of stored energy, the balance of systems costs is lower as well.

3 History of Flywheels

Flywheels have been used as a component of manufacturing equipment since their application in ancient potter's wheels. Potter's wheels first appeared in China and Mesopotamia between 6,000 BCE and 2,400 BCE.[4] In their earliest form they were turntables mounted on a pivot to allow the potter to view his work from all around. These evolved into more complex assemblies with two disks connected by a vertical axle. The upper disk was the turntable on which the work piece rested. The lower disk was a heavy kick wheel located near the floor. The disks were typically made from wood, stone or clay. This configuration allowed the potter to form the clay with both hands while keeping the wheel rotating with his foot.[5] Long before the advent of carbon and glass fibers, at least one type of potter's wheel is known to have been constructed out of a composite material using unidirectional hoop-wound bamboo embedded in a clay matrix.[6]

It is also known that small flywheels were fitted to spindles used in the making of thread.[7] In this process, the raw material from a fiber crop such as flax is placed on a holder known as a distaff. Fiber is drawn from the distaff onto a spindle. In the earliest embodiment, the spindle was a straight stick or bone that was held in one hand while the other hand was used to twist the fiber as it was drawn from the distaff. The addition of a small stone flywheel to the base of the spindle sped up the process considerably.[8]

Starting around 1200 AD the distaff and spindle system began to give way to the spinning wheel. Numerous types of spinning wheel exist including the hand-turned great wheel and the foot-powered treadle wheel. In all cases, the operator turns a flywheel that is generally known as a drive wheel. The drive wheel in turn is connected to a much smaller flyer or bobbin via a drive band. The bobbin spins at a much higher speed than a hand-held spindle. The spinning wheel with its larger flywheel improves on the productivity of the distaff and spindle by an order of magnitude or more.[9]

Flywheels remained small and human-powered until the industrial revolution. The Newcomen engine invented in 1712 was the first practical machine to use steam to perform useful work. It was a linear engine typically used to power a water pump.[10] Even though rotative power was highly sought as a means to drive mills and other manufacturing, the conversion of reciprocating power to rotary power remained out of reach for nearly 70 more years.

In 1780, James Pickard patented an arrangement of cranks and ratchet gears enabling a reciprocating steam engine to perform rotative work. Concurrently, James Watt and Matthew Boulton, stymied by the Pickard patent, developed a workaround using sun and planet gears. Work on the Watt and Boulton invention began in the mid-1770s and their invention was patented in 1781. Both machines used flywheels to smooth out the flow of energy.[11]

The "wagon wheel" configuration found in these early engines remained the most common flywheel shape for another 150 years and is still in use today. In the embodiment of this era, flywheels used heavy rims built from cast iron and later steel to damp pulsations in reciprocating engines or reciprocating loads.

Throughout the 1800s, machines of all types grew in power and size culminating in the massive Corliss engines of the late 1800s such as the Centennial Engine shown in Fig. 1.[12] The largest engines produced 1,400 hp, stood more than 40 ft tall, and employed flywheels 30 ft in diameter.

These massive flywheels stored relatively little energy by today's standards. For example, a large flywheel from the turn of the last century could weigh 50 tons and operate at surface speeds around 5,280 ft/min.[13] Conservatively, assuming that all of the mass of the rotor is concentrated at

Fig. 1. The centennial steam engine.

the periphery, the total stored energy of such a wheel would be less than 5 kWh. In comparison, a modern flywheel in use today for stabilizing the electric grid weighs about 1 ton.[14] This wheel is made from carbon and glass composite. It is less than 1 m in diameter and 2 m long. Because it spins with a surface speed in excess of 600 m/s (compared to 20 m/s for the Corliss flywheel), the modern flywheel stores more than 25 kWh of usable energy.

At the height of their use in large stationary engines flywheel bursts were fairly common, occurring several times per year. These events were quite serious and often the result of a failure of a governor.[15] A particularly large failure such as a 50,000 lb flywheel coming apart at 75 RPM would sometimes result in the destruction of the building in which it was housed.[16]

Today, the flywheel is a basic component of the internal combustion engine. Approximately one billion engine-powered vehicles exist in the world today[17] and nearly every one employs a flywheel. Here they damp out torque pulses caused by the periodic firing of cylinders. In this application, as in the steam engine, energy is stored very briefly before it is used, namely, for less than one revolution of the wheel itself.

4 Advanced Flywheels for Energy Storage and Power Management

A new class of standalone flywheel energy storage systems has emerged and is referred to here as Advanced Flywheels. These systems have a number of defining attributes.

(a) *Standalone flywheel module*: The flywheel is housed in an enclosure that separates it from the application or machinery to which it is connected.
(b) *Electrically connected to application*: With few exceptions, the flywheel is electrically connected to the application through a motor that is built into the flywheel module and a variable speed drive that sources or sinks power from or to the motor. The exceptions to this rule are rotors that are mechanically connected to the application via magnetic couplings that transmit torque into and out of the vacuum housing and variable speed transmissions that match spin speed of the flywheel to the shaft speed of the application.
(c) *Operation in vacuum*: Generally, the flywheel module is evacuated to minimize aerodynamic drag on the rotor. In some applications where higher losses are tolerable, developers use helium-filled enclosures running at reduced pressure.
(d) *High-speed rotor*: The rotor operates at speeds limited by the allowable stress of the rotor material and the end of design life ranging from 200–400 m/s for metal rotors to 500–1000 m/s for carbon composite rotors.
(e) *Long-life bearings:* Potential advantages of flywheels over other forms of energy storage are high cycle life and long calendar life. In order to fully realize these benefits in stationary applications requiring continuous operation, high-speed bearings may be expected to last 20 years or more without maintenance or replacement.

The power at which the system operates and the energy stored by the flywheel are selected independently. Since the ratio of energy to power has units of time, it is common to express the capability of a flywheel in terms of output power provided for a specified duration. In one example, a flywheel system designed to serve a ride through application may provide 1 MW for 3 s. This system provides 0.8 kWh of extractable energy. In a second example, a flywheel system designed to provide frequency regulation services may provide 100 kW for 15 min. This system provides 25 kWh of extractable energy. These two systems will be very different. The machine of the first example will have a large motor and relatively relaxed requirements for rotor design as it stores relatively little energy. The machine of the second example will require a much smaller motor than the machine of the first example and a much more energetic rotor.

The ability to select power and energy completely independently sets flywheel energy storage apart from the electrostatic and electrochemical energy storage of capacitors and batteries. While battery design may be tailored to short or long discharge times, typically expressed in C-rate (units of 1/discharge time in hours), the flexibility in selecting flywheel discharge time spans several orders of magnitude.

Flywheels have inherently long cycle life. The material properties of the metals and composites used in high-speed flywheels are sufficiently well understood to allow for a design life exceeding 10^6 cycles.

The state of charge of a flywheel and its availability are known with high precision and accuracy. Individual modules in use today range in energy capacity from a fraction of a kWh to 50 kWh.

The total cost of a flywheel system comprises three distinct cost centers.

(a) *Elements that scale with stored energy*: For a particular geometry and rotor material, rotor weight scales with stored energy. Components and subsystems that scale with rotor weight include the bearings, the housing, and structural hardware affixing the flywheel assembly to the application or to its foundation.
(b) *Elements that scale with power*: For a flywheel system with an integral motor-generator, elements that scale with power include the motor itself, the motor drive, cabling, switchgear, and feedthroughs.
(c) *Balance of System*: The remaining elements of the flywheel system scale weakly if at all with power or energy. These include the vacuum system, sensors, telemetry, diagnostics, and controls.

The cost for a complete flywheel system may be expressed as follows:

$$C_{\text{system}} = (A * \text{Power}) + (B * \text{Energy}) + C_{\text{BOS}},$$

where elements that scale with power have a cost factor of A with units of \$/kW, elements that scale with stored energy have a cost factor B with units of \$/kWh, and balance of system costs C_{BOS} have units of dollars.

Flywheel systems in service today have costs spread across all three cost centers. There appears to be no reported instance of an existing system in operation where the cost of the rotor exceeds 20% of the cost of the system.

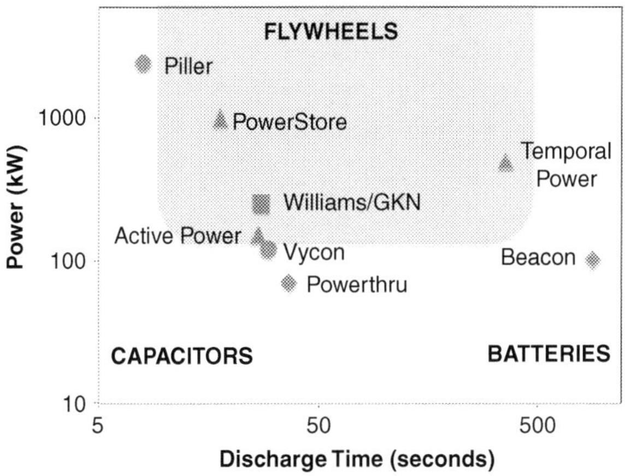

Fig. 2. Cost competitive parameter space for flywheels.

It is important to consider these aspects of flywheel system cost when comparing flywheels to other energy storage technologies. It is convenient to compare energy storage technologies on the basis of \$/kWh at the module or system level. This approach is valid for batteries and capacitors since, once \$/kWh has been ascertained at the cell level, system cost may be estimated by simple scaling. However, while a value for \$/kWh may be calculated for a flywheel system at the system level, scaling is not valid absent consideration of power, energy, and balance of system costs.

The competitiveness of ultracapacitors, batteries, and flywheels may be expressed in terms of power and discharge time. Figure 2 shows regions where flywheels, capacitors, and batteries are most cost effective. Also shown are the ratings of standalone flywheel energy storage systems from a number of current manufacturers. The shaded area indicates the region of the parameter space where flywheels are particularly advantageous.

Conceptually, with an incremental energy cost having decreased to \$20,000/kWh,[18] cycle life as high as 10^6 and moderate balance of system costs, ultracapacitors should be cost-competitive at any power level for discharge times up to several seconds. However, current applications requiring short duration discharge (3 s) in excess of 1 MW such as electromagnetic aircraft launch and ride through backup power are presently served by rotary systems. For applications requiring discharge times of 15 min or more batteries are a widely used solution.

Flywheels are a cost-effective solution for applications requiring power for more than several seconds and up to several or tens of minutes, particularly when high cycle life is required. For applications requiring less than 100 kW, balance of system costs make flywheels less cost competitive.

In order to fully realize the potential advantages of standalone flywheel energy storage systems, the systems must have acceptable acquisition and operating costs as well as long calendar and cycle life. Flywheel technology development focuses primarily on three areas: the rotor, the bearings, and the integral motor-generator.

4.1 *Rotor Design*

As mentioned above, the three prevalent forms for a rotor are disk-like, solid cylinder, and drum like, meaning a thick- or thin-walled cylinder.

Here it is useful to define terms. "Rotor" refers to the entire rotating assembly comprising the mass that stores the kinetic energy and the rotating portions of the bearings and motor. For a disk or solid cylinder the mass storing the energy makes up the bulk of the rotating element to which bearing and motor components are affixed. For a drum-like rotor the definition is a little more complicated. The rotor comprises the thick walled-cylinder that stores most of the energy as well as a distinct shaft and one or more hubs in addition to the motor and bearing elements. The term "rim" applies to the drum-like part of the rotor that stores most of the energy.

Ideally, the most effective use of rotor material would be to concentrate the mass at the periphery where the velocity is the highest. This suggests that the most cost-effective rotor configuration will be a thin-walled rim where all of the mass is at the periphery.

At first glance this appears obvious. However, for an isotropic material the stress in a thin-walled cylinder will be about twice as high as the stress in a disk or solid cylinder when both are spinning with the same surface speed. When normalized to comparable peak stresses, the solid cylinder will store more energy than a thin cylinder of comparable mass.

For example, Eqs. (2) and (3) may be compared for a thin rim and a disk made from the same material having Poisson ratio of 0.3.Setting peak stresses equal and solving for surface speed gives $V_{\text{thin-rim}} = 0.64 V_{\text{disk}}$. Applying this result in Eq. (5) shows that the specific energy of the disk will be 25% higher than the specific energy of the thin rim.

A disk or solid cylinder is much simpler to construct than a rotor assembly using composite materials. While a rim made from high strength carbon fibers offers much higher specific energy than a solid cylindrical metal rotor and will be much lighter when storing a comparable amount of energy, the simplicity of the solid cylinder makes it more economical in certain applications.

Large steel disk or solid cylinder flywheel rotors may be built to operate at surface speeds up to 200–300 m/s with readily available grades of steel. However, a large cylindrical steel rotor intended for operation at very high speed (>400 m/s) must address an additional engineering challenge. In order to operate at such high speed, exceptionally high strength steel is required. These steels tend to be brittle and have poorer fatigue and fracture behavior than mild steel. The strength required for the highest strength steel rotors can only be attained through forging.

Forging is a process through which metal, steel in particular, is deformed plastically under extremely high pressure to reorient the grain structure of the metal thereby improving its strength. Steel forging almost always involves the application of high temperature as well. The process is performed using massive hammers, presses, and rollers and is broadly used in the production of high strength plates, fittings, rings, and engine components. The influence of forging on material properties decreases with thickness and forged steel parts thicker than several inches are relatively uncommon. This presents a concern for large forged steel cylindrical flywheel rotors since the highest stress in a solid cylindrical flywheel rotor occurs along the centerline. Material properties on the centerline are critical yet difficult to control and impossible to inspect in the finished part.

It should be noted that the inner portion of the disk or solid cylinder is traveling slower than the outer portion with a velocity that is proportional to radius. Since energy scales with the square of speed, the inner portion stores much less energy than the outer portion. The mass of the rotor could be reduced substantially if the inner portion could be removed without structurally compromising the rim. This is the objective in the use of oriented materials for the thick or thin-walled rim.

Oriented materials, such as carbon fiber or even drawn steel wire are extremely strong in the fiber direction. In the fiber direction, mid-grade carbon fibers have twice the strength of high-grade forged steel. The potential life of composite rims far exceeds the potential life of steel rims. Presently, composite centrifuge rotors are used on a large scale to enrich uranium. Approximately 500,000 composite centrifuge rotors have been spinning continuously for more than 20 years. These rotors are several meters long and operate at surface speeds in excess of 1,000 m/s. The design life of these

Fig. 3. Advanced composite flywheel rotor. Courtesy Beacon Power LLC.

rotors is 35 years. Flywheel rotors derived from centrifuge technology are expected to be capable of 10 million deep discharge cycles.[19]

Various winding processes are used to construct flywheel rims using oriented material, the most common of which is wet filament winding of carbon or glass fibers. Rims constructed using oriented material, such as the advanced composite rotor shown in Fig. 3, are highly complex and pose a number of engineering challenges. A typical composite flywheel rotor will be constructed by wet filament winding glass and carbon fibers onto a mandrel. In this process a "tow" or thin tape comprising thousands of fibers is drawn through an epoxy bath and wound onto the mandrel where it cures into a solid cylinder.

Filament wound composites are highly anisotropic. The stiffness and strength of the material are very different in the circumferential, axial and radial directions. Circumferential strength and stiffness may be 1,000 times greater than strength and stiffness in the radial direction.

These thick composites are engineered materials meaning that the highly anisotropic material properties such as strength and stiffness in various directions depend not only on the type of fiber and matrix material used but also on details of the manufacturing process. Fibers with different properties may be used in different regions of the rotor or co-mingled to tailor strength, density, and modulus to the local need at the lowest cost.

An engineering challenge unique to the composite flywheel rims is radial tensile stress that can lead to delamination. The outer layers of a composite flywheel are traveling at higher speed than the inner fibers and experience greater centrifugal acceleration. The outer layers will tend to dilate more than the inner layers. A rim constructed from a single type of fiber will experience radial tensile stress throughout its thickness having a peak value near mid-thickness. In this example, delamination or cracking is resisted only by the strength of the matrix material. For this reason, radial delamination is the limiting failure mechanism for thick composite rims.

Various approaches to composite rim design are used individually or in combination to suppress delamination. These include press-fit rings, graduated modulus, graduated density, mass loading, bore loading, and winding under tension.

(a) *Press-fit rings*: The thick composite rim is constructed from a number of thinner concentric rings. The rings are produced with a shallow taper and are sized so that an interference fit exists between the rings. Pressing the rings together imposes a radial compressive stress that offsets the radial tensile stress caused by spinning at high speed.
(b) *Graduated modulus*: The composite rim is constructed of rotor materials with varying moduli of elasticity or stiffness. When lower modulus or softer fibers are used at the bore, they will dilate more than the stiffer outer layers reducing radial tensile stress.
(c) *Graduated density*: The composite rim is constructed using lower density outer fibers such as carbon and higher density inner fibers such as glass. The higher density inner fibers experience higher centrifugal acceleration and greater radial growth than lower density fibers. The greater radial growth helps to reduce radial tensile stress. Using carbon outer fibers and glass inner fibers has the benefit of using both higher modulus and lower density in the outer fibers.
(d) *Mass loading*: A layer providing weight but no strength is installed along the bore of the composite rim. At high speed the weight creates an internal pressure on the bore of the rotor, much like high hydrostatic pressure in a thick walled pipe. This introduces a compressive stress that counteracts the radial delamination stress. Here both the compressive stress and the radial stress increase with speed.

(e) *Bore loading*: Similar to press fitting rings, the composite rim is pressed or thermally assembled onto a metallic hub with large interference. Like mass loading, this introduces a radial compressive stress. Unlike mass loading, compressive stress from bore loading decreases at higher speed.
(f) *Winding under tension*:[20] During the winding process it is possible to control the tension applied to the fiber as it is wound. By varying winding tension it is possible to build significant radial compressive stress into the part.

Using graduated modulus, graduated density, and mass loading together has the drawback of increasing circumferential tensile stress along the bore of the rim.

An interesting attribute of the relative weakness in the radial direction of a thick composite rim is its positive impact on safety. The most destructive type of flywheel failure is one in which a solid rotor breaks up into three large pieces. Containment of such a failure requires a structure many times more massive than the rotor itself. The use of such a massive containment structure would render any flywheel energy storage system uneconomic. In contrast, a thick, filament wound rim with significant radial tensile stress cannot fail in this way. Irrespective of the details of construction, thick filament wound rims tend to fail by crashing and abrading composite over an extended interval.[21]

The hub represents a particularly difficult challenge in the design of a thick, filament wound flywheel rotor. The purpose of the hub is simply to attach the composite rim to a shaft. The problem arises in matching strain of the hub to the rotor. The composite rim experiences the highest centrifugal acceleration and therefore the highest stress and the highest strain. The hub experiences centrifugal acceleration ranging from zero along the axis of rotation up to the centrifugal acceleration equal to that experienced by the rim. Unless care is taken, the rim can simply grow away from and fall off of the hub.

This is especially important when a metal hub is used with a composite rim. Carbon composite may strain as much as 2% before failing while steel or aluminum will fail at a strain of less than 1%. Metal hubs can be used with composite rims but doing so requires careful consideration of the hub geometry. Often a very high stress press fit is used to drive the hub into a state of compression during assembly.

A composite hub may be constructed in such a way that the engineered material properties of the hub allow it to strain along with the rim. For example, a composite hub constructed with the following properties will support a composite rim at high speed. The hub must have a strain to failure equal to the strain to failure of the rim, the density of the hub equals the density of the rim, and the isotropic stiffness of the hub is less than or equal to half of the circumferential modulus of the rim.[22] Other approaches such as dished or spoked hubs have also been used successfully.

No blanket statement can be made asserting the cost-effectiveness of one configuration over another without careful consideration of the specific requirements of the application and the impact of rotor shape and material selection on balance of system costs. Both the isotropic disk and the oriented cylinder have their advantages.

4.2 *Rotor Dynamics*

Irrespective of size or construction, flywheel rotors tend to share a common set of rotor dynamic traits, in that inertial effects are especially important.

It is highly desirable for any rotating machine not to operate continuously at a speed that excites vibrational modes. Peak vibration occurs when spin speed equals a "critical frequency". These frequencies depend on the mass of the rotating element, the moments of inertia of the rotating element, the stiffness of the rotating element, and the stiffness of the bearing mounts.

Gas turbine engines are often accelerated through a "first critical" and then operate at a nearly constant speed between the first and second critical frequencies. A flywheel, on the other hand, will operate over a speed range of 2:1 or more so a goal of flywheel rotor dynamic engineering is to locate the solid-body modes below the operating speed range and the flexural modes above the operating speed range.

Here, a Jeffcott[23] rotor model is used to illustrate the interesting rotor dynamic behavior exhibited by flywheels. The Jeffcott rotor is a lumped parameter model where the rotor has a concentrated mass m, a polar moment of inertia about the spin axis I_p, and a moment of inertia transverse to the spin axis I_t. In the most fundamental case, the shaft is treated as rigid and the shaft runs on a pair of compliant bearings on either end of the rotor with a combined stiffness k.

The solid body modes are referred to as "whirl". The two main types are cylindrical and conical where these terms describe the shape of the orbit of the principal moment of inertia of the rotor about the spin axis. At critical speeds, energy is transferred from the spin axis to the whirl

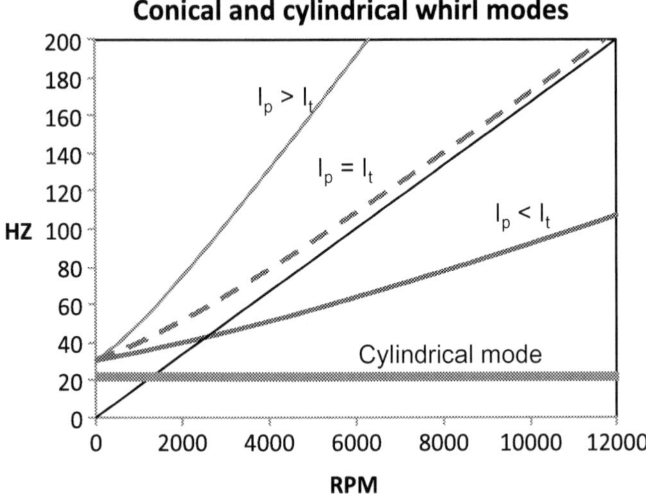

Fig. 4. Whirl plot for fly wheel rotors.

mode, sometimes with disastrous results. For the simple model here, the cylindrical mode is a function of only rotor mass and bearing stiffness and has a frequency f given by:

$$f = \sqrt{\frac{k}{m}}.$$

This frequency is constant and does not vary with the spin speed of the rotor. The frequency of the tilt mode behaves somewhat differently as gyroscopic stiffening plays an important role. Figure 4 below shows a whirl map for three different rotors, each having the same mass and bearing stiffness but differing in ratio of I_p to I_t.

The x-axis gives the spin speed of the rotor. The y-axis gives the frequency of the whirl modes. A line with a slope of 1:1 is the locus of points at which spin speed equals whirl speed. If these points cannot be avoided altogether it is desirable to cross them quickly.

Conical whirl modes are not constant. As a rotor speeds up, gyroscopic stiffening causes the whirl frequency to change. Because all three rotors have the same mass and the same bearing stiffness, they all experience the same cylindrical mode which they will traverse at a low speed. For a drum-like rotor $I_p < I_t$. The conical whirl frequency increases gradually and the rotor traverses this mode easily at low speed. A disk-like rotor has $I_p > I_t$

which causes the conical whirl frequency to rise so rapidly that this mode is never excited.

A special case is where $I_p = I_t$. In this case, the conical whirl speed asymptotically approaches the spin speed. The practical impact of this phenomenon is that as the rotor approaches critical speed, vibration gets worse until damage occurs as it is impossible for the rotor to traverse this mode. It is a fundamental attribute of the mass properties of the rotor and is to be avoided in design as it cannot be fixed or "balanced out" in operation.

As stated previously, another objective of flywheel rotor design is to drive free-body natural frequencies, like the pitch at which a bell rings, above the operating speed range. Applying damping to reduce peak vibration would appear to be an obvious solution but it is problematic. Adding damping to the rotating assembly introduces a mechanism that may transfer spin energy into whirl.[24] Large vibration can result. This phenomena has been the cause of a number of spectacular rotating machinery failures over the years. This caveat pertains to the use of damping in the rotating frame of reference only. The use of damping in the stationary frame of reference is important in bearing mount design and is also used in conjunction with flexible shafting.[25]

Flexural modes can be avoided by making the rotating assembly so rigid that the lowest free-body natural frequency occurs above the highest operating speed. For drum-like rotors this is accomplished by making the rim fairly thick. For drum-like and solid cylinder rotors this is accomplished by avoiding high ratios of length to diameter. In either case, the avoidance of flexible modes of vibration constrains the overall size of the rotor, the aspect ratio, also referred to as the l:d ratio, and the ratio of rim thickness to rotor diameter.

When rotors are designed to reach the maximum speed allowed by the strength of material, the maximum rotor size is constrained by flexural modes. The current practical limit for the maximum amount of energy that can be stored in a single stress-optimized rotor appears to be 40–60 kWh for carbon composite rotors and 50–80 kWh for steel rotors. Although it is a less cost-effective use of rotor material, a rotor may be designed to operate well below its stress limit. A rotor operating well below its stress limit will generally operate well below flexural mode limits as well. In this case, much larger rotors are possible.

The rotor dynamic behavior of flywheel rotors is influenced by spin stiffening, strain softening, and a wide variety of other nonlinear phenomena. Even so, the basic principles outlined here dictate the shape of all flywheel rotors in use today. The confluence of steps taken to manage rim stresses, avoid or traverse whirl modes, and avoid flexural modes of vibration drive flywheel designs into a relatively small number of practical form factors.

4.3 *Bearings*

Flywheel bearings support the rotor while allowing it to spin freely. Certain attributes of flywheel operation place significant demands on the bearings and the bearings are typically the life-limiting component of a flywheel system.

Flywheel rotors tend to be fairly heavy when compared to other rotating machines operating at comparable spinning speed. The load carrying capacity of the bearing must be sufficient for the weight of the rotor. Flywheels operate at high surface speed which, depending on the size of the rotor, can translate to spin speeds of more than 60,000 RPM (1,000 revolutions per second). In order to minimize operating cost it is desirable to use bearings that have very long life so that they will not have to be replaced over the operating life of the system. Bearing drag causes a freely spinning rotor to decelerate. Power is required to overcome this drag to maintain the flywheel at a constant speed. This incurs an ongoing energy cost that should also be minimized. Operation in vacuum constrains options for lubrication.

The two most common types of bearings found in flywheels are ball bearings and active magnetic bearings. Permanent magnets or solenoid coils can provide passive lift and are sometimes used in conjunction with ball bearings or active magnetic bearings to reduce the load on the bearings.

(a) Ball Bearings

Ball bearings are a type of rolling element bearing with low drag and the potential for very long life. Ball bearings can support radial and axial loads. The main types of radial ball bearing are angular contact and deep groove.

Figure 5 shows the cross-section of an angular contact ball bearing. The inner ring or race of the bearing is mounted on the shaft. The outer ring or race is mounted in a bearing holder. The bearing holder may include features such as compliance, damping and cooling. When multiple bearings are used, precise alignment of the shaft and the bearing holders is crucial.

As shown in Fig. 5, the balls run in grooves that have a slightly larger radius of curvature than the balls. An axial load such as the weight of the rotor takes out the play or looseness of the bearing. The balls contact the races in regions referred to as a Hertzian contact patch.[26] The load passes from the inner race to the outer race through the contact patches. The ball

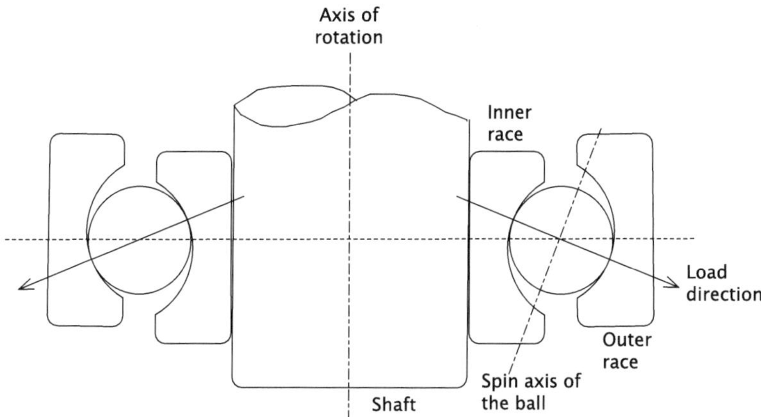

Fig. 5. Ball bearing cross-section.

spins about an axis that is perpendicular to the load line. At rest and at low speed the center of the ball also lies on the load line.

Minute details have a profound impact on bearing life. The Hertzian contact patch has a finite extent. At the center of the contact patch pure rolling occurs and there is no sliding between the ball and the race. Away from the center of the patch sliding does occur. In order to minimize wear due to sliding contact, lubricant is required. A layer that is only microns thick will suffice but cleanliness is critical. Any contaminant or surface aberration larger than the thickness of the lubrication film will result in wear.

A bearing is said to operate at high speed when the DN number of the bearing exceeds 800,000 where DN is the product of bearing diameter in mm multiplied by spin speed in RPM. For instance, a 25-mm bearing operating at 40,000 RPM would have a DN of 1,000,000. At high speed, dynamic forces within the bearing itself become important. In this case, the centrifugal force acting on the ball causes the load line to flatten out. At high speed, a line through the two contact patches will no longer pass through the center of the ball. This introduces out-of-balance forces that are caused by the balls themselves that increase stress at the contact patches.

For a clean bearing that is well lubricated, properly loaded and precisely aligned, life will be determined by fatigue limits of the balls and the race. Bearing life is forecast in terms of number of revolutions. The classic approach to bearing life calculations was formulated by Lundberg and Palmgren in 1947 using the Weibull probability distribution of metal fatigue to establish the basic theory.[27] In practice, this has been reduced to a simple calculation:

$$L_{10} = \left(\frac{C}{P}\right)^p,$$

where L_{10} is number of cycles (rotations) that will occur before 10% of the bearings have failed, C is a factor called the dynamic load rating, P is the equivalent bearing load, and p has a value of 3 for ball bearings. In theory, infinite life is possible when the equivalent bearing load is vanishingly small. In practice, extremely light loads result in balls skidding along the races with unacceptable wear. Bearing life forecasts beyond 10^{10} cycles are rare and beyond this the Lundberg–Palmgren approach becomes less reliable.

This represents a problem when predicting bearing life for a flywheel. For example, a 16,000 RPM flywheel operating continuously for 20 years will accumulate 1.7×10^{11} revolutions. Fortunately, recent advances in ball bearing theory suggest that there is a surface contact pressure below which conventional contact fatigue limit theory no longer applies. In this regime, much longer cycle life is possible.[28] Recent data indicate that when peak surface contact pressure is maintained below 300,000 psi (2,000 MPa) cycle life exceeds conventional theory by more than an order of magnitude.

Flywheel developers currently apply the following practices as means of attaining long bearing life.

Supporting most of the weight of the rotor with permanent magnets greatly reduces the load on the bearings. Without the use of passive magnetic assistance, the ball bearings must be sized so that Hertzian contact stress is within acceptable levels. Having a large contact area means using a large bearing with a correspondingly high DN. Removing the load due to rotor weight from the bearings allows the use of much smaller bearings operating at much lower contact stress with much lower DN values. This has a geometric impact on bearing life in that life improves with the cube or greater of the ratio of full load to reduced load.

So-called ceramic or hybrid bearings use Silicon Nitride (Si_3N_4) balls running in steel races. Silicon nitride balls are lighter than steel reducing inertial effects when run at high DN. They are also stronger and have much better fatigue properties than steel balls. A preferred configuration found in a number of current machines is angular contact bearings installed as duplex pairs. In this configuration matching bearings are clamped together giving good control of local alignment and contact stress. Best practices for lubrication, load management, and alignment are well know and critically important.

(b) Active Magnetic Bearings (AMB)

As shown in Fig. 6, active magnetic bearings systems levitate rotors and maintain shaft position by sensing rotor position and controlling forces applied in the radial and axial directions. Two types of actuators are used: radial and thrust or axial.

The radial magnetic bearing stator comprises electromagnets with a number of radially-oriented poles. The electromagnets are coils of wire wound around salient poles where all of the radial poles share a common back iron. The rotating portion of the radial bearing is usually made from a stack of thin laminations of magnetic steel that are mounted on the shaft. The radial bearing stator surrounds the rotating lamination stack. Current is applied to all of the coils. The current magnitude is monitored and adjusted continuously in order to control the position of the shaft.

The rotating portion of a radial magnetic bearing experiences time varying magnetic fields as it travels past the stator poles. This changing magnetic field induces recirculating electrical current within the bearing rotor known as eddy currents. Laminations have thin insulating layers on their surfaces inhibiting the formation of eddy currents and thus reducing loss to very low levels.

The stator of a magnetic thrust bearing comprises a coil of wire in a u-shaped channel. The thrust disk is mounted to the rotor shaft. Magnetic fields in a magnetic thrust bearing are more or less constant so these parts may be made from solid steel and do not require laminations.

Any object has 6 degrees of freedom of movement: x, y, and z translation and rotation about the x, y, and z axes. By convention the z-axis is the spin axis and spin speed is controlled by the motor. The active magnetic bearing controls the translation of the rotor in three axes and tilt in the other two spin axis and is sometimes referred to as a 5-axis system.

Thrust-bearing control and radial-bearing control are decoupled. Axial (z) position of the rotor is measured and the thrust bearing current is adjusted to control the axial position of the rotor. The thrust bearing almost always uses upward and downward facing actuators as this increases stiffness and controllability when compared to using a single thrust actuator.

Radial bearings are usually used in pairs that are well separated. The x and y position of the shaft is measured at each bearing. A control algorithm

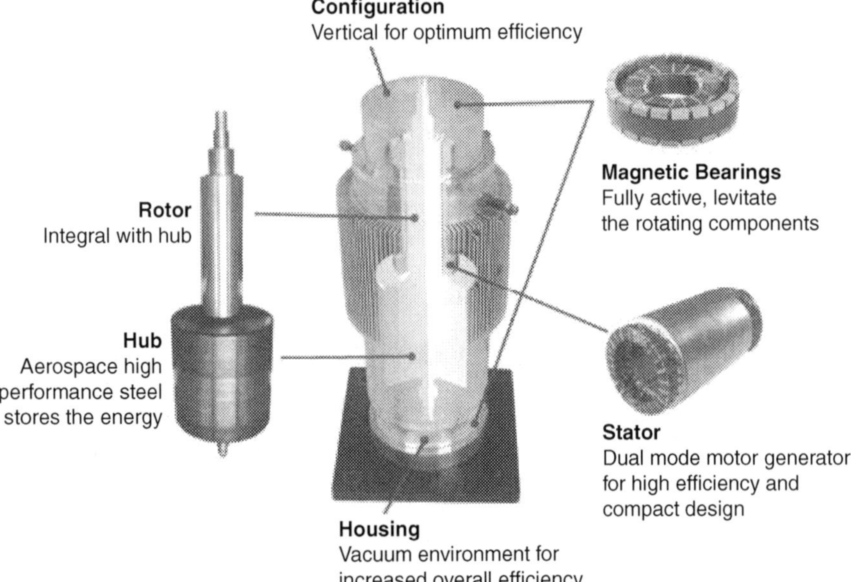

Fig. 6. Flywheel system using active magnetic bearings Courtesy Calnetix, Inc.

uses these displacement measurements to control the x and y translation of the center of gravity of the rotor and the x and y tilt.

AMBs have a number of attributes that are very useful in flywheel systems. First and foremost, because there is no contact between the rotor and the stator, there is no wear. Since the clearance between the stator and the rotor may be as much as 1 mm, the requirements for alignment of radial magnetic bearing are much less stringent than alignment requirements for ball bearings. Drag losses occurring within the bearing and power consumed in operating the bearing system are very low. AMBs run well in vacuum as they do not use lubricants. They are used extensively in turbomolecular vacuum pumps.

The continuous sensing of rotor position provides information about rotor health. A change in balance of the rotor may indicate that the structural integrity of the rotor is compromised. The magnetic bearing controller continuously monitors rotor balance and can be set to indicate a fault or shut down the machine when or if the rotor unbalance grows beyond a specified value.

It is possible to configure AMBs to be silent and vibration free. A rotor will have some residual unbalance so running on hard ball bearings transmits a once per rev force to the housing. This may result in an audible vibration. For instance, a flywheel spinning on ball bearings at 15,696 RPM will have a once per rev frequency of 261.6 Hz and may emit a tone at the pitch of Middle C.

An AMB can completely eliminate the transmission of once per rev forces to the housing by using a process called synchronous disturbance rejection.[29] A notch filter or other control strategy is implemented so that the bearing has zero stiffness precisely at the spin speed and relatively high stiffness at any other frequency.

Flywheels running on AMBs also use touchdown or backup bearings. Touchdown bearings may include heavy-duty ball bearings or bushings that are installed so that there is a clearance between the touchdown bearing and the shaft. Under normal conditions the touchdown bearing does not spin. The purpose of the touchdown bearing is to support the rotor in the event of an AMB failure to provide time for a safe discharge of the stored energy.

4.4 *Motor-Generator*

Standalone flywheel systems are generally charged and discharged using an integral motor-generator, sometimes just referred to simply as the motor. The motor may be an integral part of the inertial mass or it may be attached to the rim via a hub and shaft. A wide variety of motor types have been deployed in flywheel systems including induction, homopolar, synchronous reluctance, and switched reluctance, as well as many different configurations of permanent magnet machines.

The suitability of a motor type depends on the cost, efficiency, and thermal requirements of the application. Key determinants in motor selection are duty factor and rotor temperature limits. These factors determine the limit to rotor heating and the efficiency required to stay below this limit.

Rotor heating is caused by the circulation of electrical current on the rotor portion of the motor-generator. While the stator may be cooled through immersion or by being mounted to a cooled housing, a rotor operating in vacuum can transfer heat to its surroundings through radiation alone. Radiation heat transfer is an extremely inefficient mechanism until a large temperature difference exists between the rotor and its surroundings. Induction motors have the highest on-rotor currents and permanent magnet motors have the lowest. Switched or synchronous reluctance motors have significant on-rotor heating but may be turned off completely when not in use. In contrast, the field of a permanent magnet rotor is always exciting the stator.

Flywheels used in uninterruptible power supplies (UPS) tend to idle at full charge until called upon to support a load during a power outage. The flywheel supports the load until a backup diesel generator is started and synched. The fastest starting backup generators can come up to power in 3 s so relatively little operating time at high power is required from the flywheel.

Steel rotors are typically used in this application. Steel can handle local temperatures of several hundred degrees without compromise to structural integrity. The duty factor for this application is relatively low meaning that if discharge events subject the rotor to significant heating, there is sufficient time for the rotor to cool down by radiating heat to the housing. In this application, induction and homopolar machines are useful.

Steel flywheels used for longer duration energy storage (tens of minutes) will be used in applications with more frequent cycling than UPS flywheels and will be subjected to greater heating. There is no opportunity to cool down between discharge events as the motor may reach thermal equilibrium during a single discharge. Lower loss motor configurations are preferred although at least one developer uses an induction motor with a steel rotor where a coolant is sprayed into the bore of the motor shaft and then collected in a sump.[30]

Thermal constraints are most severe for composite rims operating at high power and high duty factor. This means that the flywheel is charged and discharged at high power often continuously. For composite rotor flywheels operating at high duty factor, it is critically important to minimize rotor heating to the greatest extent possible. Heating of the rotating portion of the flywheel motor, amounting to less than 0.1% of total rated power, can conduct heat into the composite rim and result in excessive rotor temperature. The most effective motors for these systems have proven to be permanent magnet machines where special care is taken to avoid subjecting the rotor to a time-varying magnetic field.

4.5 *Safety*

Potential flywheel failure mechanisms depend a great deal on the configuration of the rotor and the materials used. Structural failure of the rotor is the most serious potential hazard. Steel rotors are susceptible to a tri-hub burst where the rotor fails catastrophically by breaking into three pieces.[31] Composite rims are not susceptible to this failure mechanism but they do have the potential to transfer destructive amounts of kinetic energy into the housing or surroundings. Containment structures have been explored and tested but due to the substantial energy released in a flywheel failure, successful containment requires a structure many times more massive than the rotor itself. Consequently, irrespective of the specific rotor material or configuration, there will exist a speed that causes failure of the rotor and it is necessary to establish a margin between maximum operating speed and failure speed.

A peer-reviewed standards certification process created expressly for flywheels would consolidate approaches to safety that now vary from one developer to another. Absent standards developed specifically for flywheels, standards from related fields may be applied.

Military specifications provide detailed guidelines for establishing such margins for composites.[32] Through the rigorous application of these standards, it is possible to establish a design margin of 1.25 as has been done for some aerospace applications. This process requires substantial materials property and process validation[34] and yields a design margin that may be highly specific to the implementation of a particular composite rim design. As an alternative, it is useful to consider system level, design independent criteria for developing safety margins.

Various standards and practices for related structures report design margins in the range of 2.0–2.4. In particular, certain National Highway Traffic Safety Administration standards for hydrostatic testing of a Compressed Natural Gas fuel container using carbon composites call for a ratio of burst pressure to service pressure of 2.25.[33] This is relevant to carbon composite flywheels as they are also thick cylinders produced by filament winding. A safety margin for stress can be established by applying the square law relationship between stress and speed then limiting maximum operating speed to a corresponding fraction of a demonstrated failure speed.

Margin = 2.00 => max. operating speed = 70.7% of failure speed
Margin = 2.25 => max. operating speed = 66.7% of failure speed
Margin = 2.40 => max. operating speed = 64.5% of failure speed

A number of system faults have the potential to put the flywheel rotor at risk. These include:

- Loss of vacuum
- Loss of coolant (for liquid cooled systems)
- Failure of the motor drive
- Failure of the magnetic bearing controller

- An electrical short which could cause overcurrent in the flywheel stator
- Loss of control signals (e.g. phase angle, bearing temperature, etc.)

Any flywheel system must be proven to withstand all of these faults prior to installation in a commercial or industrial setting.

An important class of flywheel system hazards is the occurrence of an unsupported rotor within the vacuum housing. Depending on the specific configuration of the system, this may be due to a ball-bearing failure, magnetic-bearing failure, or failure of the hub or shaft. This condition is referred to here as a loose rotor event. The hazard associated with a loose rotor event is the possibility of high speed rubbing between the outside of the flywheel and the inside of the enclosure. In order to avoid the need for a containment structure, the system must have features that address this hazard.

An effective method for preventing high speed rub of the rotor during a loose rotor event is to apply a physical restraint to the flywheel spindle, if the configuration includes a spindle, or to the interior of the rotor if the rotor is annular and does not have a spindle. In the case of a spindle, the restraint would be a bushing or bearing. In the case of a rotor without a spindle, this structure could be a stationary post.

In either case, the restraint must prevent contact between the exterior of the flywheel rotor and the enclosure. Management of clearances between the rotor, the restraint, and the housing the must take into consideration compliance of the bearings and the bearing foundations, dilation of the rotor at its maximum operating speed, etc.

5 Applications for Advanced Flywheels

Standalone flywheel systems are in use today around the world. Applications are as varied as racecars and aircraft carriers. The flywheel systems in these applications provide electrical power for durations ranging from 2 s to 15 min. Output power from individual units ranges from tens of kilowatts to more than 500 MW. This section reviews some of the applications where standalone flywheel systems are currently in service.

5.1 *Frequency Regulation*

A large electrical grid has numerous generators and numerous loads. The grid must operate at a nearly constant frequency in order for the generators to remain synchronized. When the amount of electricity consumed differs from the amount that is produced, generator output must be controlled to minimize the difference. If this difference is not minimized quickly, the frequency of the grid may change and it may become unstable. For instance, many generators use a governor to maintain a constant speed. If the load increases faster than the governor can respond, the generator will slow down, momentarily operating at lower frequency. If the load changes are severe enough, or if there are large changes in generator output, generators may not remain synchronized and a wide spread power outage may occur.

Frequency regulation is used to improve the stability of the grid. Usually generators provide frequency regulation as an ancillary service. This means that a power plant may sell the frequency regulation service to the grid operator in addition to selling electricity. In order to provide this service effectively, the power plant must be able to ramp up and down quickly, often responding to a control signal from the grid operator that may change every few seconds or less.

Flywheels are ideally suited to this application as they are capable of millisecond response times and nearly constant cycling. Beacon Power LLC was the first company to provide frequency regulation services using flywheel energy storage. Their first energy storage plant, shown in Fig. 7, was commissioned in Stephentown, NY in 2011. The plant comprises 200 advanced composite rotor flywheels each having the ability to source or sink 100 kW for 15 min. The total output of the plant is 20 MW. The plant operates in response to commands from the New York Independent System Operator (NY ISO) and provides approximately 10% of New York's overall frequency regulation needs. In 2013, a second Beacon flywheel energy storage plant began operation. The second plant is located in Hazle Township, PA where it provides service to the local grid operator, PJM Interconnection LLC.

5.2 *Ramping*

For large grids, variations in load and generation are addressed through frequency regulation ancillary services. Islands and isolated grids are even more susceptible to instability but markets for frequency regulation services do not exist in these areas. Here the problem manifests itself as excessive ramping of the output of conventional generators that are used in conjunction with renewable energy sources.

Wind and solar are intermittent energy sources with output that cannot be controlled to match demand. Other generating assets must be

Fig. 7. Stephentown frequency regulation plant. Courtesy Beacon Power LLC.

used along with renewables in order to make up the difference between electricity from renewable sources and the load. When renewables provide more than 20–30% of the total supply, the other generating assets are subjected to severe ramping resulting in inefficient operation and damage. Island grid operators are beginning to address this problem. As an example, Puerto Rico is requiring that new solar installations limit ramp rates to 10% per minute.At the 20–30% level of renewable penetration, the impact on the grid becomes increasing difficult to manage without storage.[35] Some very large islands are targeting wind penetration exceeding 40% on a capacity basis and storage will be an essential element in implementation.[36]

The fluctuations in power produced by wind and solar vary considerably in frequency, severity, and duration. Variations in solar energy are usually gradual and occur over the course of a day. Wind, on the other hand, can have frequent variations of ±20% lasting less than 2 min.[37,38] Flywheels are particularly well suited for smoothing out the frequent, short duration variations in electricity produced from wind.

There are 50 islands with a combined average power consumption of 53 GW where the potential application of flywheel energy storage would amount to 5% of this value.[39,40] Flywheels are presently being tested as a solution in Scotland where systems produced by GKN (formerly Williams) are installed.[41]

5.3 *Mining*

Open-pit mines around the world use draglines to excavate material. A dragline comprises a large bucket suspended from a boom with wire ropes. A hoist rope supports the bucket from the boom. A dragrope draws the bucket assembly horizontally. Both ropes are powered using electric motors. The machines are often so massive that they must be built on site. Some draglines are grid connected but most are in remote locations and are powered by diesel generators on-site. The load profile of a dragline is cyclic and highly non-uniform. During the lifting phase, peak loads of 6 MW are typical. Lowering the load into a conveyance regenerates as much as 3 MW. This cycle repeats approximately once per minute continuously.

For the dragline to function, multiple generator sets (gensets) with a combined capacity equal to the peak power, run continuously. The generators therefore operate at part load and off of peak efficiency most of the time. This causes a host of problems including high fuel cost and by-products of incomplete combustion known as "wet stacking" which leads to high maintenance costs.

Displacing gensets with flywheel energy storage reduces operating expense in three ways: (1) The reduced number of gensets will operate closer to full power, thus improving efficiency. (2) Energy regenerated during the lower phase is recovered and reused. (3) O&M cost is reduced.

The Usibelli Coal mine in Healy, Alaska operates a 6 MW dragline that is fully electric and is connected to the Golden Valley Electric Association (GVEA) grid. The impact of the fluctuating load was so severe that routine dragline operation caused the lights of other GVEA customers to flicker. In 1982, the Usibelli mine purchased and installed a flywheel to stabilize the load produced by the dragline. The 40 ton flywheel consists of three one foot thick, eight foot diameter steel plates and is connected to the GVEA grid in parallel with the dragline. Since the installation of the flywheel fluctuation on the grid has decreased to only around 500 kW and there have been no issues for other customers.[42]

5.4 *Materials Handling*

Like the mining application, materials handling requires intermittent power to move loads at more or less repeatable intervals. In many cases the peak power required to lift a load is much greater than the average power of the process. Often there is no convenient way to regeneratively recover energy while lowering a load.

Fig. 8. Flywheels installed in a RTG crane. Courtesy Calnetix Inc.

Again, flywheels are well suited to this application as the load durations are short and are repeated frequently. As shown in Fig. 8, the first implementation of flywheel energy storage in material handling is found in Rubber Tired Gantry (RTG) cranes. Approximately 8,000 RTG cranes operate in container terminals around the world.

While ship-to-shore gantry cranes run on tracks and are grid connected, RTG cranes run on tires and are free to move about the terminal. For this reason RTG cranes are usually powered by diesel gensets mounted on the crane. Without on-board energy storage the high peak to average power operation of the diesel engine results in inefficient operation and high emissions. RTG cranes are a significant source of pollutants at sea ports in many countries.

Flywheels may be retrofitted to an existing RTG or installed in a new crane. When installed in a new crane, a smaller diesel genset may be used, further reducing fuel consumption and emissions. Flywheels produced by Calnetix, Inc. have been deployed in RTGs since 2006.

In a typical RTG, a 455 kW-rated diesel drives a 500 kVA alternator that powers the hoist motors. When a container is lifted, each hoist draws about 120 kW (240 kW total) for around 10 s. The power to perform the lift is provided by the flywheel system. During lowering, the hoist motors function as generators that return energy to the electrical system. This energy is recovered and stored in the flywheels. The flywheel system has been demonstrated to reduce fuel consumption by 32–38%,[43] nitrous oxide emissions by 26%, and particulate emissions by 67%.[44]

5.5 *Transit System Energy Recovery*

Currently 190 metro systems operate in 54 countries around the world. These systems comprise 9,477 stations and over 11,800 km of track.[45] Many systems have recently been or are currently being extended. In addition more than 30 new metro rail systems are currently under construction, nearly all of which are in Asia. This massive expansion has received relatively little notice in the US.

Trackside energy storage captures energy lost during braking and allows for heavier and longer trains without increasing transmission or distribution line capacity. In order to mitigate voltage sag or increase transit system capacity in an existing system without using energy storage, a new substation has to be installed. Flywheel energy storage installed at a transit station would provide the same mitigation of voltage sag as a new substation but in a small footprint with no new utility feed and at a much lower cost. Given the high rate of charge-discharge cycles, flywheels are particularly well suited for this application.

Using energy storage to recover energy lost in braking reduces metro rail electricity consumption on the order of 10%.[46] Energy cost savings of $50,000–$90,000 per station per year have been forecast.[47] When installed in regions where the utility tariff structure includes demand charges, additional savings of $75,000–$250,000 per station per year are attainable.[48] In studies and tests to date, trackside storage sized to provide 1–3 MW of launch power or energy recovery per station is an effective rating for the metro rail application.[49]

Presently, a 2 MW flywheel system produced by Calnetix is installed in the Los Angles Metro Red Line at the Westlake/MacArthur Park station.[50] The machines are similar to the ones used in the RTG application. From 2000 through 2004, flywheel systems developed by URENCO were demonstrated at trackside installations in New York, London, Paris, and Lyon.[51]

5.6 *Motorsport*

Since the late 2000s, hybrid propulsion systems have powered the cars in top-tier motorsport beginning with Formula 1 followed by the highest class of Le Mans series racing: the LMP1 series. In F1, hybrid power trains improve fuel efficiency so that cars can complete an entire race without refueling. In the Le Mans endurance races, reducing fuel consumption decreases the number of pit stops required to complete a 24-h race that covers approximately 3,000 miles.

Williams Hybrid Power (WHP) pioneered the use of flywheels in this application. WHP was a spin-off of the Williams Formula 1 racing team and currently operates as GKN Hybrid Power. While never deployed in F1, WHP flywheels were used successfully in the Audi R18 e-Tron LMP1s that won at Le Mans in 2012, 2013, and 2014.[52]

The flywheel system in the 2014 race winner uses a compact carbon composite rotor spinning at 40,000 RPM to retrieve or produce 0.5 MJ per lap at a peak power of 170 kW.[53] The integral flywheel motor in the GKN rotor uses a novel approach to embedding particles of magnetic material directly in the composite. This eliminates electrical conductivity of the rotating portion of the flywheel motor allowing for high power, high duty factor operation.

5.7 *UPS*

Flywheel systems from companies including Hitec, Piller, Powerstore, Powerthru, Active Power and others are in global use providing temporary backup electrical power. The purpose of the flywheel in this application is to support the load of a critical facility or system during a power outage until backup diesel generators can be brought up to speed and synchronized. Flywheels compete directly with batteries and offer the advantages of much longer service life and the avoidance of the need to periodically replace and recycle the batteries. Here flywheels are implemented in one of two ways.

When used a standalone energy storage device, the system is referred to as a flywheel UPS. The flywheel provides electrical power to a direct current (DC) bus and an inverter converts this into alternating current (AC) electricity to power the load. In this application, the flywheel replaces or augments a battery. Discharge times of ten's of seconds are typical. Rotors in flywheel UPS systems generally spin about a vertical axis in vacuum or reduced pressure.

Rotary UPS are variously known as ride through systems, engine coupled UPS, or diesel rotary uninterruptible power supplies (DRUPS). A typical rotary system comprises a diesel generator, an inductive coupling with a substantial moment of inertia, and an alternator all mounted coaxially on a common base frame. A clutch may be located between the inductive coupling and the generator. During an outage, kinetic energy stored in the inductive coupling drives the alternator to support the load while the diesel generator starts. Start times for diesel generators have decreased dramatically. Power from the generator may be available in as little as 3 s after an outage begins. Rotary UPS are large, the smallest having a rating around 1 MW.

The global market for UPS systems is on the order of \$8–\$10B per year. Rotary systems account for about 5% of the market total UPS market. However, when only large systems (>2 MW) are considered, Rotary UPS account for 35% of the market.[54] In Europe, where Rotary UPS are well established, half of all new UPS installations that are rated at more than 1 MW are Rotary UPS.[55]

5.8 *Electromagnetic Aircraft Launch — EMALS*[56]

Electromagnetic launch technology is under development for the purpose of replacing steam powered catapults on aircraft carriers. Steam catapults are large, heavy, and inefficient. They operate without control feedback and subject airframes to high transient loads. They are expensive to maintain. Each launch consumes 1,350 lb (615 kg) of steam produced by the aircraft carrier's nuclear reactor.

An electromagnetic launch system will be capable of launching heavier aircraft using much less energy. The weight, volume, and controllability of the launcher will also be significantly improved.

The launcher of the EMALS system is a linear synchronous motor also referred to as a coil gun. The linear motor is powered by alternators. Each alternator comprises an axial field permanent magnet motor with dual stators. The alternator's rotor disk serves as the energy storage component and the field source during power generation. Average power from the ship's electrical system is fed into the alternator between launch events. The system is sized to charge fully in 45 s. During a launch event the energy stored in the rotors is released in a pulse lasting from 2 to 3 s. Peak alternator output is 81.6 MW when discharged into an impedance matched load. When fully charged, the EMALS rotors store 121 MJ (33.6 kWh) of extractable energy at a maximum speed of 6,400 RPM. The total stored energy is much higher as the rotor speed only decreases by about 25% during a launch event.

5.9 *Pulsed Power*[57]

In an application somewhat similar to EMALs, flywheels provide pulsed power for large science experiments. Tokamaks and other fusion test facilities require high power to create strong magnetic fields that confine the plasma during an experiment. The largest and most powerful tokamak currently in operation is the Joint European Torus (JET) located in Oxfordshire, England. JET has been in operation since in 1983. A single plasma pulse at JET requires peak power of 1,000 MW and occasionally more. On a typical day 22 tests are conducted. These tests are short in duration and are referred to as "shots".

If JET were to draw this power from the grid, the impact would be severely disruptive. Instead, JET draws power from the grid continuously, charging two enormous steel flywheels. The flywheels provide power for each shot.

Each of the two JET flywheels has a diameter of about 9 m and weighs 775 tons. At full speed the rotors spin at 225 RPM and attain a tip speed of around 100 m/s. Between shots, the wheels are accelerated from half speed to full speed over a period of 9 min using 8.8 MW motors. During a 20-s shot, each flywheel can discharge 700 kWh of energy at a peak power of 500 MW.[58]

6 Conclusions and Outlook

Mechanically connected flywheels have long been a component of any machine where reciprocating movement is transformed into rotation. In this application, the continued use of flywheels is assured.

As electrically connected energy storage systems, flywheels must compete with batteries and ultracapacitors on the basis of cost where cost is evaluated over the life of a system. For low cycle applications, such as electric vehicles, battery prices are already nearing the long sought goal of $100/kWh.[59] Flywheels are unlikely to achieve this incremental energy cost using materials and subsystems available today.

However, applications requiring 10^6 cycles and a calendar life of decades will continue to be well served by flywheels as battery cycle life remains at least two orders of magnitude lower than this. In these applications, flywheels compete with ultracapacitors on the basis of the cost per unit energy delivered.

Cost reduction is perhaps the most important objective in the development of any type of energy storage. The extent to which the use of flywheels will expand or decline will depend on trends in cost reduction for flywheels and for the various competing technologies.

Cost drivers for flywheel systems are spread out over a number of subsystems including the rotor, bearings, power electronics and the balance of system. Flywheels are not deployed in numbers sufficient to drive down the cost of these subsystems through high volume manufacturing. However, flywheels will benefit as other industries drive both increasing performance and declining cost in several key areas.

Rotor material performance has a geometric impact on the cost of a flywheel system. Steel, carbon composites, and glass composites are all fairly mature and order of magnitude improvements in performance or cost are unlikely. The development of potentially transformative materials, perhaps carbon nanotube composite, aims to increase strength and stiffness by an order of magnitude over existing composites. If it were available, the use of such a material in a flywheel would not only substantially improve the energy per unit mass of the rotor but would also lead to much smaller and less costly bearings and housings.

Improved performance and lower cost of power electronics for electric vehicles translate directly into improved performance and lower cost of the flywheel motor drive and the active magnetic bearing. Prompted by the accelerating use of electric power trains, motor drive power electronics costs for electric vehicles have dropped dramatically over the last decade and are approaching $5/kW.[60]

Given the increasing need in areas where flywheels are already in use combined with performance and cost trends in the underlying technology, flywheels should remain a competitive energy storage solution for the foreseeable future.

Acknowledgments

I would like to thank Matthew Lazarewicz, a pioneer in advanced flywheel technology who led the development of Beacon Power technology and deployment of the first frequency regulation plant, for his numerous insights and for presenting me with the opportunity to write this article.

References

1. S. Rathod Balasaheb *et al.*, A case study on design of a flywheel for punching press operation, *IJEAT*, **3**(4) (2014).
2. G. Genta, *Flywheel Energy Storage* (Butterworths, London, 1985).
3. W. C. Young and R. G. Budynas, *Formulas for Stress and Strain* (7th edition), (McFraw-Hill, London, 2002), pp. 745 (#6), 746 (16.2-3, 16.2-7).
4. Potter's Wheel, *New World Encyclopedia*. Available at: http://www.newworldencyclopedia.org/entry/Potter's_wheel. Accessed on September 25, 2014.
5. V. Bryant, The origin's of the Potter's wheel, *Ceramics Today*. Available at: http://www.ceramicstoday.com/articles/potters_wheel.htm. Accessed on September 25, 2014.
6. J. B. Chang, D. A. Christopher and J. K. H. Ratner, *Flywheel Rotor Safe-Life Technology: Literature Search Summary* (Diane Publishing, Collingdale, 2002), p. ix.
7. H. Hodges, *Technology in the Ancient World* (Barnes and Noble, New York, 1992), p. 47.
8. E. W. Barber, *Prehistoric Textiles: The Development of Cloth in the Neolithic and Bronze Ages with Special Reference to the Agean* (Princeton University Press, Princeton, 1991), pp. 41–44.
9. A. Pacey, *Technology in World Civilization: A Thousand-Year History* (First MIT Press paperback edition) (The MIT Press, Cambridge MA, 1991).
10. Science Museum, Home — atmospheric engine by Francis Thompson, 1791. Available at: www.sciencemuseum.org.uk. Accessed on September 25, 2014.
11. H. W. Dickinson, *A Short History of the Steam Engine* (Cambridge University Press, Cambridge, 2011), pp. 79–82.
12. Public Domain image. Available at: http://en.wikipedia.org/wiki/Corliss_steam_engine. Accessed on September, 2014.
13. *Industry: A Magazine Devoted to Science, Engineering, and the Mechanic Arts, Especially on the Pacific Coast* (Vol. 5) (Industrial Publishing Company, Detroit1892), p. 776.
14. Carbon Fiber Flywheels, Beacon Power Corporation. Available at: http://beaconpower.com/carbon-fiber-flywheels/. Accessed on October 2, 2014.
15. *Insurance Engineering* (Vol. 10) (Insurance Press, 1905), pp. 384, 579.
16. Available at: http://www.farmcollector.com/steam-traction/100-years-ago-in-american-machinist.aspx#axzz3DIFKslUY. Accessed on September 27, 2014.
17. J. Sousanis, World Vehicle Population Tops 1 Billion Units. *Ward AutoWorld*, August 15, 2011.
18. Pricing information for Maxwell and Ioxus cells and modules obtained by the author, March 2014.
19. G. Gardiner, Composite Flywheels: Finally Picking up Speed? *Composites World*, March 4, 2014. Available at: http://www.compositesworld.com/blog/post/composite-flywheels-finally-picking-up-speed. Accessed on October 6, 2014.
20. D. A. Bender *et al.*, "Flywheel System Using Wire Wound Rotor," Application US 13/222,693, Publication number US 20120062154 A1, March 15, 2012.

21. N. C. Bracket *et al.*, "Crash Management System for Implementation in Flywheel Systems," US 7,365,461 B2, April 29, 2008.
22. D. A. Bender and T. C. Kuklo, Separators for Flywheel Rotors, Patent Number 5,775,176, July 7, 1998.
23. F. Ehrich, *Handbook of Rotordynamics* (McGraw-Hill, London, 1992), Ch. 1.2.1, "The Jeffcott Rotor."
24. W. Lemahieu *et al.*, Instability due to internal damping of rotating shafts, *Sustainable Construction and Design* **3**(2) (2012), pp. 123–127.
25. E. Sonnichsen, Spin Testing, *Global Gas Turbine News*, 37(1).
26. H. Heinrich, Contact between solid elastic bodies, *Journ. Für Reine und Angewandte Math*, **92** (1882), pp. 156–171.
27. The SKF Formula for Rolling Bearing Life, *Evolution: Business and Technology Magazine from SKF*, February 16, 2001.
28. Life — A New Life Theory, Part 5 — Technical Guide, "Super Precision Bearings," NSK Motion and Controls, file e1245f.pdf, pp. 138–144. Available at: http://www.jp.nsk.com/app01/en/ctrg/index.cgi?rm=pdfView&pno=e1254f. Accessed on September 27, 2014.
29. R. Herzog *et al.*, Unbalance compensation using generalized notch filters in the multivariable feedback of magnetic bearings, *IEEE Trans. Control Syst. Technol.* **4**(5) (1996), pp. 580–586.
30. J. A. Veltri, C. MacNeil and A. Lampe, Cooled Flywheel Apparatus, US patent application US14/072,462, Publication no. US2014012472 A1, May 8, 2014.
31. M. P. Boyce, *Gas Turbine Engineering Handbook* (Elsevier, Amsterdam, 2012), p. 912.
32. Military Handbook MIL-HDBK-17-1D, *Polymer Matrix Composites* (Vols. I, II, III).
33. Department of Transportation, National Highway Traffic Safety Administration, 49 CFR Part 571, RIN [2127-AF14], Federal Motor Vehicle Safety Standards; *Compressed Natural Gas Fuel Container Integrity*, p. 69, S7.2.1.
34. Beacon Power Stephentown Advanced Energy Storage Case Study, Clean Energy Action Project. Available at: http://www.cleanenergyactionproject.com/CleanEnergyActionProject/CS.Beacon_Power_Stephentown_Advanced_Energy_Storage_Energy_Storage_Case_Study.html. Accessed on October 7, 2014.
35. International Renewable Energy Agency, Electricity Storage and Renewables for Island Power, May 2012.
36. EirGrid, System Operator for Northern Ireland, All-Island Generating Capacity Statement 2012–2021.
37. V. Gevirgian and D. Corbus. Ramping Performance Analysis of the Kahuku Wind — Energy Battery Storage System, NREL/MP-5D00-59003, November 2013.
38. *Ibid.*, p. 11.
39. Available at: http://en.wikipedia.org/wiki/List_of_countries_by_electricity_consumption. Accessed on September 27, 2014.
40. Available at: http://en.wikipedia.org/wiki/List_of_islands_by_population. Accessed on September 27, 2014.

41. R. Williams, *Flywheel turn Scottish Islands on to Stable Power* (Energy Storage Publishing, West Sussex, 2014). Available at: http://bestmag.co.uk/industry-news/flywheels-turn-scottish-islands-stable-power. Accessed on October 7, 2014.
42. Usibelli Coal Mine Flywheel — Alaska Energy Wiki. Available at: http://energy-alaska.wikidot.com/usibelli-flywheel. Accessed on September 24, 2014.
43. Flywheel Energy Storage for Rubber Tired Gantry Cranes, Green Car Congress (April 9, 2009). Available at: http://www.greencarcongress.com/2009/04/flywheel-energy-storage-system-for-rubber-tired-gantry-cranes.html. Accessed on October 8, 2014.
44. M. Flynn, P. McMullen and O. Solis, Saving energy using flywheels, *IEEE Ind. Appl. Mag.* **14**(6) (2008), pp. 69–73.
45. Available at: http://en.wikipedia.org/wiki/List_of_metro_systems. Accessed on October 7, 2014.
46. Available at: http://www.abb.com/cawp/seitp202/265455d72a797481c1257b59003b8600.aspx. Accessed on October 7, 2014.
47. Schroeder, Yu, Teumin, *Guiding the Selection & Application of Wayside Energy Storage Technologies for Rail Transit and Electric Utilities*, *Transit Cooperative Research Program*, Transportation Research Board, Contractor's Final Report for TCRP Project J-6/Task 75 Submitted November 2010.
48. APTA whitepaper, SEPTA Recycled Energy Optimization Project with Regenerative Braking Energy Storage, Jacques Poulin, Product Manager – Energy Storage, ABB Inc., Montreal.
49. P. McMullen, Green ovations: Innovations in green technologies, reducing peak power demand with flywheel technology, *Electric Energy* (January/February 2013). Available at: http://www.electricenergyonline.com/show_article.php?mag= &article=680. Accessed on October 7, 2014.
50. *Ibid.*
51. C. Tarrant, Kinetic Energy Storage Wins Acceptance, *Railway Gazette*, April 1, 2004. Available at: http://www.railwaygazette.com/news/single-view/view/kinetic-energy-storage-wins-acceptance.html. Accessed on October 7, 2014.
52. A. Cotton, Audi R18 (2014), *Racecar Engineering*, June 1, 2014.
53. GKN, GKN Helps Power Audi to Victory at Le Mans, June 6, 2014. Available at: http://www.gkn.com/media/News/Pages/GKN-helps-power-Audi-to-victory-at-Le-Mans.aspx. Accessed on October 8, 2014.
54. Kinetic Energy Storage vs. Batteries in Data Centre Applications, Hitec Power. Available at: http://www.datacentre.me/downloads/Documents/KINETIC%20ENERGY%20STORAGE%20VS%20BATTERIES%20IN%20DATA%20CENTRE%20UPS%20APPLICATIONS.pdf. Accessed on October 9, 2014.
55. G. Gagliano, Applications Director, S&C Electric Company, private communication, March 2014.
56. GlobalSecurity.Org. Available at: http://www.globalsecurity.org/military/systems/ship/systems/emals.htm. Accessed on September 24, 2014.

57. European Fusion Development Website (EFDA). Available at: http://www.efda.org/jet/, http://www.efda.org/2011/10/775-tons-of-steel/, and http://www.efda.org/jet/history-anniversaries/. Accessed on October 7, 2014.
58. M. Huart and L. Sonnerup, JET Flywheel Generators, in *Proc. Inst. Mech. Eng. A: J. Power Energy* **200**(2) (1986), pp. 95–100.
59. S. Zacharay, Are EV battery prices much lower than we think? Under $200/kWh, *Clean Technica*. Available at: http://cleantechnica.com/2014/01/07/ev-battery-prices-much-lower-think/. Accessed on October 9, 2014.
60. S. Rogers, Advanced Power Electronics and Electric Motors R&D, *US DOE Presentation* (May 14, 2013). Available at: http://energy.gov/sites/prod/files/2014/03/f13/ape00a_rogers_2013_o.pdf. Accessed on October 9, 2014.

About the Contributors

Alberto Benato was born in Padova, Italy, in 1985. He received the B.E. and M. Tech. degrees in energy engineering from the University of Padova, Padova, Italy, in 2007 and 2010, respectively. In 2011, he joined the Department of Mechanical Engineering, University of Padova, as an Assistant Researcher and started his lecturer activity for the Turbomachinery courses. Since 2012, he has been with the Department of Industrial Engineering of the University of Padova and in 2015 he received the Ph.D. degree in Industrial Engineering. From February 2015, he has worked in the Department of Industrial Engineering, University of Padova, as a post-doctoral researcher. His current research interests include thermoelectric and nuclear power plants, waste heat recovery units, and energy storage systems. His research activity is documented by more than 30 papers published in referee journals and presented in important international congresses. Dr. Benato is a member of the American Society of Mechanical Engineers (ASME) and of the Italian Thermal Machines Engineering Association (ATI). He is also a frequent speaker at national and international conferences and serves as reviewer for several international journals.

Donald Bender is a Mechanical Engineer and Principal Member of the Technical Staff at Sandia National Laboratories. He holds BSME and MSME degrees from the Massachusetts Institute of Technology. Mr. Bender is a subject matter expert in the area of flywheel energy storage with 25 years of industrial and research experience in the field. At Sandia he supports System Surety Engineering and special projects pertaining to energy storage.

Francisco Blázquez was born in Toledo, Spain in 1972. He received the Dipl. degree in Industrial Engineering and the Ph.D. degree in Electrical Engineering from the Technical University of Madrid, in 1997 and 2004, respectively. Since 1999 he has been a Professor of Electrical Machines and Drives in the Department of Electrical Engineering at the Technical University of Madrid. His current research interests include Electrical Machine Design, Electrical Machine Protection Systems and Renewable Electric Power Generation. He has worked on many research and industrial development projects and has produced numerous publications and patents in the field of electrical engineering.

Giovanna Cavazzini received her degree with honors in Mechanical Engineering in 2003 from the University of Padova and her doctorate with European Label in Energetics in 2007 from the University of Padova. In 2005, she spent a research period at the Ecole Nationale Superieure Arts et Métiers, Paris Tech de Lille with emphasis on the study of unsteady turbulent phenomena developing in turbo machines. In 2009, she received the qualification of Maitre de Conference from the Ministère de l'Enseignement Supérieur et de la Recherche in France and since 2016 she has been an

Associate Professor of Fluid Machines and Energy Systems at the University of Padova. Her main research interests include analysis and modeling of renewable energy systems, techno-economical optimization of hydropower plants and design optimization of fluid machines with particular emphasis on hydraulic turbines and pump-turbines. She is author of about 80 scientific publications, the most part of which have been published in International Journals with significant Impact Factor and in Proceedings of International Conferences. She is Deputy Chair of the joint sub-program of the European Energy Research Alliance (EERA) on "Mechanical Energy Storage".

Manuel Chazarra received the Industrial Engineering degree with specialization in electricity and the M.Sc. degree, from the Comillas Pontifical University, Madrid, Spain. He is currently working toward the Ph.D. degree at the Technical University of Madrid (UPM) with a thesis entitled "Short-term scheduling of the joint energy and secondary regulation reserves of pumped-storage hydropower plants with variable speed or operated in hydraulic short-circuit mode in the Spanish electricity system".

Paul Connor is a Senior Research Fellow and Lecturer in the School of Chemistry at St. Andrews University, Scotland. He started working at St. Andrews in 1998 researching negative electrodes for rechargeable lithium–ion batteries. He has continued looking at various aspects of Li battery development as well as researching other electrochemical energy-based devices such as Solid Oxide Fuel Cells, reversible fuel cells, electrolyser technology, and even solar collection. His Ph.D. was awarded in 1998 from the University of Otago, in his native New Zealand, for studies of the metal oxide-water interface relating to photoelectrochemical systems. He previously received 1st class BSc (Honors) degrees in both Chemistry (1991) and Physics (1992) from the University of Otago.

Gael Corre received a Chemical Engineering degree from the INSA, Toulouse (France) in 2003. After having worked on CFD projects addressing the physicochemical modeling of environmental processes, he joined Professor John Irvine's research group at the University of St. Andrews in 2005 as a research student and received a Ph.D. in 2009. His Ph.D. work addressed Solid Oxide Fuel Cells (SOFC), with a strong focus on the development of oxide anodes for SOFCs direct operation on ethanol and methane fuels. During his Ph.D., Gael actively collaborated with Professors Gorte and Vohs research group at the University of Pennsylvania, leading to the development of nickel-free oxide anodes, offering performance levels comparable to nickel anodes and able to operate on undiluted methane. After receiving his Ph.D., Gael undertook a more engineering-focused project, by working on the development of short stacks for the Hybrid Direct Carbon Fuel Cell Technology, successfully developing stack units operating on pyrolysed MDF. In 2010, Gael joined the Hydrogen Centre at the University of Glamorgan as a research fellow, where his research addresses the direct use of biogas in SOFCs.

Mark Dooner received his Master's Degree in Electronic Engineering from the University of Warwick, UK, in 2011 with first class honors. He remained at the University of Warwick to study for a Ph.D. in Automotive Mechatronics and Control. Mark was awarded his doctorate in 2016 and graduated in the summer of that year. He is currently a Research Fellow at the University of Warwick where he conducts research into combustion engine optimization and control, and generating electricity using Compressed Air Energy Storage (CAES). Mark has experience in Mechatronics modeling and control, power systems and battery management, machine control, optimization, pneumatics, engine control, electromagnetic modeling, CAES, and real-time, dynamic and finite element system modeling. Mark has also worked in industry prior to his Master's degree, designing and testing circuits for aircraft engine control systems.

Jesús Fraile-Ardanuy was born in Madrid, Spain, in 1972. He received the Telecommunication Eng. Degree from the Technical University of Madrid (UPM), Madrid in 1996 and the Ph.D. degree from UPM, Spain, in 2003. He is an Associate Professor at ETSI Telecomunicación-UPM. He is the author of 12 books, more than 50 articles in peer-reviewed journals and international congresses. His research interests include applications of intelligent control to renewable energy generation and integration of electric vehicles on the grid.

Wentian Gu graduated from Georgia Institute of Technology under the guidance of Professor Gleb Yushin. His research interest was focused on high energy fluoride-based cathode materials for lithium–ion batteries. He is currently working as a researcher at General Motors, China Science Lab, participating in projects concerning high-power batteries for low-voltage start/stop system.

Professor John Irvine is Professor of Chemistry at the University of St. Andrews, currently holds a Royal Society Wolfson Merit Award, is a Visiting Professor at Queens University, Belfast; Co-director of the Energy Technology Partnership, 1000 Talents Professor at Fujian Institute for Research in the Structure of Matter and is European Councillor for the International Society of Solid State Ionics. His first degree is in Chemical Physics from Edinburgh University and he obtained a DPhil from the University of Ulster in Photoelectrochemistry. He performed his post-doctoral studies working with Anthony West in Aberdeen and was subsequently appointed to a BP/RSC fellowship, lectureship and senior lectureship at Aberdeen University. In 1994, he was Visiting Professor at Northwestern University and then moved to the University of St. Andrews as Reader and then Professor of Inorganic

Chemistry in 1999. In 2005, he was elected a Fellow of the Royal Society of Edinburgh. In 2008, he received the Royal Society of Chemistry Materials Chemistry Award and the Sustainable Energy Award in 2015 and European Solid Oxide Fuel Cell Forum Schönbein Gold Medal, in 2016. He has over 400 publications in refereed journals including *Nature* and *Nature Materials.* He has developed new concepts in fuel cells such as the Hybrid Direct Carbon Fuel Cell, has a leading role in the field of developing redox stable, coking tolerant oxide electrodes for SOFCs and discovered the first significant interstitial oxide ion conductor. He was Chairman of the 2010 European SOFC Forum in Lucerne and Chair of the Scientific Committee for the Faraday Discussion, York 2015. His research interests include solid state ionics, new materials, ceramic processing, electrochemistry, fuel cell technology, hydrogen, photoelectrochemistry, electrochemical conversion, and heterogeneous catalysis.

Christopher Krupke received the B.E. degree in Mechanical Engineering from the University of Applied Sciences, Wolfenbüttel, Germany, in 2013. At the same time, he completed an apprenticeship to an industrial mechanic at Volkswagen AG, Wolfsburg, as part of a dual degree system. Alongside his undergraduate studies he received a second B.E. degree in Mechatronics from Coventry University, U.K., where he was awarded the best overall student. Following his degrees he studied for a Ph.D. in Mechatronics at the University of Warwick, UK, with a research focus on wind power and compressed air energy storage. During his research, Christopher was involved in a variety of research and industrial projects, ranging from compressed air energy storage applications, pneumatic and electric drives, system modeling and control as well has real time prototyping. After gaining the Ph.D. degree in 2016 he commenced work at Volkswagen AG as a test engineer in the Research and Development department. His tasks encompass whole vehicle analysis with respect to fuel consumption and driving performance.

Xing Luo received his B.E. degree from Xi'an University of Technology, China, in 2004, the M.Sc. degree from the University of Liverpool, U.K., in 2006, and his Ph.D. degree from the University of Birmingham, U.K. in 2010. Currently, he is a Research Fellow with the School of Engineering, the University of Warwick, U.K. His research interests include electrical energy storage, smart grid, engine and pneumatic system modeling and identification, hardware-in-the-loop simulation, energy efficient systems, power electronic systems, real-time control development.

Juan I. Pérez-Díaz was born in Madrid, Spain in 1979. He obtained the Ph.D. degree in Civil Engineering in 2008 from the Technical University of Madrid (Spain), where he has worked as a non-tenured Associate Professor since 2011. In 2014, he was Visiting Professor in the Department of Industrial Engineering of the University of Padova. His current research interests focus on power generation scheduling, load-frequency control in power systems and grid integration of renewable energy, with special emphasis on pumped-hydro energy storage. He has participated in and coordinated several research projects for Spanish power companies and funded under national and international calls for research proposals.

Carlos A. Platero was born in Madrid, Spain, in 1972. He obtained the Dipl. degree and Ph.D. degree in Electrical Engineering from Universidad Politécnica de Madrid, Spain, in 1996 and 2007, respectively. From 1996 to 2008 he has worked in ABB Generación S.A., Alstom Power S.A. and ENDESA Generación SA, always involved in design and commissioning of diesel, thermal and hydro power plants. In 2003, he began teaching at the Electrical Engineering Department of the Industrial Engineer School of Universidad Politécnica de Madrid, and joined an energy research group. Since 2008 he became full-time Associate Professor. During these years he has worked in the protection and diagnosis of electrical machines, especially large synchronous generators. He has been a Visiting Professor in EPFL (Swiss Federal Institute of Technology, Lausanne) and Coventry University in England.

Jose Angel Sanchez Fernandez is a Civil Engineer (1986) and holds a doctorate (1995) from the Technical University of Madrid (Spain). He has been an Associate Professor of Electrical Engineering since November 1996 at this University (UPM). He has lectured in several courses of Basic Electrical Engineering, Electrical Technology, Energy Systems and Hydropower Engineering. He has been Chair of the Institute of Electric and Electronic Engineers (IEEE) Industry Applications Society Spanish Chapter from 2000 to 2007. At the Technical University of Madrid, he has been Head of the Hydraulic and Energy Engineering Department between May 2008 and July 2014. Currently (since July 2014), he is Head of the Hydraulic, Energy and Environmental Engineering Department at this University. His main lines of research are hydropower operation and grid integration of wind energy systems. He has participated in 13 competitive research projects, being the main researcher of one of them. He has also participated in 16 non-competitive projects, being the main researcher for seven of them. He is co-author of 2 books, 3 book chapters, 3 articles in national journals, 27 articles in international journals and 31 conference papers. He is also co-inventor of four patents. He has received the Medal of the UPM (2013) and is a Senior Member of the IEEE.

Anna Stoppato graduated with honors in Mechanical Engineering at the University of Padova (Italy) and is a Ph.D. in Energetics. She is an Associate Professor in Turbomachinery and Energy Systems at the University of Padova. The main scientific field of her activity is the study and the modeling of energy conversion plants. Her research activity is documented by more than 90 scientific publications and is mainly focused on the simulation of thermal plants operated at different loads. She implemented a modular code (DIMAP) for energetic, exergetic, and exergoeconomic analysis of energy conversion plants, at both design and off-design operation. An original exergetic–exergoeconomic technique for plants diagnostics is also implemented in the code. She is involved in studies about the development of techniques aimed at evaluating the influence of power plant management strategies on its components residual life. She is also involved in research activities on waste heat recovery both for CHP purposes and electric production and on Energy storage, with a particular focus on mechanical and thermal storage. Anna Stoppato is a member of ATI (the Italian Thermal Machines Engineering Association) and of ASME (American Society of Mechanical Engineers). She is a Co-winner of the 2002 ASME Edward F. Obert Award.

Jihong Wang (Senior Member of IEEE09, Fellow of IET) received the B.E. degree from Wuhan University of Technology, Wuhan, China, the M.Sc. degree from Shandong University of Science and Technology, Shandong, China, and the Ph.D. degree from Coventry University, Coventry, U.K., in 1982, 1985, and 1995, respectively. She is currently a Professor of Power Systems and Control Engineering within the School of Engineering, the University of Warwick, Coventry, U.K. She was a Technical Editor to *IEEE Trans. on Mechatronics*. Her main research interests include compressed air energy storage, nonlinear system control, system modeling and identification, power systems, energy-efficient actuators and systems, and applications of intelligent algorithms.

Lu Wei received her Ph.D. degree from the Georgia Institute of Technology (USA) in the School of Materials Science and Engineering with Professor Gleb Yushin as her advisor. She received a B.S. in Materials Science and Engineering from Zhengzhou University (China) in June 2005. Upon graduation she joined Northwestern Polytechnical University (China) as a graduate student in the National Key Laboratory of Thermostructural Composite Materials. At Georgia Institute of Technology, she has been engaged in the synthesis of innovative nanostructured materials for energy storage applications, such as supercapacitors and Li-batteries. Her research interests are focused on the investigation of low cost nanoporous carbon electrodes with controlled pore size from environmentally friendly precursors, and the manufacturing of on-chip micro supercapacitors consisting of single particle electrodes with precisely defined dimensions.

Gleb Yushin is a Professor in the School of Materials & Engineering and the Director of the Center for Nanostructured Materials for Energy Storage at Georgia Institute of Technology. Since 2011, Dr. Yushin has served on the Board of Directors at Sila Nanotechnologies, Inc., an engineered materials company focused on dramatically improving energy storage technologies. Since 2016, Dr. Yushin serves as an Editor-in-Chief for *Materials Today*, the flagship journal of the *Materials Today* family, dedicated to covering the most innovative, cutting edge and influential work of broad interest to the materials science community. Dr. Yushin received a Ph.D. degree in Materials Science from North Carolina State University (NC, USA) and a M.S. degree in Physics from the Polytechnic Institute (Saint-Petersburg, Russia). The current research activities of his laboratory are focused on synthesis and characterization of nanostructured and nanocomposite materials for use in advanced lithium and lithium–ion batteries, supercapacitors, and lightweight structural materials.

Professor Xiaoxiang Xu received his Bachelor's degree from the University of Science & Technology of China (USTC) in 2006 and the Ph.D. degree from the University of St. Andrews in 2010 (Supervisors: Professor John Irvine and Professor Shanwen Tao). He continued his research at the University of St. Andrews as a post-doctorate with Professor John Irvine until 2013. He was recruited to join the 1000 Talent Plan from the Chinese government in 2014 and started his own career in Tongji University. His research interests include photocatalytic water splitting and fuel cells.

Index

D

E

Q

R

S

Printed in the United States
By Bookmasters